Wine and Oil Production in Antiquity in Israel and Other Mediterranean Countries

JSOT/ASOR Monograph Series 10

Wine and Oil Production in Antiquity in Israel and Other Mediterranean Countries

Rafael Frankel

Sheffield
Academic Press

Published by Sheffield Academic Press Ltd
Mansion House
19 Kingfield Road
Sheffield S11 9AS
England

Printed on acid-free paper in Great Britain
by Bookcraft Ltd
Midsomer Norton, Bath

British Library Cataloguing in Publication Data

A catalogue record for this book is available
from the British Library

ISBN 1-85075-519-1

Contents

Preface 15
List of Photographs 17
List of Figures 20
List of Maps 22
List of Charts 23
Introduction 25
Conventions Used for Data on the Disc 31
Abbreviations 34

Preamble A
Grape and Wine Ecology: Origins and Methods of Cultivation 35
1. The Vine 35
2. The Olive 36

Preamble B
Wine and Oil: Their Importance in Antiquity—Aspects of Modern Research 38

Preamble C
Wine and Olive Oil Production: Basic Processes 41
1. Processes of Wine Production 41
   Treading 41
   Pressing and Varieties of Must 42
   First Must 42
   Must of the Second Pressing 42
   After-Wine 42
   Lees-Wine 43
   Boiled-Down Must 43
   Fermentation and Types of Wine 43
   Pre-Treatment of Grapes 43
2. Olive Oil: Its Uses and Importance 43
   Cosmetics 43
   Light 44
   Industrial Purposes 44
   Olive Wood 44
   Ritual 44
   Food 44
   The Uses of Oil Lees 45
3. Olive Oil: Processes of Production 46
   Crushing 46
   Olive Oil Extraction and Oil Quality 46
   Trees and Olive Groves 46
   Stages of Ripeness of Olives 46
   Treatment between Picking and Processing 47
   Oil Extraction: Pressing and Other Methods 47
   Oil Separation 48

Preamble D
History of Previous Research 49

Chapter 1
Simple Installations (T1) 51
1A The Simple Treading/Crushing Installation (Ts111, 121, 131) 51
1A1 Rock-Cut Simple Treading Installations (T111) 51
1A1.1 Character and Date 51
1A1.2 Simple Treading Installation (T111): Size and Form 52
1A1.3 Cutting Simple Rock-Cut Installations 52
1A2 Built and Free-Standing Simple Treading Installations (Ts121, T131) 54
1B Sub-Types of Simple Treading Installation 54
1B1 Small Crushing/Pressing Installations (Ts112, 113) 54
1B2 Pairs of Installations 54
1B3 Small Hole in Centre of Treading Floor (T171) 54
1B4 Two Small Cup-marks on Either Side of the Treading Floor (T172) 55
1B5 Stone-Cut Tethering Rings (T173) 55
1C Simple Installations from the Iron Age 56
1D Topographical Distribution of Simple Installations (Ts1): Simple Wine Husbandry 56
1E Oil Production in Simple Installations 56
1F Simple Wineries in Other Regions 58
1F1 Countries South of the Mediterranean 58
1F1.1 Egypt 58
1F1.2 North Africa 58
1F1.3 Syria 58
1F1.4 Turkey 58
1F2 Countries North of the Mediterranean and Mediterranean Islands 59

Chapter 2
The Simple Lever and Weights Press (T2) 61
2A The Principles of the Lever Press 61
2B Simple Lever and Weights Press: Israel and Environs 62
2B1 Date of Introduction 62
2B2 Typology and Distribution 62
2B2.1 The Anchoring Point (Fulcrum) 62
2B2.2 Pressing Point (Load Point): Lateral Collection and Central Collecton 62
2C The Simple Lever and Weights Press in Other Countries 66
2C1 Syria 66
2C2 Cyprus 67
2C3 Crete 67

Chapter 3
Olive-Crushing Devices (T3) 68
3A Crushing Devices Associated with Simple Lever and Weights Press T2 (Ts1152, 1153, 126, 133, 134) 68
3B The Round Rotary Olive Crusher (T3) 68
3B1 The Round Rotary Olive Crusher in Israel and Environs 68
3B2 Crushing Installations in Other Countries 72
3B3 Crushing Installations in Ancient Literature and Historical Conclusions 74

Chapter 4
The Improved Lever Press (T4) 76
4A Improved Lever Presses in Israel and Environs 76
4A1 The Maresha Press: Lever Press, Central Vat, Plain Piers (T40111) 76
4A2 The Zabadi Press: Lever Press, Lateral Collection, Slotted Piers (T40222) 77
4A3 Perforated Piers (T423) 82
4A4 Guider Mortices (T44) 83
4A4.1 Two Mortices below Beam (T441) 83

4A4.2	Two Mortices below Free End of Beam (T4481)	83
4A5	Other Types of Improved Lever Press from Israel and Environs	84
4A5.1	Lateral Collection	84
4A5.1.2	The Qedumim Press-Bed and Vat (T47114)	84
4A5.2	Central Collection	84
4A5.2.2	Central Vats with Radial Grooves (T4612)	84
4A5.2.3	Central Vats for the Production of First Oil	84
4A5.3	Press-Bed with Radial Grooves (T4624)	85
4A6	Improved Lever Presses in Israel and Environs: Conclusions	85
4B	Descriptions of Presses in Classical Literature	86
4B1	Cato the Elder	86
4B2	Vitruvius	86
4B3	Pliny the Elder	86
4B4	Hero of Alexandria	87
4C	Lever Presses in Other Countries	88
4C1	The Eastern Mediterranean	88
4C1.1	North Syria	88
4C1.2	Greece, Crete, Delos, Kalymnos	90
4C1.3	Cyprus	90
4C1.4	The Crimea	91
4C2	The Northern and Western Mediterranean	91
4C2.1	Italy	91
4C2.2	Yugoslavia	93
4C2.3	France and Germany	93
4C2.4	Spain	94
4C3	The Southern Mediterranean: North Africa and the Islands	94
4C3.1	Libya–Tunis: Perforated Pier Press (Ts4022, 4032, 424)	94
4C3.2	Algeria	95
4C3.2.1	Dovetail Beam Niche Press (Ts4044, 416)	95
4C3.2.2	The Madaure Oil Separator (Ts47132)	96
4C3.2.3	T-Shaped Pier Mortice Press (Ts4071, 435)	96
4C3.2.4	The Taourienne Pier Base; The Double Dovetail Mortice Pier Base (T4315)	97
4C3.2.5	Single-Socket Pier Base (T436)	97
4C3.3	Morroco: The Four-Mortice Pier-Base Press (Ts4061, 433)	97
4C3.4	Lever Presses in North Africa: Conclusions	98
4C3.5	Malta	98

Chapter 5

Beam Weights (T5)		99
5A	Simple Beam Weights in Israel and Other Countries: Unworked Field Stones (Ts5110), Weights with Horizontal Bore (Ts5112, 5113), Weights with Vertical Bore (Ts521)	99
5B	Complex Weights in Israel, the Southern Levant and Cyprus: The Reversed-T Weight (T53), Weight with Hook (T54), Large Weight with Horizontal Bore (T512)	100
5C	The Semana Weight, Rectangular Beam Weights with External Mortices (Ts55, 56)	102
5D	Cylindrical Weights with External Mortices	105
5D1	Italy: The Francolise Posto Weight (Ts55110, 55190, 5541)	105
5D2	Morocco: The Volubilis and Cotta Weights (Ts55111, 55412, 56911)	105
5E	Beam Weights with Two Vertical Bores (Ts57)	105
5E1	The Methana Weight, Greece (T5712)	105
5E2	The el-Beida Weight, Libya (T5732)	105
5F	The Taqle Weight with Reversed-T Bore and External Mortices, Syria (Ts5352, 535202, 5362)	105
5G	Beam Weights: Conclusions	106

Chapter 6
The Lever and Screw Press (T6) 107
6A The Introduction of the Screw Press 107
6A1 Principles and Chronology 107
6A2 Technology: Lever and Screw, Intermediate Piers (Ts441-446) 107
6B Models of the Lever and Screw Press Based on Pre-Industrial Presses, Classical Written Sources and Ancient Artistic Representations: Models 6A–6M 108
Model 6A: The Grand Point Press 108
Model 6B: The Pivoted Screw Press 108
Model 6C: The Tied Screw Lever Press 108
Model 6D: The Taissons Press 109
Model 6E: The Fixed-Block Press 109
Model 6F: The Raised Press-Bed Press 109
Model 6G: The Languedoc Press 109
Model 6H: The Arginunta Press 110
Model 6J: The Box Press 110
Model 6K: The Perforated Weight Press 111
Model 6L: Hero's Press B 111
Model 6M: The Fenis Press 111
6C Archaeological Evidence for Lever and Screw Presses: Screw Weights (T62): Israel and Environs 111
6C1 Screw Weights with Central Sockets 111
6C1.1 The Samaria Screw Weight (T6211) 111
6C1.2 The Dinᶜila Screw Weight (T622) 113
6C1.3 Screw Weight with Central Socket, Square Channel and External Mortices (T623) 113
6C1.4 The Kasfa Screw Weight (T624) 114
6C2 Screw Weights without Central Socket 114
6C2.1 The Bet Ha-ᶜEmeq Screw Weight (T625) 114
6C2.2 The Luvim Screw Weight (T626) 116
6C2.3 The Miᶜilya Weight (T6271); The Midrasa Weight (T6272) 117
6C3 Chronology of Screw Weights in Israel 117
6C4 Quarrying and Dressing Screw Weights 117
6D Screw Weights in Other Countries (Ts62) 117
6D1 North Syria and Lebanon 118
6D2 Greece, the Pontus, Kalymnos, Lesbos, Cyprus 119
6D3 Italy 119
6D4 France 120
6D5 Spain and Portugal 121
6D6 North Africa 121

Chapter 7
Direct-Pressure Rigid-Frame Presses: Single-Screw, Double-Screw and Wedge Presses (T7) 122
7A Chronology and Typology 122
7B Models of Rigid-Frame Presses Based on Pre-Industrial Presses, Classical Written Sources and Ancient Artistic Representations (Models 7A–7G) 122
Model 7A: Single Rotating-Screw Press 122
Model 7A1: Single Rotating-Screw Press, Handles at Lower End of Screw 122
Model 7A2: Single Rotating-Screw Press, Handles at Lower End of Screw, Vertical Auxiliary Drum 123
Model 7A3: Single Rotating-Screw Press, Handles at Lower End of Screw, Horizontal Auxiliary Drum 123
Model 7A4: Single Rotating-Screw Press, Handles at Upper End of Screw 123
Model 7B: Press with Two Fixed Screws 123
Model 7C: Press with Two Rotating Screws 123
Model 7C1: Press with Two Rotating Screws, Handles at Upper End of Screw, Female Threads in Lower Press Frame 123

	Model 7C2: Press with Two Rotating Screws, Handles at Upper End of Screws, Female Threads in Upper Press Frame	124
	Model 7C3: Press with Two Rotating Screws, Handles at Lower End of Screws, Female Threads in Upper Press Frame	124
	Model 7D: Press with Two Anchored Rotating Screws, Female Threads in Pressing Board	124
	Model 7D1: With Frame and Handles at Upper End of Screws	124
	Model 7D2: Without Frame and with Handles at Upper End of Screws	125
	Model 7D3: Without Frame and with Handles at Lower End of Screws	125
	Model 7E: Wedge Presses	125
	Model 7E1: Horizontal Wedge Presses	125
	Model 7E2: Horizontal Wedge Presses with Hinged Hammers	125
	Model 7E3: Vertical Wedge Presses	125
	Model 7F: Screw and Weight Press	126
	Model 7G: Drum and Weight Press	126
	Models 7A–7G: Summary	126
7C	Archaeological Remains of Rigid-Frame Presses in Israel and Environs	126
7C1	The Grooved-Pier Press (T711)	126
7C2	The Cross Press (T72)	130
7C3	Screw Press Bases	130
7C3.1	The ʿEin Nashut and Ṣippori Presses (Ts73-74)	130
7C3.2	The Tabgha and Weradim Presses: Screw Press Bases with Central Collection (Ts7313, 73221, 73222, 7323, 751)	133 133
7C3.3	The Rama and Mishkena Presses: Screw Press Bases with Mortices and Insertion Channels (Ts742, 743)	135
7C3.4	The Manot Press: Screw Press Bases with Mortices with Horizontal Bores for Pins (T744)	136
7C3.5	Miscellaneous Screw Press Bases (Ts745, 751, 752)	137
7D	Screw Press Bases from Other Countries	137

Chapter 8

Improved Wineries: Ancillary Installations (T8)		138
8A	Differences and Similarities between Wineries and Oil Presses	138
8B	Wineries: Typology	138
8C	Compartments Adjacent to Treading Floor (T80)	139
8C1	Small Compartments Adjacent to Treading Floor (T801)	139
8C2	Large Compartments Adjacent to Treading Floor (T802)	139
8C3	Auxiliary Treading Floor and Collecting Vat Connected to Main Treading Floor (T803)	139
8D	Wine Presses	140
8D1	Mortices of the Single Fixed-Screw Press (Ts81-83)	140
8D1.1	The Ayalon Press: Single-Fixed-Screw Press with Square Mortice (T81)	141
8D1.1.1	Geographical Distribution of the Ayalon Press	141
8D1.1.2	The Form of the Press-Bed of the Ayalon Press	141
8D1.1.3	The Form of the Mortice of the Ayalon Press	142
8D1.1.4	Connecting Channels (Ts81---1, 831--1)	143
8D1.2	The Ḥamad Press: Single Fixed-Screw Press, Mortice in Form of Socket and Internal Mortices (T82)	144
8D1.3	The Ḥanita Press: Single Fixed-Screw Press, Closed Dovetail Mortice (T831)	144
8D1.3.1	Geographical Distribution of Ḥanita Press	144
8D1.3.2	Form of Press-Bed of Ḥanita Press	145
8D1.4	Mortices of Single Fixed-Screw Presses: Conclusions	145
8D2	Lever and Weights Press in Wineries (Ts85)	145
8D2.1	Simple Lever and Weights Presses in Wineries	145
8D2.2	The Marj el-Qital Beam Niche (T851)	146
8D2.3	Large Rock-Cut Tethering Ring as Device to Operate Lever Press (T855)	146
8D3	Other Presses in Wineries	146
8E	Rollers in Wineries (Ts861, 862)	146

8F Devices for Containing the Material to be Pressed 147
8F1 Frails: *ʿql* (Hebrew); *fiscus, fiscina, fiscellus* (Latin) 147
8F2 *Regulae, galeagra*: Wood Frames and Ropes 148
8F3 Devices for Containing Grape Skins and Stalks in the Single Fixed-Screw Wine Press (Ts81-83) 148

Chapter 9
Plans of Improved Wineries (T9) 149
9A Improved Wineries in Israel and Environs 149
9A1 Wineries in the 'Four-Rectangle Plan' (T91) 149
9A2 Wineries in the 'Composite Plan' (T92) 150
9A3 Wineries in the 'One-Axis Plan' (T95-96) 152
9A4 Galilean Wineries with Two Collecting Vats (T97) 153
9A5 Winery with Bell-Shaped Collecting Vat, Ledge for Lid and Bore (T94) 153
9A6 Installations with Collecting Vat in Centre of Treading Floor (T93) 153
9B Wineries in Ancient Artistic Representations and Classical Literature (Ts02) 154
9C Wineries in Other Countries (Ts98) 154
9C1 Wineries in Egypt 154
9C2 Wineries in North Africa 155
9C3 Wineries in Syria 155
9C4 Wineries in the Crimea 155
9C5 Wineries from the Greek Mainland, Mikonos, Delos, Crete and Cyprus 156
9C6 Wineries from Italy 156
9C7 Wineries from Yugoslavia 157
9C8 Wineries from France and Spain 157
9C9 Wineries from Germany 157
9D Improved Wineries: Conclusions 157

Chapter 10
Pre-Industrial Installations (Ts01-02) 159
10A Presses for Purposes other than Oil and Wine Production 159
10A1 Clothes Presses 159
10A2 Cider Making 159
10A3 Cheese Presses 159
10A4 Printing Presses 159
10B The 'Mystic Press' 159
10C Pre-Industrial Presses: A Regional Survey 160
10C1 Pre-Industrial Presses from Israel and Environs 160
10C2 Pre-Industrial Presses from North Africa 161
10C3 Pre-Industrial Presses from Europe 161

Chapter 11
Conclusions 164
11A Israel and Environs: A Regional Technological Survey 164
11A1 Judaea: The Southern Culture 164
11A2 Phoenicia: The Northern Culture 165
11A3 Samaria and the Sharon Coastal Plain 167
11A4 The Carmel, Lower Galilee and Eastern Upper Galilee 167
11A5 The Golan Heights 169
11A6 The Jerusalem Area 169
11A7 Israel and Environs: Summary 169
11B Oil and Wine Presses in the Mediterranean Countries 170
11B1 Beam-Anchoring Devices 170
11B2 Methods of Applying Force to the Press-Beam 171
11B2.1 Lever and Weights Press: Beam Weights 171

11B2.2 Lever and Drum Press 171
11B2.3 Lever and Screw Press 172
11B3 Oil Separation 174
11B3.1 Skimming 174
11B3.2 Overflow Decantation 175
11B3.3 Underflow Decantation 175
11B3.4 Combined Overflow and Underflow Decantation 175
11C Spatial Patterns and Chronological Sequences 176
11C1 Technical Continuity, Regional Diversity and Cultural Identity 176
11C2 Integrated Cultures and Eclectic Cultures 178
11C3 Central Sophistication and Peripheral Diversification 179
11C4 Slow Development and Fast Development 179
11C5 Cultural Diffusion 180

Appendix 1
Wine and Oil Production in Ancient Hebrew Literature 185
A1.1 Terms in the Hebrew Bible 185
A1.1.1 Installations in the Hebrew Bible 185
A1.1.1.1 *yqb* 185
A1.1.1.2 *gt*, *gnt* 185
A1.1.1.3 *pwrh* 185
A1.1.1.4 *yqb*, *gt* and *pwrh*: Conclusions 185
A1.1.2 Types of Oil in the Hebrew Bible 186
A1.2 Installations in Talmudic Literature 186
A1.2.1 Installations Connected to Wine Only 186
A1.2.1.1 *gt*, winery (Hebrew) 186
A1.2.1.2 *ʿṣrh*, *mʿṣrʾ*, *mʿṣrtʾ*, winery (Aramaic) 187
A1.2.1.3 *zyyrʾ* (*mʿṣrʾ zyyrʾ*), press; *mkbš*, press 187
A1.2.1.4 *ʿgwly* (*ʿygwly*, *hgt*) beam weights 187
A1.2.1.5 *tpwḥ*, cup-mark in centre of treading floor; *lḥm*, press cake 187
A1.2.1.6 *mštyḥ* (*mšṭḥ*) *šl ʿlym*, auxiliary treading floor; *mštyḥ* (*mšṭḥ*) *šl ʾdmh*, compartments 188
A1.2.2 Installations Connected to Oil Production 188
A1.2.2.1 The Oil Press of the Mishnah 188
A1.2.2.1.1 *byt bd*, oil press; *qwrh*, press-beam 188
A1.2.2.1.2 *ym*, collecting vat; *mml*, press-bed 188
A1.2.2.1.3 *btwlwt*, slotted piers 189
A1.2.2.1.4 *glgl*, winch 189
A1.2.2.1.5 *ʿkyrym* (*ʿkydym*), beam weights 189
A1.2.2.1.6 The Oil Press of the Mishnah: Conclusions 190
A1.2.2.2 The Oil Press of the Mishnah: Additional Texts 190
A1.2.2.2.1 The second list 190
A1.2.2.2.1.1 *nsrym*, *ʾswrym*, *yṣydyn*, *ysyryn* = *syrym*, pots/vats 190
A1.2.2.2.1.2 *yqbym*, vats 190
A1.2.2.2.1.3 *mprkwt*, upper and lower *ryḥym*, crushing devices 190
A1.2.2.2.1.4 (*ʿbyrym*) *sqyn*, *mrswpyn*, sacks, bags 190
A1.2.2.2.2 The oil press of the Mishnah: Additions—the third list 190
A1.2.3 Texts Concerned both with Oil and Wine Production 191
A1.2.3.1 Wine and oil production in the Sabbatical year 191
A1.2.3.1.1 *bwdydh* 191
A1.2.3.1.2 *qwṭb*, *qwṭby* (*qynby*, *qtkw*) 191
A1.2.3.2 The Baraitha of the Lulabim: *lwlbyn*, screw; *dpyn*, pressing board, *ʿdšyn*, press-bed; *ʿqlym*, frails 192
A1.2.4 Installations in Other Texts 193
A1.2.4.1 *ʿwqh*, collecting vat 193

A1.2.4.2	*ʿqrb*, beam clamp; *ʾnqly*, wall hook	193
A1.2.4.3	*ʾwllh*, batch? press-bed?	193
A1.2.4.4	The seat attached to the press-beam	193
A1.2.5	The Oil for the Temple Offering	193
A1.2.5.1	The places from which the oil was brought	193
A1.2.5.2	Unsuitable oils	194
A1.2.5.3	Three types of olive	194
A1.2.5.4	The three types of oil	194
A1.2.5.5	The nine grades of oil in the Mishnah	195
A1.2.6	*mwḥl, my zytym*, olive lees	195
A1.2.7	Variations in Terminology in Talmudic Literature	195
A1.2.7.1	The complete oil press	195
A1.2.7.2	Press-bed	195
A1.2.7.3	Collecting vat	195
A1.2.7.4	Press-beam	196
A1.2.7.5	Winch, slotted piers, screw	196
A1.2.7.6	Beam weight	196
A1.2.7.7	Crushing equipment	196
A1.2.7.8	Frails	196
A1.2.7.9	Wooden components	196

Appendix 2

Types of Wine in Ancient Hebrew Literature		198
A2.1	Terms for Wine in the Hebrew Bible and Related Epigraphic Sources	198
A2.1.1.1	*yyn*	198
A2.1.1.2	*tyrwš*	198
A2.1.1.3	*ʿsys*	198
A2.1.1.4	*škr*	198
A2.1.1.5	*ḥmr*	198
A2.1.1.6	*ḥmṣ*	198
A2.1.1.7	*mz*	198
A2.1.1.8	*smdr*	198
A2.1.1.9	*msk, mzg*	198
A2.1.2.0	*yyn ʾgnt*	198
A2.1.2.1	*yyn hrqḥ, yyn ṭwb*	199
A2.1.2.2	*yyn yšn*	199
A2.1.3	Wines Designated by Provenance	199
A2.1.4.1	The *lmlk*, seal impressions	199
A2.1.4.2	Other seal impressions and jar inscriptions	199
A2.2.1	Types of Wine in Talmudic Literature	199
A2.2.1.1	Wine named according to age	200
A2.2.1.1.1	New wine	200
A2.2.1.1.2	Old wine	200
A2.2.1.1.3	Vintage wine	200
A2.2.1.2	Wines named by colour	200
A2.2.1.2.1	Red wine	200
A2.2.1.2.2	Dark wines	200
A2.2.1.2.3	White wines	200
A2.2.1.2.4	Clear wine	201
A2.2.1.3	Wines named by quality, taste and aroma	201
A2.2.1.3.1	*yyn yph*, fine wine	201
A2.2.1.3.2	Aromatic wine	201
A2.2.1.3.3	*yyn qšh*, strong wine	201
A2.2.1.3.4	*tylʾ hrypʾ*, acrid tillia	201

A2.2.1.3.5	*yyn ḥd*, sharp wine	201
A2.2.1.3.6	*(yyn) mr*, bitter wine	201
A2.2.1.3.7	Sweet wine	201
A2.2.1.3.8	*yyn qwss*, pungent wine	201
A2.2.1.4	Wines named according to provenance	201
A2.2.1.4.1	*yyn hšrwny*, Sharon wine	201
A2.2.1.4.2	*yyn krmly*, Carmel wine	201
A2.2.1.4.3	*yyn ʿmwny*, Ammonite wine	201
A2.2.1.4.4	*yyn hʿyṭlqy (bʾtlqy)*, Italian wine	201
A2.2.1.4.5	*yyn pwrgyytʾ*, Phrygian wine/Plugata wine	202
A2.2.1.5	Names of wine related to the production process	202
A2.2.1.5.1	*(yyn) ʿylwyʾ*, improved wine	202
A2.2.1.5.2	*(yyn) mʿšn*, smoked wine	202
A2.2.1.5.3	*yyn ṣymwqyn*, raisin wine	202
A2.2.1.5.4	Boiled wine	202
A2.2.1.5.5	*yyn šmrym*, lees-wine	202
A2.2.1.5.6	*tmd*, after-wine	202
A2.2.1.5.7	*yyn qrwš*, coagulated wine	202
A2.2.1.5.8	*yyn mrtp*, wine from the wine cellar	202
A2.2.1.6	Diluted and undiluted wine	203
A2.2.1.6.1	Diluted wine	203
A2.2.1.6.2	Undiluted wine	203
A2.2.1.6.3	*ḥmrʾ mrqʾ*, yellowish wine? pale wine, diluted wine	203
A2.2.1.7	Wines with additives	203
A2.2.1.7.1	*yyn mbwsm*, spiced wine	203
A2.2.1.7.2	*ḥmr bsym, bsymʾ*, well-seasoned/sweet/over-fermented wine	203
A2.2.1.7.3	*yyn mṭwbl*, spiced wine	203
A2.2.1.7.4	*yyn qprysyn*, caper wine	203
A2.2.1.7.5	*hmr wplplyn*, wine with pepper	203
A2.2.1.7.6	*yrnqʾ, yrqwnʾ*, vegetable wine	203
A2.2.1.7.7	*ʾsprgws*, asparagus beverage	203
A2.2.1.7.8	*qwryyṭy*, date wine	203
A2.2.1.7.9	*yyn tpwḥym*, apple wine	203
A2.2.1.8	Wines the names of which derive from Greek or Latin	203
A2.2.1.8.1	*ʾynwmylyn, yyn ymylyn*, honey wine	203
A2.2.1.8.2	*ʾlwnṭyt*, an aromatic wine	204
A2.2.1.8.3	*hylysṭwn*, sweet wine made of grapes left in sun	204
A2.2.1.8.4	*qwndytwn*, spiced/peppered wine	204
A2.2.1.8.5	*ʾpsyntyn, psynṭṭwn*, absinth (wormwood) wine	204
A2.2.1.8.6	*ʾynmrnynwn*, myrtle wine?	204
A2.2.1.8.7	*qrynʾ*, an Asiatic sweet wine	204
A2.2.1.9	Other types of wine	204
A2.2.1.9.1	*tylʾ, tlyyʾ*, a bitter white wine	204
A2.2.1.9.2	*ḥmrʾ d'qrym*, partly fermented wine	205
A2.2.2	Products Containing Wine	205
A2.2.2.1	*mwryys*, salted fish sauce	205
A2.2.2.2	*ḥylq/ḥyqh*, fish sauce made of the *ḥylq* fish	205
A2.2.2.3	*ʾnygrwn, ʾngwrnwm*, sauce of wine and garum	205
A2.2.2.4	*ʾksygrwn, ʾnsygrwn*, sauce of vinegar and garum	205
A.2.2.3	Jar Inscriptions and Other Epigraphic Evidence from the Roman Period	206
A2.2.4	Conclusions	206

Bibliography 207

Index 224

Computer Disc
List A
List of Sites and Installations
Installations in Sample Square 16/26 (100 Sq km)

List B
List of Installations according to Type

List C
Alphabetical List of Sites (Site Index)

Preface

My interest in the subject of agricultural installations arose from participation in surveys of the Israel Archaeological Survey, that of the Rosh Ha'ayin Map under the leadership of Moshe Kochavi in 1972–73 and that of Western and Upper Galilee led by myself starting in 1975. As a result my doctoral dissertation, of which this book is a revised version, was on this subject (Frankel 1984). It was written in Hebrew at Tel Aviv University and supervised by Moshe Kochavi and and Shmuel Avitsur. With the help of Seymour Gitin it was offered for publication to the American Schools of Oriental Research and after the Hebrew manuscript was read by Baruch Halpern, who suggested some important changes, it was accepted for publication by that body who later arranged that it would be published by the Sheffield Academic Press. I wish to thank Moshe Kochavi, Shmuel Avitsur, Seymour Gitin and Baruch Halpern for their help and encouragement.

Some of the ideas presented in the book have appeared in other publications, especially Frankel 1987, 1988–89, 1992a, Frankel, Patrich and Tsafrir 1990, Amouretti and Brun 1993: 107-18, 477-81, Frankel, Avitsur and Ayalon 1994: 19-89 and in a lecture given in a Symposium at Dumbarton Oaks, Washington, DC, in April 1995 which will appear in due course in Dumbarton Oaks Papers.

I can thank only some of the many others that helped in bringing this book to the final stage of publication: first of all the team of the Western Galilee section of the Israel Archaeological Survey and especially Shmuel Beer, may his memory be blessed, Shlomo Grotkirk, Nimrod Getzov and Yigael Tepper who was the first to show us the importance of the subject of agricultural installations; the archaeologists and others who showed me their finds and put them at my disposal David Amit, Eitan Ayalon, Motty Aviam, David Ben Ami, Chaim Ben-David, Yoram Ben Meir, Ariel Berman, Dave Davies, Zvi Eilan, may his memory be blessed, David Eitam, Emanuel Eisenberg, Claire Epstein, Israel Finkelstein, Nurit Feig, Benny Frankel, Danny Friedman (Sion), Zvi Gal, Yoseph Gat, may his memory be blessed, Zeev Greenhot, Zeev Herzog and the Tel Michal Excavation Team, Yaakov Kaplan, may his memory be blessed, Amos Kloner, Micah Livneh, Zvi Maoz, Yitshak Magen, Shelomit Nimlich, Eleazer Oren, Avner Raban, Vasilius Tzaferis, Gladys Weinberg, Sam Wolf and the Gezer Hebrew Union College Excavation Team, Yigael Yadin, may his memory be blessed, Adam Zartal, John Zeligman and Zvi Zuck; the teams of the Field Schools Carmel and Ophrah of the Society of the Protecton of Nature that did the same and especially that of Western Galilee that helped in excavations; those who allowed me to use archival material, the former Head of the Israel Department of Antiquities Avi Eitan and Ronnie Reich; the Archives of the Present Department and those of the Mandatory Period and Zeev Yeivin; the Archives of The Israel Archaeological Survey, Shulamit Nimlich; the Archives of the Katserin Museum and Shimon Gibson; the Photographic Archives of the Palestine Exploration Fund; those who searched for me for material abroad, Claudine Dauphin, Pirhiah Beck, Steve Israels, and Paul Rivlin; Moshe Asis who helped with some of the Talmudic sources although he is certainly not responsible for the relevant sections in the book; Aliza Braun the secretary of the Research Authority of Haifa University who helped and encouraged me at all stages of preparing the manuscript, the typists Heather Karnoff, Angela Greenson and Danielle Friedlander for the arduous work of typing it, Hagit Tahan for drawing the endless number of illustrations and maps with such patience and care and Eli Zefadia of the Geography Department who carried out the scanning of the figures for the computer disc.

The research and publication were funded in part by Doctoral Scholarships of the Memorial Foundation for Jewish Culture, The David Ben Gurion Research Fund of the Histadruth, a Doctoral Research Grant from the Centre for the Research of Eretz Israel and the Yishuv—Yad Izhak Ben Zvi, and Research Grants from the Israeli Academy of Sciences.

Finally I want to thank my Kibbutz and particularly my wife and family who accompanied and helped at every stage in the very long and often difficult process of producing the dissertation and then this book.

I am, of course, solely responsible for any errors or omissions.

The book was completed in 1992 but the bibliography includes later publications and also items that came to the notice of the writer at a later date; these have been marked with an asterisk. Similarly in exceptional cases additional information has been included in the notes.

I thank the following for allowing material to be reproduced in this volume:

The American School of Classical Studies: photograph 38 from Frantz 1985
Eitan Ayalon: photograph 79
Jean-Pierre Brun from Brun 1986: photographs 25, 26, 40, 41
École Française de Rome: photograph 42 from Leveau 1984
Eli Elgart, Eretz Israel Museum: photograph 20
Shlomo Grotkirk: photograph 43
Sophocles Hadjisavvas: photograph 39 from Hadjisavvas 1992
Amos Kloner: photograph 27
Libraire Gunegard publishers of Humbel 1976: photographs 50a, 50b, 50c, 86 and Figures 16, 17, 20
Carter Litchfield: photographs 50, 61, 62, 67
Adam Zartal: photograph 68

List of Photographs

Photo

1	Rock-cut simple treading installation/winery (T111): (1626)50/007(02)	51
2	Unfinished winery (T03): Yirka East	52
3	Unfinished winery (T03): ʿAmqa South	52
4	Small rock-cut crushing/pressing installation (T112): (1626)74-009	54
5	Pear-shaped small crushing/pressing installation with central groove (T113): Abu Sinan North (see also Photo 7)	54
6	Winery (T111) with two cup-marks, one on either side of treading floor (T172): (1626)63/010(01)	55
7	Winery (T111) with two cup-marks, one on either side of treading floor (T172): Abu Sinan North (see also Photo 5)	55
8	Winery (T111) with two cup-marks, one on either side of treading floor (T172): (1626)75-004(01)	55
9	Rock-cut tethering ring (T173): (1626)50/009(01)	55
10	Winery at Iron Age site (T111): (1626)95-004(01)	56
11	Rock-cut press-bed with small collecting cup-mark (T114): ʿAmqa East Inst. 03	57
12	Two connected crushing and collecting cup-marks (T117): Bat el-Jebel East Insts. 3 and 4	58
13	Free-standing central vats with groove and bore (T26111110): Bet Mirsham Inst. 07 (Albright 1941–43: Pl. 53A)	63
14	Rock-cut lever and weights press with central vat (T26112116): Qurnat el-Ḥarmiya	63
15	Free-standing central vat without connecting device (T2612110): Bet Shemesh Inst. 01 (Grant and Wright 1938: Pl. 20.4)	64
16	Free-standing round press-bed (T262211): Rosh Zayit Inst. 06	66
17	Free-standing round press-bed (T26221): Shiqmona	66
18	Rock-cut lever and weights press with central vat (T26122100?): ʿAvdon Inst. 02	66
19	Free-standing central vat, square with groove and bore (T26111210): Megiddo Inst. 03	66
20	Pre-industrial olive crusher (T01712): Dura el-Qarʿa (Photo: Eli Elgart, Eretz Israel Museum)	68
21	Round crushing basin with sunken socket (T31): Ṣur Natan Inst. 02	70
22	Round crushing basin with sunken socket (T31): Ṣur Natan Inst. 03	70
23	Round crushing basin with raised socket (T32): Tabgha Inst. 02	71
24	Round crushing basin, concave crushing surface, central protusion, no socket (T341): Zabadi Inst. 02	71
25	Horizontal oil mill (T37): Volubilis, Morocco (Brun 1986: fig. 24)	72
26	Round crushing basin, concave crushing surface, raised socket, the *trapetum* (T331): from one of the villas at Boscoreale, today at Pompei (Brun 1986: fig. 18)	73
27	Improved lever and weights presses (T401110002): Maresha Inst. 15 (Photo: Amos Kloner)	76
28	Improved lever and weights presses (T402221203): Zabadi Inst. 02, from south	77
29	Improved lever and weights presses (T402221203): Zabadi Inst. 02, from north	78
30	Improved lever and weights presses (T402221203): Zabadi Inst. 02, western press from east	78
31	Improved lever weights and screw press (T402221203/8): Karkara Inst. 01	78
32	Slotted piers (T42120): Karkara Inst. 03	79
33	Slotted piers (T42123): Shubeika Inst. 01	79
34	Slotted piers (T42113): Kafr Yasif Inst. 01	79
35	Lever and screw press with central collection and slotted piers: Quṣeir Inst. 03, press B, stage 1 (T402260047); stage 2 (T403160047), complete press	80
36	Slotted niche (T4221): Umm el-ʿAmed	81
37	Lever and screw press with central collection and slotted piers: Quṣeir Inst. 03, press B, stage 1 (T402260047); stage 2 (T403160047), central vat	84
38	Central vat for first oil (T46124): Quseir Inst. 02, press A2	84
39	Pear-shaped press-bed with splaying spout (T4632): Laureatic Olympus, Greece (Frantz 1986; Pl. 76e)	90

40	Cypriot slotted pier (T427) (Hadjisavvas 1992: fig. 191)	91
41	The Tivoli pier base (T43111): Sanary Saint Ternide (Brun site 91), France (Brun 1986: fig. 179)	92
42	North African press with perforated piers (T424311000): Bir Sgaoun, Algeria (Brun 1986: fig. 48)	94
43	Examples of the T-shaped pier base (T435) (Leveau 1984: fig. 232)	97
44	Four-mortice pier base (T433): Volubilis, Morocco (Photo: Shlomo Grotkirk)	97
45	Bell-shaped weights with horizontal bore (Iron Age T5113): Bet Shemesh Inst. 13 (Grant and Wright 1938: Pl. 21.4)	100
46	Doughnut-shaped weights (Iron Age T5212): Bet Shemesh Inst. 14 (Grant 1934: Pl. 18)	100
47	Weight with hook (T5411): Karkara Inst. 01 (see also Photo 34)	100
48	Reversed-T weight (T5310): Kurqush Inst. 03	101
49	Reversed-T weight with large upper opening (T5323): Khawka	101
50	Pre-industrial lever and weights press (T01114), raising the weight: Bet Guvrin (Photo: American Colony, Jerusalem)	102
51	Model 6A. The grand point press: Chateau de Vinzelles, France (Humbel 1976: Pl. XX)	108
52	Model 6D. The Taissons press: 'Maison Rustique 1749' (Humbel 1976: Pl. IX)	109
53	Model 6E. The fixed-block press: Zürich, Switzerland (Humbel 1976: Pl. XVIII)	109
54	Model 6J. The box press: 'Maison Rustique 1749' (Humbel 1976: Pl. X)	110
55	Model 6K. The perforated weight press: Elche, Spain (Photo Carter Litchfield)	111
56	The Samaria screw weight (T62121): Tiberias, Hot Springs	112
57	The Dinᶜila screw weight (T62211): Shubeika Inst. 03	113
58	The Kasfa screw weight (T62411): Jerusalem Notre Dame Monastery	114
59	The Kasfa screw weight (T62411): Jerusalem Notre Dame Monastery	114
60	The Bet Ha-ᶜEmeq screw weight (T62511): Quṣeir Inst. 04, press C	115
61	The Luvim screw weight (T26611): Safti ᶜAdi	116
62	The Luvim screw weight (T26611): Safti ᶜAdi	116
63	The Miᶜilya screw weight (T6271): Quṣeir Inst. 01, press A	117
64	Quarrying and dressing cylindrical press component (T03): ᶜAmqa Central	118
65	Quarrying and dressing cylindrical press component (T03): ᶜAmqa Central	118
66	Screw of pre-industrial press Model 7A1 (T0131): Bet Ha-ᶜEmeq	123
67	Grooved-pier press (T711): Bet Natif (Photo: Carter Litchfield)	127
68	Grooved-pier press (T711): Bet Natif (Photo: Carter Litchfield)	127
69	A grooved pier, a southern variant (T7121): Rafat	129
70	A grooved pier, a northern variant (T7131): Kefar Naḥum	130
71	Ṣafṣafot, three screw presses and crushing basin—complete installation from north	130
72	Ṣafṣafot, three screw presses and crushing basin—eastern press from west (T7132)	130
73	The ᶜEin Nashut screw press base (T732): Fakhura	131
74	The Ṣippori screw press base (T741): Majarbin (Photo: Adam Zartal)	133
75	The Tabgha screw press base (T73222): Tabgha	134
76	Manot screw press base (T744): Manot Lower (sugar press)	136
77	Manot screw press base (T744): Mana East (attached to winery)	136
78	Manot screw press base (T744): Judeida (attached to winery)	136
79	Ayalon screw mortice, free-standing round press-bed (T8163): Karak Inst. 05	141
80	Ayalon screw mortice, rock-cut square press-bed and connecting channel (T81121): Usha	142
81	Ḥanita screw mortice, free-standing rectangular press-bed (T83112): Habay	145
82	Ḥanita screw mortice, free-standing irregular press-bed (T83110): Musliḥ	145
83	Simple lever and weights press in winery (T85): Kenisa East	146
84	Pre-industrial olive pulp frail: western Galilee 1980	148
85	Winery, four-rectangle plan (T9111), small compartments (T8010), Ayalon screw mortice (T81221): Tel Aviv, Eretz Israel Museum Inst. 02 (Photo: Eitan Ayalon)	149
86	Winery, four-rectangle plan (T9112), compartments (T8012), screw mortice (T81301): Burak Inst. 01	149
87	Winery—one-axis plan with intermediate cavity (T951): Liman West	152
88	Winery—one-axis plan with intermediate cavity (T951): Liman West	152
89	Winery—one-axis plan with intermediate cavity (T951): Yaᶜara West	152

90 Winery—bell-shaped collecting vat, ledge for lid, bore connecting treading floor to vat (T94): Yirka Central 153
91 Pre-industrial lever and drum press—'Casse-Coue' (T0115): Saint Lauren de la Plaine, France (Humbel 1976: Pl. LXIV) 162
92 Pre-industrial oil separator (T0191): Elche, Spain (Photo: Carter Litchfield) 176

List of Figures

Figure

1 Vintage scene: Tomb of Nakht, Thebes (New Kingdom) 58
2 Vintage scene: Attic black-figure amphora: Wurzburg 265 (18-902) 59
3 Lever and weights press, square central vat with radial grooves (T2613121): Gezer Inst. 01 (Macalister 1912: 2, fig. 257) 62
4 Horizontal oil mill (T37) 72
5 Improved lever and weights press, southern type: Maresha Inst. 09 (Bliss and Macalister 1912: fig. 92) 76
6 Improved lever and weights press, northern type: Zabadi Inst. 02; plan and section 78
7 Reconstruction of southern Maresha press and northern Zabadi press—comparison 78
8 Cato's lever and drum press (Drachman 1932: fig. 12) 86
9 Hero of Alexandria's press A: the lever, drum and weight press (Drachman 1932: fig. 20) 87
10 Model 6L: Hero of Alexandria's press B (Drachman 1932: fig. 23) 88
11 Winery from Beḥyo, Syria (Inst. 50) with typical monolithic beam niche with dovetail base (T41410001): note plain rollers (T861) (De Vogue 1865–77: Pl. 118a) 89
12 North African dovetail beam niche (Ts416) 96
13 Pre-industrial lever and weights presses from ᶜAjlun and el-Ṭafile (Dalman 1928–42: figs. 55, 56) 102
14 The Semana weight (T55121) 103
15 Libyan press with perforated piers (Ts424) and Semana weight (T55121)—reconstruction 103
16 Model 6B: the pivoted screw press (Drachman 1932: fig. 14) 108
17 Model 6C: the tied screw lever press (Humbel 1976: fig. 15) 108
18 Model 6F: the raised press-bed press (Humbel 1976: fig. 11b) 109
19 Model 6G: the Languedoc press (D'Alembert and Diderot 1762: Pl. 1) 110
20 Model 6H: the Arginunta press (Paton and Myers 1898: fig. 1) 110
21 Model 6J: box press—variant that lifts stones placed on board (Humbel 1976: fig. 17) 110
22 Model 6M: the Fenis press (Drachman 1932: fig. 13) 111
23 Screw weights from the Pontus (Anderson 1903: 15) 119
24 Model 7A1: single rotating-screw press, handle at lower end of screw 122
25 Model 7A4: single rotating-screw press, handle at upper end of screw 123
26 Model 7B: press with two fixed screws 123
27 Model 7C1: press with two rotating screws, handles at upper end of screws, female threads in lower press frame 123
28 Model 7C1: screw press from Fayum (Billiard 1913: fig. 157) 124
29 Model 7C2: press with two rotating screws, handles at upper end of screws, female threads in upper press frame 124
30 Model 7C3: press with two rotating screws, handles at lower end of screws, female threads in upper press frame 124
31 Model 7C3: Fuller's press—Frescoe from Pompei 05 (Billiard 1913: fig. 156) 124
32 Model 7D1: press with two anchored rotating screws, female threads in pressing board, with frame and with handles at upper end of screws 125
33 Model 7D2: press with two anchored rotating screws, female threads in pressing board, without frame and with handles at upper end of screws 125
34 Model 7D3: press with two anchored rotating screws, female threads in pressing board, without frame and with handles at lower end of screws 125
35 Horizontal wedge press 125
36 Model 7F: screw and weight press 126
37 Model 7G: drum and weight press 126

38 Grooved-pier press (T711): reconstruction 127
39 Cross press (T72)—reconstruction 130
40 Depiction of single fixed-screw press: mosaic, Tyre, Qabr Hiram 140
41 Fixed-screw wine press (Ts81-83)—reconstruction (Frankel, Avitsur and Ayalon 1994: fig. 82) 140
42 The *galeagra*, press frame, two types as described by Hero of Alexandria and reconstructed by Drachman (1932: figs. 18, 19) 148
43 Winery—composite plan (T92): ʿAvedat South-West. A. Treading floor; B. Collecting vats: C. Screw press mortices with connecting channel (T82601); D. Large compartments (T802) 151
44 Winery of Roman Period from Meinarti, Egypt, based on Adams (1966: fig. 2). A. Treading floor; B. Intermediate vat; C. Collecting vat 154
45 Winery from Tipasa, Algeria, based on Gsell (1984: fig. 55). A and B. Treading floors; C. Collecting vat; D. Draining sump 155
46 Triple winery with beam weight: Tiritake Inst. 07, Crimea, based on Gaidukevych (1958: fig. 68). A. Treading floors; B. Intermediate vat: C. Collecting vats; D. Fragment of raised working area; E. Press-bed; F. Press weight 156
47 The Madaure oil separator (T47132), based on Christofle 1930b 175

List of Maps

Map

1 Central collection: simple lever and weights press (Iron Age) (T261) 64
2 Free-standing central vats (Iron Age) (T261-1) 65
3 Crushing basin, sunken socket (T31) 69
4 Crushing basin, raised socket (T32) 70
5 Crushing basins without socket (Ts 34) 71
6 Crushing basins, Mediterranean Basin (Ts3) 73
7 Plain piers 77
8 Slotted piers with hole and notch 80
9 Slotted piers with upper hole and notch 80
10 Slotted piers with insertion channel 80
11 Slotted piers with hole and angular channel 80
12 Slotted piers (T421) 81
13 Perforated piers 82
14 Central collection in improved lever presses 85
15 Simple beam weights (Ts51, 52) 99
16 Weights with reversed-T bore (T53) 102
17 Semana weight—main sub-types (T55) 104
18 Developed beam weights—main types: Mediterranean Basin 106
19 Samaria screw weight—socket and external mortices (T621) 112
20 Dinᶜila screw weight—socket and open dovetail channel (T622) 113
21 Kasfa screw weight—socket and internal mortices (T624) 115
22 Bet Ha-ᶜEmeq screw weight—no socket, central dovetail mortice (T625) 116
23 Luvim screw weight—no socket, square mortice (T627); Miᶜilya weight—closed dovetail channel (T627) 117
24 Samaria screw weights (Ts621): Mediterranean Basin 120
25 Kasfa, Sarepta and Ponteves screw weights (Ts624) 121
26 Grooved-pier press (T71) 128
27 Grooved piers without mortices (T4618) 129
28 Cross-screw press (T72) 131
29 Rectangular screw press base, open mortices (T731) 132
30 Ṣippori screw press base, closed mortices (T741) 133
31 ᶜEin Nashut, Tabgha and Veradim screw press bases, rounded corners, open mortices (T732) 134
32 Mishkena screw press (T742); Rama screw press (T743) 135
33 Manot press (T744) 136
34 Ayalon press, square mortice for single fixed-screw press: shape of press-bed (T81) 141
35 Ayalon press: shape of mortice (T81) 142
36 Screw press mortices of single fixed-screw press with channels (Ts81-1, 81-3, 831121) 143
37 Ḥanita screw press, central dovetail mortice for single fixed-screw press: construction and shape of press-bed (T831) 144
38 Crushing rollers in wineries (T86) 147
39 Wineries in 'four-rectangle plan' (Ts911) 150
40 Various types of winery (Ts9) 151
41 Pre-industrial presses (Ts01) 160
42 Closed dovetail mortice (T041) 166
43 Central vats with radial grooves (T042) 168
44 Beam presses—anchoring devices: Mediterranean Basin 170
45 Lever and drum presses 172
46 Screw weights without mortices (Ts629) 174

List of Charts

Chart

1	Area of collecting vat in wineries from sample square 16/26	53
2	Area of treading floor in wineries from sample square 16/26	53
3	Shape of treading floor and collecting vat in wineries from sample square 16/26	53
4	Quantities of rainfall in summer months at places in the Mediterranean (millimetres)	60
5	Wineries—Dimensions	181
6	Beam length and mechanical advantage of beam presses	184
7	Terms used for installations in Talmudic literature	197

Introduction

This book is concerned with the installations for wine and oil production used in antiquity in Israel and other lands of the Mediterranean region, their character, technology and history. The main theme of this study is the regional diversity of these installations, an aspect of the subject that has been noted in the past but has until now not been examined in depth. A detailed analysis will be presented showing both the differences between installations found in the various regions of Israel, and the differences between these and those from other countries. This analysis is also the basis for wider cultural and historical conclusions. The data used derive largely from archaeological surveys and this book is therefore also a case study in the use of such surveys in spatial analysis in general and in the study of regional diversity in particular.

The study of agricultural installations has not been in the mainstream of modern archaeological research although the technology and history of these devices have been described and discussed from earliest times, and have also benefited from some important contributions in recent years (Drachman 1932; Camps-Fabrer 1953; Rossiter 1978; Mazor 1981; Callot 1984; Brun 1986; Mattingly 1988a).[1] One of the reasons that agricultural installations found in excavations have usually received little attention is because archaeologists from the beginning of this century onwards have been concerned mainly with trying to construct chronological sequences of facets of material culture, artefacts, types of buildings, etc. based on stratified assemblages. However, while many other types of artefacts such as pottery or objects of artistic significance can, in this way, be dated stylistically to within half a century, the differences between agricultural installations of different periods are rarely marked. Even after new techniques were introduced, the old methods continued in use together with the new. This is clearly demonstrated by the fact that almost all known ancient techniques survived and have been recorded in pre-industrial installations still in use in recent times. As a result it is not possible to construct a clear chronological sequence of types of installations. However, although the chronological diversity of these installations is often negligible, regional diversity is frequently great. Techniques continued to be practised and types of devices used for long periods in one region while completely different methods were found at the same time in other regions, often very close by. These regional differences can often be traced for thousands of years. Drawing up distribution maps of the various types of installation has made it possible to understand the complicated dynamics of development and diffusion of the different types of installation and techniques. It has enabled us to define seperate regional cultures, often to connect these to ethnic and/or political entities, and to trace them from period to period. It has also been possible to follow the development of specific devices over long periods, and to understand the manner in which devices diffused from region to region and thus how new techniques were absorbed into existing technical cultures.

There are three aspects of this study that are fundamental to the research strategies followed but are to some degree unusual in modern scholarship and therefore demand explanation. The first is the wide chronological range examined, from the beginnings of history or even before until modern times. This was deemed necessary because certain types of installations were in use for very long periods and the processes of development and diffusion of the various techniques could only be fully understood when viewed in a wide chronological perspective. The second unusual aspect is the wide geographical area covered. Although stress was laid on Israel and its immediate environs, the study includes the whole of the Mediterranean Basin and the surrounding countries. This is for two reasons. First, in general, it is only possible to assess the significance of the culture of a particular region by comparing it to those of other areas and defining in which ways it differs from neighbouring cultures and in which ways it is similar to them. Secondly, many of the installations under discussion are found in various parts of the Mediterranean world and one of the important questions that must be tackled is their region of origin and the routes by which they were diffused. The third unusual aspect of this study is the fact that the catalogue of installations on which the research is based is not limited to those from clearly dated stratified contexts but includes many undated finds from surveys. This was considered essential because the dated installations, while being of utmost importance, were in most cases insufficient in number to allow us to reach coherent conclusions while the remarkable regional diversity made it possible to map clearly defined distribution patterns even when undated items were included.

The study includes both wine and oil because the installations used to produce these two substances are very similar and it is therefore often difficult to distinguish them. Similarly, identical or similar terms are often used for these installations in ancient literature. Thus separating the two subjects was fraught with difficulties and the study therefore discusses both.

The research here presented draws on four main types of sources: first and foremost the archaeological finds; secondly, pictorial depictions such as wall paintings from places as far apart geographically and chronologically as Pharaonic Egypt and Roman Pompeii, classical Athenian vase paintings, mosaics, reliefs from Roman and Byzantine periods and mediaeval illustrated manuscripts and altarpieces; thirdly, pre-industrial installations that were in use until recent times, both the actual installations and ethnological descriptions of how these were operated; and fourthly, written sources, primarily Hebrew and Aramaic biblical and Talmudic texts, but also classical texts: mainly Latin but also Greek. The main aim has been to use the latter three sources in order to understand the technology and history of the first, the archaeological evidence. The wooden parts of the ancient devices rarely survive and it is only with the help of the pre-industrial installations and the pictorial depictions that attempt can be made to reconstruct them in their original form while the ancient written sources provide invaluable historical, cultural and chronological data. These four sources are, however, complementary and each aids in understanding the full significance of the other three. For example, the beginning of modern research in the subject in the eighteenth century centred on trying to reconstruct Cato's famous press in the light of contemporary presses that today we would call pre-industrial (Meister 1763).

As the study is primarily a geographical and typological analysis of installations, the data are presented in a catalogue on a computer disc in which the sites are numbered according to their geographical position and the installations are classified according to a detailed numbered decimal typology that distinguishes not only between different types of complete installations but also between all the various types of their component parts, weights, piers, press-beds, vats etc. The catalogue includes not only the archaeological finds but also the pictorial depictions and the pre-industrial installations mentioned above, the latter two being grouped together in Type 0.

However, the catalogue includes only those installations the descriptions of which allow for clear typological definition. Thus the vast majority of installations for which only a verbal description is available, such as those appearing in the classical surveys, the Survey of Western Palestine (Conder and Kitchener 1881–83) and that of Guerin (1868–80) had to be omitted.

In the catalogue on the computer disc there are three interrelated lists:

List A: Sites and Installations

In this list each site appears with its installations, each of these designated by type number, a short description and bibliography. The sites are arranged according to site number in two corpuses: the first, those from Israel and environs; and the second, those from other countries. In the one from Israel and environs the site number is based on the Israel–Palestine Grid and the corpus includes almost all available data, not only published reports and information from the archives of the Mandatory Department of Antiquities of Palestine (ADAP) and those of the Israel Archaeological Survey (IAS) but also from many other unrecorded sites visited by the author.

List A does not include all the examples of the very simple installations classified as Type 1 (see Chapter 1). A complete list of these, however, from a sample area of 100 sq km, Map 16/26 Israel Grid, has been presented in a chart at the end of List A.

In the second corpus, the one of sites and installations from other countries, the site number is based on country and longitude and latitude (the countries are designated and divided as they were before the division of the USSR, Yugoslavia and Czechoslovakia into smaller independant states, and each island is regarded as a separate country). As regards these regions, the corpus is of published sites only, is clearly far from including all of these and is partly based even on secondary sources.

List B: Installations according to Type

This list is arranged according to type number. The typology is based on a decimal system and at the beginning of each main type the principles on which the division into sub-types is based is explained in a chart. Each type is described and appears with an illustration. As the installations of each type are arranged according to site number, they automatically appear in geographical order which enables easy spatial analysis, the drawing-up of maps, etc.

List C: Alphabetical List of Sites

In this list all the sites appear by name in alphabetical order together with their site numbers. As in List A, this list is divided into the two corpuses: those from Israel and environs and those from other countries. The purpose of this list is to facilitate finding, in List A and thus also in List B, sites the names of which are known (in the text of the book site numbers were omitted). The bibliography of the catalogue is combined with that of the book.

The discussion in the book itself is not according to chronological periods but according to the types of installations. The types are arranged approximately in the order in which they were probably first introduced. The book opens with four preambles that present briefly some basic background subjects: (A) the ecology and the beginnings of the cultivation of the olive and the vine (it should be stressed that while the olive is indigenous to the Mediterranean forest, the vine apparently originated in northern Anatolia and was introduced to the southern Levant probably in the Early Bronze Age); (B) the importance of the olive and the vine in light of current research; (C) the basic techniques of wine and oil production; and (D) the history of research of the installations. Each of the following chapters is devoted to one type and numbered accordingly.

Chapter 1: Simple Installations (T1)[2]

The most important of these is the simple treading installation or winery consisting of treading floor and collecting vat. This was the *gt* or *yqb* of the Bible, terms usually translated somewhat misleadingly as 'wine press'. Rock-cut examples of these are found in very large numbers throughout the region, usually outside settlement sites. This shows that they were normally placed in the vineyard, and the written sources show that the first fermentation took place in the vat. This type of winery is rare in Europe, probably because of the difference in climate. In Europe rain often falls in the vintage season. As a result, in Europe fermentation was usually carried out in ceramic vessels, the *pithoi* or *dolia* of the classical periods.

Chapter 2: The Simple Lever and Weights Press (T2)

This press is first found in Israel in the Iron Age. By Iron Age II a distinction can already be made between a southern type with central collection, the Bet Mirsham press, and a northern type with lateral collection, the Rosh Zayit press. The lever and weights press is already found in Syria and Cyprus in the Late Bronze Age. It is not completely clear whether the lever and weights press was in use in the Aegean in the Bronze Age.

Chapter 3: Olive-Crushing Devices (T3)

In the Iron Age olives were crushed by rollers or in mortars. By the Hellenistic period, however, the rotary olive crusher (T3) was almost universal. The regional variation of this type is considerable. The date and place of origin is, however, still not known. An unusual type of olive crusher is the horizontal oil mill (T37) found in North Africa and Spain. It is very similar to the rotary donkey mill used for grinding grain and is almost certainly the *mola olearius* referred to by Collumella.

Chapter 4: The Improved Lever Press (T4)

In Israel in the Southern Maresha press the collection is central as in the previous period, the additional element being plain piers placed on either side of the central collecting vat. In the north, in the Zabadi press, lateral collection continues and the main innovation is the appearance of slotted piers at the fulcrum. This is the Mishnaic Press (*B. Bat.* 4.5) and the slotted piers are the *btwlwt* of the Mishnah and the *arbores* of Cato's press, in the latter case of wood. Wooden pre-industrial examples are common in western Europe.

Several classical writers describe or refer to oil or wine presses. Cato the Elder (early second century BCE) describes an oil and a wine press that are both lever and drum presses with slotted piers. Vitruvius (first century BCE?) refers apparently to both a lever and drum press and a direct-pressure screw press. Pliny the Elder (first century CE) gives an interesting survey of the history of the press, referring to a lever and drum, a lever and screw and a direct-pressure screw press. Hero of Alexandria (first century CE) gives instructions for constructing four types of press: lever, weight and drum; lever and screw; double-screw direct-pressure press and single-screw direct-pressure press.

Different types of press are found in different parts of the Mediterranean Basin, the main distinguishing feature being the form of the fulcrum. In North Syria the characteristic press has a large impressive beam niche with a dovetail base and grooves (Ts413–415), in Cyprus a unique type of slotted monolith appears (T427) and in North Africa regional differences conform almost exactly to the Roman provinces but probably precede them chronologically. Perforated piers (T424) are found in the region of the Roman Province of Africa, former Carthage, the dovetail beam-niche (T416) in Numidia, the 'T-shaped pier mortice' (T435) in Caesarean Mauritania and the 'four-mortice pier base' (T433) in Tinjanian Mauritania. In Italy, Spain, southern France and Yugoslavia the Tivoli pier press (T4311) is found which clearly secured wooden slotted piers as described by Cato and similar to the stone examples found in the Phoenician Zabadi press. The characteristic of the Greek press was the beautifully shaped pear-shaped press-bed (T463). In most countries these presses were lever and weights presses, but in Italy and adjacent countries (Yugoslavia and parts of Provence) lever and drum presses are found as described both by Cato and Pliny. In Provence, however, there are also lever and weights presses that pre-date the Roman occupation and evince Aegean influence.

Chapter 5: Beam Weights (T5)

The simple beam weight with one bore is found in many countries. In Israel, in the Iron Age regional variation can be discerned in the types of simple weights used: in the north naturally perforated field stones; in the south fashioned weights; and in the southern mountains round 'doughnut-shaped weights'. By the Hellenistic period two main types of developed weights had evolved, that with reversed-T bore (T53) is found in the southern Levant and Cyprus. The other type, the Semana weight (Ts55, 56), was rectangular in shape with mortices on the short sides, and usually with a groove on the upper surface joining them. It is found in North Africa, Spain, southern France, Cyprus and in the Aegean area, where it apparently originated, developing perhaps from the Methana weight with two vertical bores (T5712) found in that region in even earlier periods. One of the main differences between these two types is that the weight with reversed-T bore was raised using a drum attached to the beam while the Semana weight was raised with the aid of a drum attached to the weight. The Methana weight probably operated in a similar manner. In North Syria a weight was found, the Taqle weight (Ts535-536), that apparently combined the characteristics of the reversed-T weight and the Semana weight.

Chapter 6: The Lever and Screw Press (T6)

This press is described both by Pliny and Hero and there are many different types of pre-industrial presses that work on this principle (Models 6A–6M) The main archaeological evidence for this press is the screw weight (Ts62) that appears in a great variety of forms. In most cases there was a socket in the middle of the top of the weight in which the screw rotated. However, those weights without sockets and with central mortices (Ts625, 626, 627) that were used in presses in which the nut and not the screw was turned are found only in Israel and environs.

Chapter 7: The Direct-Pressure Rigid-Frame Press: Single-Screw Presses, Double-Screw Presses and Wedge Presses (T7)

There is evidence, both pre-industrial and pictorial, for a great variety of rigid-frame presses (Models 7A–7G). From the archaeological remains, however, it is in most cases not possible to distinguish the exact original form of the press so that the archaealogical remains have here been classified according to other criteria. In Israel and environs these presses are found mainly in those regions where screw weights are not found. The grooved-pier press (T71) is the main type in Judaea and probably developed from the southern lever and weights press, the Maresha press. Other types are found on the Carmel, in Lower and eastern Upper Galilee and on the Golan. Few examples appear in other countries, although Vitruvius refers to, and Pliny and Hero describe, presses of these types and pre-industrial examples are common. The presses of this type in other countries were probably usually completely of wood and therefore left no remains.

Chapter 8: Improved Wineries—Ancillary Installations (T8)

There is a variety of compartments round the pressing floor but the main device is the single-fixed screw wine press (Ts81–83). This press, which consists of a mortice in the centre of the treading floors of wineries, is an additional type that appears only in the southern Levant, features in mosaics in this region only (in Jordan and Lebanon) and is apparently referred to in the Jerusalem Talmud but not in classical literature. This type of press was particularly suited to the large wineries characteristic of the region. It appears in two main sub-types: the southern Ayalon press (T81) with square mortice, and the northern Ḥanita press with central dovetail mortice (T831).

Chapter 9: Plans of Improved Wineries (T9)

The three main components of wineries—the treading floor, the collecting vat and the intermediary vat (which was sometimes a settling tank and sometimes a container for sieving material)—and other components, compartments, presses, etc. were arranged in different characteristic plans. In Israel three main plans were discerned: 'the four-rectangle plan' (T91) in the centre of the country; 'the composite plan' (T92) in the Negev Highlands; and 'the one axis plan' (Ts95–96) in Galilee. In Egypt a winery with a raised collecting vat appears in Byzantine sites and pictorial depictions from the Pharaonic periods and the Hellenistic period suggest that it existed also in much earlier periods. In the Crimea characteristic local types were also recorded.

Chapter 10: Pre-Industrial Installations (T10)

This chapter opens with a survey of the uses of presses for purposes other than wine and oil: clothes presses, cider presses, cheese presses and printing presses. The distribution patterns of pre-industrial presses are then examined. There are marked regional differences in pre-industrial installation which are in almost all cases a continuation of the situation in late antiquity. Striking examples are the survival of a press in North Africa with a beam weight similar to the ancient Semana weight, the survival of lever and drum presses in France and Switzerland similar to Cato's press and the survival of central oil collection, all three types of beam weights and the most common of the ancient screw weights, the Samaria weight (T6211), in pre-industrial presses in Israel and its environs.

Chapter 11: Conclusions

In this chapter three aspects of the subject are presented.

(A) *Israel and Environs: A Regional Survey.* Two integrated regional cultures are defined which can be followed at least from the Iron Age until the Byzantine period and to some extent even later: that of Judaea in the south characterized by central collection and square mortices; and that of Phoenicia in the north-west characterized by lateral collection and dovetail mortices. In each case, when the screw is introduced it is adapted to the press found in the region in the previous period so that in Judaea the typical screw press is a direct-screw press (T71) and the lever and screw press does not appear, while in western Galilee (Phoenicia) the reverse situation is found. The typical press is a lever and screw press and the direct-pressure frame screw press does not appear. The technical cultures of other regions—Samaria, Lower Galilee, Golan and Jerusalem—are then related to these two main technical cultures.

(B) *Oil and Wine Presses in the Mediterranean Countries.* The distribution and diversity of the different types of devices are examined in turn throughout the region showing the differences between the various areas and the relationships between them. A subject examined in detail is the relationship between the various devices in which the beam was anchored, and it is suggested that the slotted piers originated in the Levant. It is also suggested that the Roman lever and drum press, Cato's press, probably developed from the press that used the Semana weight, but in this case it is not clear where this development took place. The relationship between the various types of screw weights and the different types of the single fixed-screw wine press are examined and separate and parallel lines of development suggested at least one in the west and one in the Levant. Different methods of oil separation are noted. The simplest method, skimming, by which the floating oil was skimmed from the lees, was probably universal but other more sophisticated methods are shown to vary in different areas. Overflow decantation was usual in the Levant and underflow decantation in the Aegean, while in North Africa and the western Mediterranean there is archaeological evidence for a sophisticated technique combining both methods. Pre-industrial examples of this method are found in the same region.

(C) *Spatial Patterns and Chronological Sequences.* In this section an attempt is made to explain why regional diversity is more marked and persistent in agricultural industries than in other facets of human material culture. It is suggested that the fact that these installations were large, often of stone and almost unbreakable, the fact that they were rarely transported far, and the way of life of the people who made and used them, were all factors that led to one type being used for long periods and only rarely being affected by methods used in other regions. The relationship of these local technical cultures to ethnic identity is also examined showing that, although they often coincided geographically with ethnic units, they were rarely regarded as culturally identifying attributes by the people who used them or by their neighbours.

Different types of technical cultures are defined examined and explained. Integrated cultures such as those of Judaea and Phoenicia are compared to the eclectic culture of the Jerusalem area. Different types of distribution patterns are also noted, such as central sophistication and peripheral diversification. The possibilities of distinguishing the spatial patterns of slow and fast development are examined on the basis of the distribution of various types of press piers and screw weights in western Galilee. Finally, a summary is presented of the conclusions reached as regards the origins of some of the main types of installations, the lever and weights press, the Semana weight, the lever and drum press, the rotary olive crusher, the screw and the various types of screw weights.

The works of the Greek and Roman engineers and agronomists, such as those of Cato, Vitruvius, Pliny the Elder, Hero of Alexandria, Varro, Columella, Palladius and the Geoponica, are referred to and discussed in the relevant sections in the book. Where relevant, the biblical and Talmudic sources are also referred to in the text, but as the book centres on Israel it was felt to be in place to discuss these sources in greater detail and this discussion, because of its length, was relegated to two appendices.

Appendix 1: Wine and Oil Production in Ancient Hebrew Literature

Terms for installations and for types of oil used in the Hebrew Bible are examined. In Talmudic literature terms used for installations are also examined and the texts discussing the oil for the temple offering are discussed in detail. It was shown that a variety of terms was used for same parts of presses and other devices. It is difficult, however, to ascertain if these semantic differences imply differences in the form of the device concerned or whether they show regional linguistic diversity.

Appendix 2: Terms Used for Wine in Ancient Hebrew Literature

Types of wine mentioned in the Hebrew Bible and related epigraphic texts are discussed and types of wine mentioned in Talmudic literature are enumerated according to wines named by age, colour, quality (taste, aroma), provenance, production process, names deriving from Greek and Latin, and other names. Jar inscriptions and other epigraphic evidence from the Roman period are then summarized. Differences and similarities to wines in classical literature are noted.

Notes

1. Before translation and revision were complete, two doctoral dissertations on similar subjects were published, that of Callot (1984) on oil presses in North Syria and that of Brun (1986) on oil presses in the Var district of Provence, France. These not only provided much new data on presses in these regions, but the latter especially opened up a large bibliography not previously known to me. In my dissertation there was only a catalogue of sites and installations in Israel and environs, and only these were given type numbers. Resulting from the additional data these publications provided, in this book an additional catalogue of presses from other countries has been added. *Giving these installations type numbers has made it necessary to revise the numbering system slightly.* Therefore readers should note that, where scholars have in the past referred to type numbers based on the dissertation, these will not be identical to those in the catalogue of the book.

2. T is used for 'type number', and Ts for 'type numbers'. This distinguishes them from site numbers.

Conventions Used for Data on the Disc

Site Name

All sites except for one group have been given names—where necessary by the addition of a point of the compass to the name of a nearby site or geographical feature.

The exceptional sites which have not been given names are the installations of Type 1 which have only been presented in full from the sample 100 sq km area of Map 16/26. These appear, with shortened site number only, in the chart at the end of List A on the disk.

In the first list (Israel–Palestine Grid) attempt has been made to reach a uniform English transliteration of Hebrew and Arabic names, that used in the Schedule of Monuments. In the case of the latter, the definite article has been always given as *el*. Terms such as *Ḥ.*, *Kh.* and *el* have been placed after the name to make the alphabetical order more logical. Occasionally two names have been given for a site; the second is usually the historical one. In the second list (other countries) the name has been given as in the source.

Site Number

For sites in the first list, the method of numbering is that used by the Israel Archaeological Survey with the addition of three noughts at the beginning.

The first four digits after the noughts designate the 10 × 10 km square (which is also the area of the 1:20,000 maps) and the next two digits the square km in which the site is found. The final digits are an arbitrary survey site number of sites within each square km.

Structure of Site Number of Sites in the Israel–Palestine Grid Area

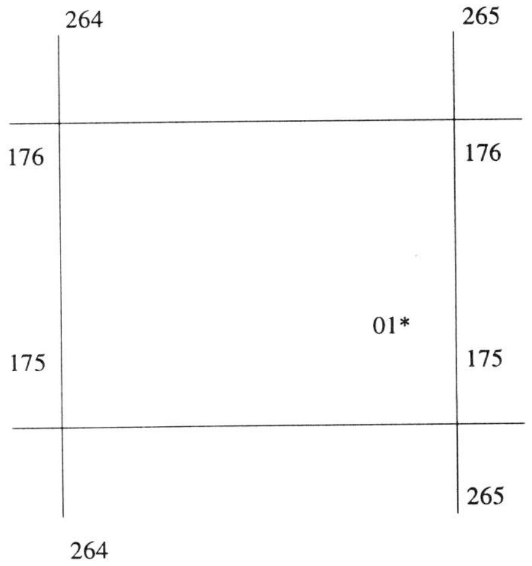

Map Reference: (MR):**26**_4_9.**17**_5_1
Site Number: 000-**2617**-_45_-001
(site 1 in square km)

As only the Israel Survey can give a site number, in site numbers numbered here the last three digits are in bold print.

In the site numbers of sites from the sample square that appear in the chart after List A on the disc the first four digits 1626 denoting the 10 × 10 km_a square, the map, have been omitted as they are all from the same map.

For sites in the second list (other countries) the first two digits designate the countries according to present political borders (1989), except islands, each of which receives a separate number code. If the third digit is 1, the site is west of Greenwich; if 0, it is east of Greenwich. A '9' designates data for which geographical position is not available or not relevant. The fourth and fifth digit represent degrees of longitude and the sixth and seventh degrees of latitude. The final digits are again an arbitrary site number of sites within the square of one degree of latitude and longitude—a 'degree square'. As far as possible, the site numbers (the final digits) and installation numbers within a site are those given by the excavators or surveyors.

Structure of Site Number of Sites outside the Israel–Palestine Grid Area

			14	
01-	1-	05	34-	00-001
Morocco	West of Greenwich	°West	°North	Site number in °square

The number codes used for the countries are:

–1. Morocco	–2. Algeria	–3. Tunis
–4. Libya	–5. Egypt	–6. Sudan
–7. Ethiopia	–8. Lebanon	–9. Syria
10. Turkey	11. Iran/Persia	12. Iraq
13. Afganistan	14. Russia	15. Bulgaria
16. Rumania	17. Czechoslovakia	18. Greece
19. Albania	20. Yugoslavia	21. Austria
22. Switzerland	23. Germany	24. Belgium
25. Holland	26. Luxemburg	27. Italy
28. France	29. Spain	30. Portugal
31. Hungary	32. Great Britain	33.
34.	35.	36.

37.		38.		39.	
40.		41.		42.	
43.		44.		45.	
46.		47.		48.	
49.		50.	Cyprus	51.	Crete
52.	Rhodes	53.	Sardinia	54.	Corsica
55.	Sicily	56.	Malta	57.	Manorca
58.	Majorca	59.	Ibiza	60.	Formantara
61.	Karpathos	62.	Casos	63.	Nitros
64.	Andros	65.	Tilos	66.	Samos
67.	Leros	68.	Khios/Chios	69.	Lesbos
70.	Skiros	71.	Limnos	72.	Jasos
73.	Naxos	74.	Milos	75.	Kalimnos
76.	Amorgos	77.	Delos/Dhilos	78.	Myconos/Mikonos
79.	Corfu/Kerkira	80.	Guernsey	81.	
82.		83.		84.	
85.		86.		87.	
88.		89.		90.	
91.		92.		93.	
94.		95.		96.	
97.		98.		99.	

Locating sites in other countries raised considerable difficulties. In many cases, the places were not to be found on the maps available and in others there was more than one place with the same name. In all cases where the location could be ascertained to the accuracy of a degree (i.e. the possible mistake being not greater than one degree of latitude and/or longitude), a complete geographical site number was given; in the few cases where this is not so, only the country code was used so that the unlocated sites are at the begining of each country in the list of sites.

The Maps

The site numbers are structured to enable mapping to be based directly on them; the maps are drawn directly from List B.

The site numbers include sufficient information to map sites in the Israel–Palestine Grid to the nearest km and other sites to nearest degree. The two standard maps used include 10 km and 10° grids respectively, which allow for identification of sites without great difficulty. The maps of the Mediterranean Basin are only sufficiently detailed to recognize large-scale special characteristics. However, only in a few cases (e.g. southern France, North Africa) is the data sufficient for more detailed mapping to be of value and in most cases the information did not allow for greater geographical accuracy in site numbers and maps.

Installation Type Numbers

The type numbers are based on a hierarchical system similar to the Dewey book classification system. The detailed explanations for numbering of each type is incorporated in the list of installations according to type, List B, but a short explanation of the ten main divisions follows.

1		Simple installations
2		Simple beam and weights presses
	20	Press as a whole
	21–29	Press components
3		Olive-crushing devices
4		Complex beam presses
	40	Press as a whole
	41–49	Press components
5		Beam weights
	62	Screw weights
7		Frame presses (wedge or screw)
8		Component parts of wineries
9		Wineries as a whole (architectural plan)
01		Pre-industrial examples and artistic representation of presses and crushing devices
02		Pre-industrial examples and artistic representation of wineries and wine making
03		Installations in stages of production
04		Characteristics common to several types of installation
09		Unclassified installations

Each additional digit usually represents a specific characteristic e.g. shape, type of mortice etc. and therefore if only some of the characteristics are referred to hyphens will appear at the other digits. In those cases where the relevant definitive digits are all at the beginning of the type number the hyphens at the end of the number are dispensed with.

An attempt has been made to base the numbering system on typology only and to distinguish oil from wine production. However, it was found impracticable to conform to these principles completely, and the reader is alerted to the main deviations from these guiding principles. The distinction between Types 2 and 4 is, in effect, chronological. Type 2 refers to lever and weights presses dating from the Iron Age or earlier (earlier than 600 BCE) and Type 4 to those from later periods. Types 01 and 02 include all installations that are complete and in which the wooden and other components of perishable materials appear, thus allowing for a typology based on different principles from that used for the archaeological remains. Types 01 and 02 include not only pre-industrial installations but also ancient representations—reliefs, mosaics, tapestries, paintings, manuscript and book illustrations, and also the rare ancient press that has survived complete. Weights from pre-industrial installations, however, that can be treated similarly to other weights have been included in the general categories of Types 5 and 6. Types 8 and 9 consist only of installations that are in our opinion for wine production; however, Types 4, 5, 6 and 7 include presses that were definitely or probably used for wine production, but the similarity in type to oil presses necessitates classifying them together. The same is true of complete presses. Type 02 are wineries but Types 01 include presses used for wine, for oil and also for other purposes—cheese, etc.

Sites

Sites can be located in List A by ascertaining the site number in the site index (List C).

Dating

In Middle-Eastern dating, the system used in the *Encyclopaedia of Excavations* (Avi-Yonah and Stern 1978) has been followed, except for the Iron Age. Albright's simpler method has been retained dividing Iron Age I from II at 925 BCE. As for other regions, the local terms have been retained except that the West European Iron Age is termed Pre-Roman.

Disc Format

The CDRom contains lists A, B and C in PDF format and can be accessed on a PC or Macintosh using Adobe Acrobat Reader.

Abbreviations

ADAP	Archives of the Department of Antiquities of Palestine (Mandate)—housed in the 'Rockefeller Museum' Jerusalem
AIDA	Archives of the Israel Department of Antiquites
AQM	Archives of the Qaserin Museum
BIBL	Bibliography
BL	Beam Length
Ca	Circa
Byz	Byzantine
CB	Crushing Basin
CHALC	Chalcolithic Period
CS	Crushing Surface
CV	Collecting Vat
D	Depth
DIAM	Diameter
EB	Early Bronze Age
EXT	Exterior
F	Fulcrum
E	East
F	Force
H	Height
H	Horvat
HEL	Hellenistic Period
IAS	Israel Archaeological Survey
INST	Installation
INT	Interior
IV	Intermediate Vat
Kh	Khirbat
L	Length - Late - Locus
LB	Late Bronze Age
M	Middle
MA	Mechanical Advantage - BL : PAF - F
MB	Middle Bronze Age
MR	Map Reference - Israel (Palestine) Grid
N	North
PAF-F	Distance between point of applied force and Fulcrum
Pers Com	Personal Communication
Pers	Persian Period
PB	Press Bed
Rom	Roman Period
S	South
SM	Screw Mortices (unless otherwise stated, diameter or width of opening of mortice)
SW	Screw Weight
SV	Separating Vat
T	Type/Tel/Tell
TF	Treading Floor
W	West - Width

Preamble A

Grape and Wine Ecology: Origins and Methods of Cultivation

1. The Vine

The grape-vine grows best in regions with warm-to-hot and dry summers and cool winters but cannot survive temperatures below 0°F (-18°C). The wild grape (*Vitis Vinifera Silvestris*) is found today in southern Europe in the region between the Caspian and Black Seas, the Danube Basin, mainland Greece, northern Italy, southern France and parts of Spain, and also in the coastal regions of North Africa from Morroco to Tunis and of southern Anatolia (see map in Zohary and Spiegel-Roy 1975: 322). Although in some cases, as with the olive, it is difficult to distinguish between escapees of the cultivated grape (*Vitis Vinefera L*) and the wild variety, there is no doubt that the wild grape was indigenous to the Caspian region and the Balkans. It is, however, not found at all in the southern Levant or in eastern North Africa, Libya and Egypt.

The earliest evidence for the presence of the vine in the region are grape pips found in several Early Bronze I sites (Zohary and Spiegel-Roy 1975: 321; Stager 1985: 188, fig. 1). It was in this period also that we have the first written evidence for vine growing and wine producing in Egypt (Stager 1985: 174). The vine is indigenous to northern Anatolia and it was almost certainly introduced from that region, a region from which several other cultural influences are felt at this period, for example grey burnished ware (Hennesey 1967: 35-36; Amiran 1970: 47-49). The similar names for wine in Ugaritic (*yn*), Hebrew (*yyn*), Greek (*[w]oinos*), and Cypro-Syllabic (*wo-i-no*) almost certainly derived from the Hittite *wiyanas*, making the probability that the vine and wine of the Levant and Greece originated from this region even greater (Brown 1969: 147-51).

The classical sources describe in great detail the different ways in which the vine was grown and show these to be very similar to those practised today. The Hebrew and Aramaic texts reflect a similar picture. Pliny (17.35.164-166) describes five methods of growing the vine:

1. allowing it to trail on the ground;
2. growing as a small bush that supports itself;
3. being trained on vertical posts (*adminiculum*);
4. on simple horizontal 'yokes' (*jugum*) clearly in rows; and
5. trained to grow overhead 'on four bars in a rectangle' (*compluvium*).

To these should be added the method of training the vines on trees known as an *arbustrum* and specially planted for the purpose (Columella 5.6; Varro 1.8.3; see White 1975: 19-23 for full discussion of methods and variations in terminolgy).

The *gfn srḥt* of Ezek. 17.6 is clearly the vine trailing on the ground, Pliny's first type, and the *gfn ʾdrt* of 17.8 probably one trained on a fence. In the Mishnah (*Men.* 8.5) it is stipulated that wine for the temple offerings should be taken from the *rglywt*, clearly the later Hebrew term for vines on the ground, and not from the *dlywt*, those trained on a fence. In the parallel text in the Tosefta (*Men.* 9.10) the term *krmt* is added, apparently another type of trellis and in other texts the term *ʿrys* is used (e.g. *Kil.* 6.6).

Yields in vineyards using traditional methods in the Hebron region were:

1. vines trailing on the ground (Pliny's first method), 3–6 metric tonnes per hectare;
2. those trained on trellises in rows (Pliny's fourth method), 15 tonnes per hectare; and
3. those trained overhead (Pliny's fifth method), 30 tonnes per hectare (Ben Yaakob 1978).

Pliny (17.35.166) also tells us that vines trained to grow overhead give the highest yields, but he and other ancient writers, while also specifying very similar yields and similar variations in yield to those mentioned above, do not state which type of vineyard they are referring to. Varro (1.2.7), Pliny (14.5.14) and Columella (3.3.2; 3.9.3), all quoting Cato directly or indirectly, speak of exceptional yields of 10–15 *cullei per iugerum*, approximately the equivalent of 20,000–30,000 litres per hectare (c. 25–37 tonnes). Both Columella and Pliny (14.5.14), however, regard the yields of 7 (8) *cullei per iugerum* of Seneca's vineyard (15,000 litre/17.5 tonnes per hectare) as exceptional, while Columella (3.3.33) mentions his own vineyards which gave 7 and 5 *cullei*, but recomends that a vineyard giving less than 3 *cullei* (6000 litres/7.5 tonnes per hectare) should be uprooted (3.3.11).

The vintage season varies according to the variety of grape and even more so according to the region. In Israel today vintage commences in the hotter regions in

June and in colder areas, such as the mountains of Upper Galilee, continues until October. There are hints of a vintage festival in the Bible (Judg. 9.27; 21.20-21) and in the Mishnah (*Ta'an.* 4.8) there is a description of how on the 15th of Ab 'the daughters of Jerusalem went forth to dance in the vineyards'. In the *Temple Scroll* (11QT; Yadin 1977) 19.11–21.10 there is a description of the Feast of the First Fruits of Wine, which was to be celebrated fifty days after the Feast of the First Fruits of Grain, a festival that clearly corresponds to biblical Pentecost. Reeves (1992) has shown that this festival of the Dead Sea Sect is closely connected to two descriptions very similar to each other of a vintage festival celebrated by Noah, one in the book of Jubilees and the other in the so-called *Genesis Apocrypha* (1QapGen), both texts also associated with the sect. Furthermore, he suggests that this festival derives from one described in Ugaritic tablets dating much earlier, to the Late Bronze Age, which was to be celebrated 'in the month "First of the Wine" on the day of the new moon' which Reeves shows to have been a vintage festival celebrated at the new year in the autumn.

2. The Olive

The Olive demands warmer conditions than the grapevine. Although the adult tree can usually withstand temperatures of 12°F (-11°C) a young tree will die at 20°F (-7°C). Thus in southern France, for example, olive groves were destroyed by severe frosts four times in the century ending in 1985 (1885, 1929, 1956, 1965). On the other hand, the olive needs between 600 and 2400 hours a year below 45°F (7°C) in order to give fruit. As a result, the belt in which the wild olive (*Olea Europaea L. var Oleaster*) is found, while to some extent overlapping with the southern margin of that of the wild grapevine, reaches further south and includes the southern Levant, although not the hottest parts of this region (map in Zohary and Spiegel-Roy 1975: 321).

Olive stones have been found in stratified contexts in a Natufian site on the Carmel and in the Naḥal Ḥemer cave, a site that dates to the pre-pottery Neolithic B and also in many sites of the Chalcolithic period (Liphschitz *et al.* 1991: 447; Stager 1985: Table 1). An exceptional discovery was that of a large quantity of olive stones found in a round pit in an underwater site south of Haifa dating to the Pottery Neolithic (c. 5500 BCE; Galili and Sharvit 1994–95; Kislev 1994–95; Carmi and Segal 1994–95).

What has been considered of particular significance is the finding of a large number of olive stones at Teleilat Ghassul, the large Chalcolithic site in the dry and hot Jordan Valley. This has been considered evidence for olives having been cultivated in the region under irrigation (Zohary and Spiegel-Roy 1975: 319) or for their having been cultivated in other regions of the country and brought to the site (Kenyon 1976).

Clearly the olive stones found in archaeological contexts could have been of wild olives collected for food. It was thought that the stones of wild olives could be distinguished from those of the cultivated variety (Hjelmquist 1973: 236-37, 252-55), but several scholars are today of the opinion that it is virtually impossible to distinguish between the wood, stones or pollen of the wild and the cultivated olive (Runnels and Hansen 1986: 301-302; Liphschitz *et al.* 1991). This has led the latter to opine that the only dependable criterion for the cultivation of olives is the percentage of olive wood found in archaeological contexts as compared to that of other types of wood. The evidence shows the percentage of olive wood to have risen sharply during the Early Bronze Age and suggests this as being evidence for the beginning of olive cultivation at this period. In their article Liphschitz *et al.* (1991) bring evidence for a Chalcolithic site in the Golan region, Rasam Harbush, with 27 samples of olive wood. As they regarded it as exceptional, they chose to ignore the site when reaching their final conclusions. The question arises, however, as to whether the site should not be regarded as 'the exception that proves the rule' showing that olives were already cultivated in the Chalcolithic period. Trough-like crushing basins (T136) and pottery oil separators (T182) found at Chalcolithic sites in this region were clearly used to produce olive oil (Epstein 1993). Recent evidence from additional sites in Jordan suggest that the olive was cultivated during the Chalcolithic period in this region also (Neef 1990). Thus, olive cultivation almost certainly started before the Early Bronze Age. This apparently was nevertheless the first period when it reached significant proportions (Stager 1985; Liphschitz *et al.* 1991).

An important question that arises is whether cultivation of the olive originated in one area or whether this step was taken independently in different countries. The identification in Crete of what were considered to be olive stones of forms transitional between those of the wild and the domesticated varieties was seen as evidence for independent domestication of the olive in that island (Hjelmquist 1973: 237). If it is accepted, however, that the stones of the cultivated and the wild olive cannot be distinguished (Runnels and Hansen 1986: 301-302; Liphschitz *et al.* 1991), this evidence can no longer be considered as proof.

It is, however, perhaps relevant that, as opposed to the vine and wine in which case the terms used in the Semitic and the Indo-European languages clearly derive from a common source, in the case of the olive and of oil this is not so. The Semitic root for oil is *šmn* (Ugaritic, Hebrew, *šmm*; Accadian, *šamnu*) and for the olive *zt* (Ugaritic, *Zt*; Hebrew, *zyt*; Accadian, *ze/irtu* [*se/irdu*] probably connected and possibly derived from the West Semitic *Zt/zyt* [Malul 1987]; Egyptian, *ḏt* [Stager 1985: 174]), whereas the Greek for olive is ἐλαία (Latin, *olea*, *oliva*) and for oil ἔλαιον (Latin, *oleum*).

As today, in the past the olive was grafted (Cato 40–43; Columella 5.11). A method still practised in Israel is to use wild olive saplings taken from the forest as root stock and grafting them with scions from a particular tree in the olive groves of the village known to be fruitful. This method is apparently hinted at in the New Testament (Rom. 11.17), although there the wild olive is the scion. In Latin literature this method is first attested to only in the fourth century (Palladius 3.18.6).

The olive lives to a great age. In old trees the centre of the trunk decays and finally becomes hollow, such trees often reaching a girth of over 1.5 m. The exact age of these ancient trees is difficult to determine, however, as the tree rings cannot be counted in the olive. Traditionally in Israel the ancient trees are called 'Roman Olives' and in sixteenth-century Ottoman tax estimates a distinction was made between *zaytun islami*, Islamic olives, and *zaytun rumani*, Roman olives, the latter estimated at half the yield (Hutteroth and Abdulfallah 1977).

The olive gives high and low yields on alternate years, although in exceptional cases two bad years will follow each other, as happened in Israel in the winters of 1950–51 and 1951–52, and then the order is reversed. Classical writers were aware of the difference of yields in alternate years. Varro (1.55.6) and Pliny (15.1.1) explained that beating the trees while picking was the reason for it and Columella (5.9.11-12) suggested interplanting on alternate years as a remedy.

The injunction in Deuteronomy (24.20) is that olives harvested by beating the tree (*ḥbt*) are for the owner and those left to be picked by hand (*pʾr*) were to be left for the stranger. Varro's instructions (1.55.1-3) suggest the opposite procedure, first carefully picking the olives by hand and then harvesting the remainder by beating the branches. Varro stresses that this should be done using reeds not poles. The Mishnah suggests a similar method to Varro's, but the Mishnaic Hebrew terms are different from the biblical ones: picked olives = *zyty msyq*; beaten olives = *zyty nqwp* (*Ḥal.* 3.9).

In Galilee today olives are usually planted at 10 m intervals and there are therefore 100 trees per hectare. The average density of planting in Greece reaches 120 trees per hectare, in Spain 90, in Italy 85 and in Libya in desert conditions only 30. The yields of unirrigated olives in Israel today are 800 kg—3 tonnes per hectare or 8–30 kg per tree—although in exceptional cases a tree will give as much as 50 kg. Basing their calculations on classical sources, scholars have reached estimates of similar yields in ancient times: 7–30 kg per tree and between 700 kg and 3 tonnes per hectare (Frank 1932; White 1970: 392).

The olive-picking season in Israel starts in the middle of October. The villagers usually begin picking after the first rains, perhaps in order to allow the rain to wash the dust off the olives. The length of the picking season varies and depends on the size of the crop and the workforce available. The quicker the task is completed, the better, and in some cases it is finished in a month and less but in others it can take two months or more. In the texts of the Dead Sea Sect there is also a Feast of the First Fruit of Oil to be celebrated fifty days after that of Wine (Reeves 1992: 351 and nn. 6, 8).

In Europe the olive-picking season is later.

Preamble B

Wine and Oil: Their Importance in Antiquity—Aspects of Modern Research

The fruit of the vine and the olive, together with wheat, are the three traditional staple products of Mediterranean dry farming. They are all three ideally suited to the climate of the region, and their products—oil, wine, raisins and grain—could be stored indefinitely. As such, they not only comprised the basic components of the human diet and played a crucial role in the economic history of the region but were an integral part of the cultures, languages, religion, literature, art, customs and folklore of the many nations that peopled the lands of the Mediterranean through the ages.

In the parable of Jotham (Judg. 9.8-9) the olive was the first tree to be requested by the others to reign over them, and similarly Columella (5.8.1) called the olive 'queen of trees'. The oft-recurring biblical triad *dgn–tyrwš–yṣhr*, corn–wine–oil (Deut. 28.5, etc.—the Hebrew terms are poetical and not the usual ones), symbolizes the place of these commodities in the agriculture of ancient Israel and, as such, their place in the Mediterranean world as a whole. This special position still finds expression in modern Western culture. The olive leaf carried by the dove (Gen. 8.11) has come to symbolize peace and the cluster of grapes borne by the spies upon a staff (Num. 13.3) to symbolize plenty. Kings of England are still anointed with olive oil as were the kings of Israel and Judaea and as will be the Messiah (*mšyḥ* = anointed). Wine is basic both to Jewish ritual—the Kiddush, the ceremonial blessing over wine on Sabbath and festivals and the four cups of wine of the Seder night at Passover—and to Christian: the wine of the Eucharist.

Much research has been devoted to the subject of wine and oil and therefore, before continuing to the main concern of this book—the actual installations used for wine and oil production—I will first present a short summary of the reçent research in some of the many other aspects of the subject together with minimal relevant bibliography.

An aspect already referred to above is that of the date and place of the domestication of the vine and olive and their introduction to the various countries and regions. This question is closely connected to the relationship between the domesticated and the wild varieties—whether it is possible to distinguish the wild from the domesticated varieties when found in archaeological contexts; to what extent olives or vines found in the wild are truly wild plants or are hybrids with escapees from cultivated areas or with remnants of deserted groves or vineyards; and whether the percentage of wild olive trees or vines in the forest reflects that of the primaeval forest or is a result of human activity (Hjemquist 1973; Liphschitz *et al.* 1991; Neef 1990; Galili, Weinstein-Evron and Zohary 1989).

Another subject that has aroused considerable controversy is the significance of the changes in pollen count recorded in samples from various periods. As the southern part of Israel is desert, a comparatively small change in rainfall can result in a radical change in the marginal areas and in the border between the desert and the sown changing its position. The country is therefore a potential laboratory for the study of climatic change. Horowitz (1971, 1979) has interpreted changes in pollen count of the olive as being a barometer of climatic change. Baruch (1986), on the other hand, believes these changes to be the result of human activity and to represent changes in the extent of olive groves.

A similar discussion connected to the significance of the olive in the history of the eastern Mediterranean is its place in the enormous changes that took place at the beginning of the third millenium BCE—usually termed the Urban Revolution. The discussion has arisen in particular regarding the Aegean region. Renfrew (1972) suggested that a large-scale increase in the extent of the olive groves at this period was one of the main factors in this change. Runnels and Hansen (1986) have recently questioned the validity of the data regarding the olive on which this thesis was based.

Some of the most interesting attempts to investigate ancient agriculture are those that examine the actual fields. Jashemski (1972, 1979) has excavated a vineyard in Pompei and been able to identify the holes left in which the vines grew. Mattingly (1994: Fig. 1) brings the remains of what was probably an olive grove that was excavated unintentionally by Tine at Passo di Covo in Apulia while uncovering a Neolithic site. Mattingly (1994: 95-97) also brings examples of use of air photographs in investigations of planting density of olive trees in North Africa.

Considerable research has been invested in attempts to make quantitative estimates of various aspects of olive oil production in antiquity. Frank (1932) made an interesting analysis of quantities, prices, etc. based on Cato's text alone. Safrai (1996) has based similar researches on the Talmudic texts. Others have taken the actual oil presses and the land holdings as their point of departure to estimate the output of the presses, regional production, consumption, surpluses, export, etc. (Amouretti *et al.* 1984; Mattingly 1988a; Frankel, Patrich and Tsafrir 1990: 298). Others have attempted to estimate the production of whole regions. Heltzer (1987: 111), basing his calculations on written sources, has estimated that the 200 villages of Late Bronze Age Ugarit produced 5500 metric tonnes of olive oil. Mattingly (1988b) has assessed the annual production of Roman Lepcis Major, basing his calculation on data from the field, as 10,000,000 litres/10,000 tonnes.

Various corpora of written sources have been studied in detail as regards wine and oil. The wine of Mari in Mesopotamia in the Middle Bronze Age (early second millenium BCE) has been discussed by Finet (1974–77), that of the Mycenaean Aegean by Palmer (1993) and that of Pharaonic Egypt by Lutz (1922), while Heltzer (1990) has discussed the wine of Late Bronze Age Ugarit and also the oil of that kingdom (1996). Hoffner (1974), Guterbock (1968) and Singer (1987) have discussed oil in Bronze Age Hittite texts from Anatolia; Waetzold (1985) and Malul (1996) the oil from Mesopotamia; and Melena (1983) that of the Mycenaean Aegean.

Similar studies have been devoted to texts of later periods. Zapetal (1920) has discussed biblical references to wine and Weinfeld (1996) has dealt with the use of oil in the cult of ancient Israel.

The most detailed written sources for the study of ancient agriculture are the works of the classical writers Cato, Varro, Columella, Pliny the Elder, Palladius (and the later treatise in Greek, the *Geoponica*), who devoted whole books to various aspects of agricultural practices, laying particular stress on viticulture and wine making. Unfortunately, the work of Mago the Punic agronomist is lost to us, although some quotations have survived in these works. See White (1970b) for a short summary of these works and a very helpful bibliography. The vast Talmudic literature has numerous references to agricultural equipment, practices and methods; see Feliks (1963) and Krauss (1910-12) for a detailed discussion. Of great importance for the understanding of the economic aspects of the subject is Diocletian's Price Edict which provides a list of types of wine and oil as well as other products together with their prices (Lauffer 1971).

Another aspect of the subject is the study of ancient diets (Vickery 1936; Hamel 1983) and a source of particular interest is the book of Roman recipes by Apicius (see also the English translation and introduction in Flower and Rosenbaum 1958).

Billiard (1913) has discussed in detail the classical texts referring to wine, Tchernia (1986) the Roman wines of Italy and Ricci (1924) the Graeco-Roman texts relating to wine in Ptolomaic Egypt.

Amouretti (1986) has studied Greek texts relating to oil, Camps-Fabrer (1953) Latin texts and inscriptions referring to oil in Roman North Africa and Brun (1986) those referring to oil in Provence at the same period.

Of particular interest are those papers that compare terms and concepts in different languages. Paul (1975) compares different Semitic terms for types of wine while Brown (1969), in his fascinating paper, discusses what he calls 'The Mediterranean Vocabulary of the Vine' ranging from Bronze Age Hittite, Ugaritic and Cypriot texts to later Hebrew texts from the Bible, classical Greek and Latin texts, New Testament, Talmudic and Arabic texts and shows that not only do the actual terms often derive from a common source but many concepts and phrases are common to texts in different languages.

Another aspect of the subject to which much research has been devoted is that of trade in wine and oil as indicated by the presence of jars or amphorae from one region that can be shown to have derived from another. The export of jugs and jars from Israel (Canaan) to Egypt is attested to in the Early Bronze Age (Amiran 1970: 59-66), and the Late Bronze Age 'Canaanite Jar' has been found both in Greece and in Egypt (Grace 1956). From the Hellenistic period the Rhodian amphorae and those from other Aegean regions, Knidos, Chios, Lesbos and Thassos, each of a typical shape and many with stamped handles, were found throughout the Mediterranean and even further afield (Grace 1952, 1953, 1961). The trade of oil and wine in the Roman period is attested for by a wide distribution of a great variety of amphorae from throughout the Mediterranean. The basic typology is that of Dressel and the most impressive evidence that of the mountain of used amphorae at the Testaccio mound in Rome. See Peacocke and Williams (1986) for typology of Roman amphorae; Sciallana and Sibella (1991) for a similar catalogue but including some earlier types; Zemer (1977) for amphorae found in the Levant from all periods; and Remesal Rodriguez (1996) for a review and an initial bibliography of the particularly well-researched production and trade of oil from Betica in Spain.

In this short review only some of the aspects of the subject have been touched upon. For example, no mention has been made of the vine and olive in art. The truth is that, whatever facet of the ancient Mediterranean cultures is approached, some connection to the vine and the olive will be found.

NB. In recent years several congresses have taken place concerned with the subjects discussed here, the proceedings of which are particularly useful: in Madrid in 1980 (Blazquez Martinez 1980) and in Seville in 1982 (Blazquez Martinez and Remesal Rodriguez 1983), both concerned with oil only and particularly with amphorae; in Haifa in 1987 (Heltzer and Eitam 1987; Eitam and Heltzer 1996), also concerned only with oil; in Paris in 1988 (Chevallier 1990), concerned only with wine; and in Aix en Provence in 1991 (Amouretti and Brun 1993), concerned with both oil and wine.

Preamble C

Wine and Olive Oil Production: Basic Processes

We learn of the methods used in oil and wine processing from descriptions of the pre-industrial installations (Ts01, 02). Wine and olive oil are both derived from liquids expressed from fruit and, as a result, the basic processes of production were very similar.

The three main stages of processing were in each case:

Wine	Olive-oil
1. Treading the grapes to express most of the must	1. Crushing the olives to a mash
2. Pressing the remaining must from grape skins and stalks	2. Pressing the mash to extract the expressed fluid
3. Fermentation	3. Separating the oil from the heavier, dark, watery lees; 20%–30% was oil, the remainder lees

The first two stages in both processes were similar, but with two important differences. First, in the case of the grape, most of the must was extracted in the first stage, the treading, whereas in that of the olive it was only in the second. Secondly, whereas grapes were almost always trodden by foot, olives were usually crushed by mechanical means, using mortar and pestle, stone rollers or the more sophisticated rotary crusher (T3).

In spite of these differences, in the most primitive methods of production the first two stages—crushing/treading and pressing—were carried out on the same simple treading/crushing installation or simple winery. This installation (Ts111, 121, 131) consisted of a slightly sloping treading floor on which the grapes were trodden and the olives crushed and pressed, and next to it and slightly lower a collecting vat to which the expressed liquid flowed.

As techniques developed, different types of installation emerged for wine and oil production and for each of the three stages of production.

1. Processes of Wine Production

Treading

The treading of grapes or olives is mentioned ten times in the Hebrew Bible and three terms appear for the simple treading installation or winery in which this operation was carried out: *gt*, *yqb* and *pwrh*. In Talmudic literature only the term *gt* is used (see Appendix 1). These terms are usually translated as 'wine press' or 'pressoir'. In this paper the term 'press' is reserved for true presses (lever, screw or wedge presses) and the terms 'treading installation', 'treading/crushing installation' or 'winery' are used for these installations.

In the LXX both *gt* and *yqb* are usually translated as ληνός, which is clearly the term for a winery, treading floor and collecting vat. In one case, however, *yqb* is translated as προλήνιον (Isa. 5.2, literally, 'in front of the winery') and in several cases as ὑπολήνιον (Isa. 16.10; Joel 3(4).13; Zech. 14.10, literally, 'below the winery'). In these cases the term *yqb* has been understood as 'collecting vat'. In the parallel to Isa. 5.2 in Mark (12.1) the term ὑπολήνιον is used, whereas in the same parallel in Matthew (21.33) the simple ληνός appears.

In Latin a winery was known as a *calcatorium* (literally, 'a treading area'; Palladius 1.18). The more usual term was *torcular*, *torculum*, which had a much wider meaning, referring also to a press room with all its equipment (Cato 12.1) and apparently also to a press proper (Pliny 18.74.317; Varro 1.55.7). In the Vulgate *gt* and *yqb* were both without exception translated as *torcular*.

There are several contemporary descriptions of treading of grapes. Ben Yaʿakob (1978) has described how the grapes are trodden today in the Hebron area. The grapes are trodden in a simple treading installation (T111). About 100 kg are placed on the treading floor and trodden barefoot by one man. Initially the connecting channel to the collecting vat is blocked up by sprigs of *poterum* (great burnet) which act as a sieve but these are quickly blocked up. As the grapes are trodden, the grape skins and stalks are pushed to the centre of the pressing area, using either the feet or a broom or wooden shovel. There the grape skins and stalks gradually form a hard block. This is done to allow the must to flow freely and to concentrate the skins and stalks in one area for further treading or for pressing by other methods (see below). After enough must is collected, the *poterum* is removed to allow the juice to flow. The skins are then arrested by hand. One man could tread 100 kg in 45 minutes. At Masada treading of grapes by five men has been reported and there one tonne of grapes was processed in 3 hours.

Both in the Hebron area and at Masada the grapes were not trodden in order to produce wine, forbidden by Muslim and Druze precepts, but to produce 'Dibes', a sweetmeat. An interesting technical detail is that in both cases marl was added 'to make the must clear'. Grape treading has been described also from Stavros in Greece and Valenca do Douro in Portugal. For references to treading grapes in German literature see Honigsberg (1962: 70 n. 2). Detailed instructions for treading are given in the tenth-century agricultural manual *Geoponica*: 'the feet are to be washed before treading, those treading are to remove leaves, wizened grapes and clusters and are not to eat or drink during work. They should also be fully dressed and have their girdles on on account of the violent sweating' (*Geoponica* 6.11; translation, White 1970b: xv).

Scenes of treading of grapes appear on wall paintings from Egyptian tombs from the first, second and third millenium BCE, on Greek classical vases, on Roman stone reliefs, on mosaic pavements, in manuscript illustrations and tapestries from the Middle Ages and later paintings (Ts020–024). From these we learn that the treading was often accompanied by music played on various instruments, reminding one of the rejoicing reflected in biblical texts (Isa. 16.10; Jer. 48.33). From the pictorial representations we learn that the men treading kept their balance in various ways by placing their hands on each others' shoulders (T02-3), by holding on to a horizontal pole (T02-2), onto ropes (T02-1) or small sticks (T02-4).

The second stage of processing the wine is usually pressing (see following paragraph), the presses often being elaborate. These are often described in great detail without mention of treading (e.g. Cato 19; Humbel 1976), which could lead to the conclusion that the grapes were not first trodden. Plommer (1973: 8) therefore suggests that Palladius's fourth-century-CE treading installation, the *calcatorium* (see below, p. 155), was technically retrogressive and a result of the easy availability of slave labour, thus making the investment in a wine press uneconomic. However, Plommer apparently misunderstood the purpose of the wine press which was to press the grape skins and stalks after treading. Younger (1966: 22) is perhaps correct in explaining that grapes are today trodden only to produce certain types of wine. However, Gallego Gongora Diaz, the owner of a large vineyard at Villanueva Ariscal Seville, informed us that when he was young he remembered the grapes being first trodden before pressing, and nearly all the pictorial representations even of late-nineteenth-century plants in California (Baird 1979: fig. 27) show both treading and pressing or only treading. In ancient times grapes were certainly always first trodden and then pressed. Classical writers often describe only the press because treading was taken for granted.

Pressing and Varieties of Must

By mechanical means it was possible to extract additional must from the grape skins and stalks. In the Hebron area these are put into a sack and the must extracted by one or two men stamping on the sack and then by placing stones on the sack. Bag presses using torsion to express the must are depicted in wall paintings from Egyptian tombs of the second and third millenia BCE. The bags were twisted in different ways. In some cases the bag was attached between two poles (T01003), often with as many as five men operating them, two on each pole and a fifth to keep the poles apart. In other cases the bag was held in a fixed frame (T01004) or attached to a fixed pole at one end only (T01005). There are records of similar, although simpler, pre-industrial methods from Corsica and Italy (T01001).

The description of the many other types of presses used in later periods will be deferred to the relevant chapters below.

From classical and Talmudic literature, however, we learn of the way the pressing operations were actually carried out and of the varieties of must of different qualities that were produced.

First Must

'Called in Greek *prototropum*—must that flows of its own accord before the grapes are trodden' (Pliny, 14.12.85). '*Mustum lixivium*—which will be that which has flowed from the grapes before they have been much trodden' (Columella 12.41). This was the finest must.

Must of the Second Pressing

According to Varro (1.54.3), 'when the grapes have been trodden the stalks and skins should be placed under the press so that whatever remains in them may be pressed out into the same vat. When the flow ceases under the press, some people trim around the edges of the mass and press again: this second pressing is called *circumsicium*, and the juice is kept separate because it tastes of the knife.' Columella (12.36) calls this must *mustum tortivum* (literally, 'pressed must'), while Cato (23.4) combines both terms (*tortivum mustum circumcidaneum*).

After-Wine

By adding water to the pressed grape skins and pressing again, a poor type of wine was produced. It was called *lora* in Latin (Cato 25; Varro 1.54.3; Columella 12.4.40), *deuterius* in Greek (Pliny 14.12.86) and *tmd* in Hebrew (see Appendix 2). It is a poor wine and was supplied to the workers.

Lees-Wine

This was made by pressing the dregs of the wine itself (Cato 43; Pliny 14.12.86). Cato calls it *vinum faecatum*, whereas Pliny regarded it as a variety of after-wine. It was known in Hebrew as *yyn šmrym* (*Pes.* 42a; *B. Bat.* 97b), literally, 'lees-wine', but also as *tmd drwwq* (after-wine of the straining bag) as opposed to *tmd dpwrṣny* (after-wine of the kernels), ordinary after-wine (*Pes.* 42b; Schapiro 1932: 58). As regards tithes, etc., the former, lees-wine, was considered as wine, whereas the latter, after-wine, was not.

Boiled-Down Must

This concentrated must was made by boiling down fresh must. Pliny (14.9.80) distinguishes between *sapa* that was boiled down to one-third of its original volume and *defrutum* which was boiled down to one-half. Columella (12.20.2) stated that *defrutum* is boiled down to one-third. Palladius (11.18) defines *defrutum* as being of thick consistency, *caroenum/carenum* as having been boiled down to two-thirds of its original volume and *sapa* as having been boiled down to one-third. Boiled-down must was very sweet and used to enrich musts of other wines. Boiled-down must was without doubt the *yyn mbšl* (literally, 'cooked wine') of Talmudic literature that was particularly sweet (see Appendix 2). In the British Museum is a small stone relief depicting the making of boiled-down must (27902—Italy?).

Fermentation and Types of Wine

To become wine, must goes first through a short and very turbulent fermentation at a temperature of 15°–20°. According to the Tosefta (*Ter.* 7.15), it lasts three days, and according to Pliny (14.25.124) nine days. In modern Spain it lasts 10, 15 or 20 days (Moreno Novarro *et al.* 1981: 207). The second fermentation is long and slow. During this stage the preferable temperature is 6°–12° (Winkler 1949: 62-64). For this reason, wine going through the second stage of fermentation is kept in cool places. In the Tosefta (*Men.* 9.12) it is stipulated that wine was not to be brought as an offering to the temple before it was 40 days old, newer than which it was not considered to be wine.

In modern wine production, antiseptic agents are used to purify the must which is then inoculated with selected pure yeast strains which allow for a controlled fermentation (Pricket 1980: 55). In ancient times there was no way of controlling the fermentation process and neither the quality of the wine produced nor how long it would last was ever predictable. This is well illustrated by the Mishnaic injunction: 'If a man sold wine to his fellow and it turned sour he is not answerable' (*B. Bat.* 6.3). Note, however, that the injunction according to the Tosefta is stricter, the vendor being responsible for the keeping quality of the wine until Passover (*t. B. Bat.* 6.5). In Talmudic literature there are hints as to means to stop wine going sour, e.g. 'He makes a cover of shingles so that it would not go sour' (*M. Qaṭ.* 2.2). Sealing the wine in jars with oil was probably for the same reason (*Ḥul.* 94a).

In classical literature there are long and complicated instructions as to how to make a myriad variety of wines as well as a variety of processes in which many different substances were added, such as sea water, honey, and a great variety of spices. Pliny devoted all of Book 14 to the subject and Columella chapters 18–41 of Volume 12, while over 70 names of types of wine are known from biblical and Talmudic sources (see Appendix 2).

The main reason for these many additions and varied processes was without doubt to attempt to overcome the vagaries of the fermentation process, or to counteract its ill effects. That this was the current belief is illustrated by the Mishnaic injunction that completes the one quoted above: 'If he (the vendor) had said I am selling you spiced wine he is answerable for its remaining (good) until Pentecost' (*B. Bat.* 6.3).

Pre-Treatment of Grapes

The need to store the various substances that are to be added to the must are clearly relevant to the understanding of the function of parts of ancient wineries, as are the instructions for pre-treatment of the grapes. The instructions of Columella (12.27) for making sweet wine are to leave the grapes in the sun for three days before treading while keeping the first must produced during this time separate. He also tells us (12.39.2) that, according to Mago the Punic agronomist, making raisin wine also required leaving the raisins in the sun for three days. In Talmudic literature wine that had been left in the sun was called *hylstwn*, a name that derived from the Greek ἡλιαστον ('of the sun'). *hylstwn* and raisin wine are mentioned together as being forbidden in the temple ritual.

2. Olive Oil: Its Uses and Importance

Today, olive oil is used almost exclusively for culinary purposes; however, in ancient times food was only one of its many uses.

Cosmetics

The use of olive oil as an unguent or as a basis for the production of perfumes and similar cosmetics was perhaps in certain cultural contexts its main use. References to its use for cosmetic purposes appear in Mesopotamian sources (Malul 1986: 100) and in

Hittite documents (Singer 1987). Melena (1983) has shown the importance of its use for this purpose in the Mycenaean world as reflected in the tablets.

In the Hebrew Bible various types of perfumed oils are mentioned, *šmn hṭwb* (Ps. 133.2 2; Kgs 20.13), *šmn hmr* (Est. 2.12), *šmn rwqḥ* (Qoh. 10.1) (below Appendix 1). Men annointed their heads and beards with oil (Pss. 133.2; 141.5; Qoh. 9.8) and the maidens received beauty treatment with oil enriched with myrrh for six months before going in to Ahasuerus (Est. 2.12). The obvious use for olive oil was as a cosmetic and not as food (Mic. 6.15).

The fact that Joab told the wise woman not to anoint herself with oil when he wanted her to behave as if she were in mourning suggests that the use of oil as an unguent was general practice (2 Sam. 14.2). In Talmudic literature the oil made from olives picked early—*ʾnpiqnwn* (see below p. 44)—was recommended for cosmetic use.

The importance of oil as a cosmetic in the classical world is stressed by Pliny (14.29.150) who states: 'There are two liquids that are agreeable to the human body: wine inside and oil outside.' The method used to clean oneself in the Greek world was to rub olive oil on one's body and then to scrape off the oil, sweat and dust with a special sickle-shaped instrument known as a strigil (Boardman 1976: 192, pll. 4.4, 5; Reich 1991).

Light

Ceramic oil lamps are attested for from the Early Bronze Age both in the Levant and the Aegean and, although the question has been raised as to what type of oil was used (Runnels and Hansen 1986: 305), there is little doubt that olive oil was the main fuel used in the very many oil lamps found from this period onwards.

In oil presses excavated in Israel, oil lamps are often found clearly using the oil from the press. Similar oil lamps can be seen in pre-industrial presses in Spain today.

Both in the Bible and Talmudic literature the oil for light in the temple is mentioned (Exod. 25.6; 27.30; 35.8; 35.15; Lev. 24.2; *Men.* 8). In the latter much discussion is devoted to what types of oil should be used for this purpose (see Appendix 1).

Industrial Purposes

Melena (1983: 117) has shown that Mycenaean tablets provide evidence for use of oil both in textile processing and tanning.

One of the more important uses of poor-grade olive oil in the Levant in recent centuries is in the production of soap. Dalman (1928–42: IV, 273-77) describes the methods used and two pre-industrial installations have recently been described, one from Jerusalem (Cohen 1989), and the other from Lydda (Ayalon 1990). The early history of soap is apparently still not clear. The terms used in Latin and Semitic languages are without doubt connected (French *savon*, Arabic *ṣabon*) and the substance was apparently known to Pliny (27.51.191) as *sapho* and in Talmudic literature (e.g. *t. Nid.* 8.11) was *ṣpwn*. The classical evidence suggests it was usually produced from animal fats and both classical and Talmudic texts suggest that it was not an important product in Roman and Byzantine periods (Blumner 1921; Krauss 1910–12: I, 155, 578).

Olive Wood

It is worth noting that *ʿṣ šmn* (tree of oil) (1 Kgs 6.31, 32, 33; Isa. 41.19; Neh. 8.15) used as building material in the temple and often translated as 'olive wood' is clearly not olive wood as is shown by its appearance together with olive wood in Neh. 8.15.

Ritual

In the same way as wine, oil also played an important part in many ritual acts and rites. The Bible shows that kings were anointed with olive oil (1 Sam. 16.1-13; 2 Kgs 9.1-3), as were priests (Exod. 29.7), the word 'Messiah' deriving from the Hebrew *mšyḥ*, 'anointed'. Not only people, but stelae were sanctified by being anointed with oil (Gen. 28.18; 35.14) as was the tabernacle and all its contents (Exod. 30.23-29; 40.9).

Oil was used in purification ceremonies (Lev. 14.12-32) and was an integral part of the various types of meal offerings that were offered in the temple, either alone (Lev. 2) or together with animal sacrifice (Num. 15.1-16; Weinfeld 1996). The Talmudic discussion as to which oils were fit to be used in these ceremonies are an important source for understanding ancient oil-extracting processes (see Appendix 1).

Hittite sources point to similar ritual traditions: the use of olive oil in purification rites, in the anointing of kings and in the anointing of the bones of the deceased after cremation (Singer 1987: 185-86).

In later Christian practice, holy relics were anointed with oil which was poured into the reliquaries. In some cases, apparently, the oil was sanctified in this way and then poured out of the reliquary through a small pipe into the vessels of pilgrims (e.g. Aviam 1990b: 360-61).

Food

Olive oil is still a staple element in the diet of Mediterranean peoples. In the Arab villages of Judaea and Samaria the consumption per capita per annum is 17–20 kg (50–55 g per day; Dar 1986) and similar quantities are attested for in Greece (Amouretti 1986: 287).

Both the written records and the archaeological evidence show that great quantities of olive oil were produced in the Roman Empire. Mattingly (1988b) suggests that millions of litres of oil were produced in North Africa. The stamped handles from Spanish oil amphorae are evidence for the scale of the production in that region and the mountain of discarded amphorae at Mount Testaccio in Rome hint at the destination of a large part of this production (Remesal Rodriguez 1996). There is no doubt that this massive production was used largely for consumption as food, as is illustrated by the degree to which oil features in Roman cooking recipes (Apicius; Flower and Rosenbaum 1958).

The question that arises, however, is whether oil was such an important element in diet in earlier periods. It has been usual to regard the triad of *dgn–tyrwš–yṣhr* (corn–wine–oil) that appears in the Bible 18 times as representing the three staples of Mediterranean agriculture and diet. Renfrew (1972) suggested that the main factor in the big changes that took place in the Aegean world in the third millenium BCE during the Early Bronze Age was the introduction of the vine and olive to that region. However, in recent years several studies have been published suggesting that even as late as the Late Bronze Age olive oil was used in the Aegean mainly for cosmetic purposes and was extracted largely from the fruit of the wild olive (Melena 1983). In conjunction with this new approach, other models were suggested to explain the social and economic changes during the Aegean Early Bronze Age (Runnels and Hansen 1986).

The biblical evidence is also not unequivocal. The triad *dgn–tyrwš–yṣhr* appears mainly, if not only, in late biblical books and there are few specific references to oil as food. The references in Ezek. 16.13, 19 to fine flour, honey and oil classes oil as a sumptuous food. Miracles both of Elijah (1 Kgs 17.8-16) and Elishah (2 Kgs 4.1-7) are connected to increasing a quantity of olive oil suggesting it to be a valuable product but also showing it to be expected that a simple family would have some oil for food in their home.

There is, however, also considerable ancient epigraphic evidence that allows for some degree of quantitative analysis.

In the Arad letters from the seventh century BCE are consignment orders, found in a fort in southern Judaea that gave orders to supply military units with rations, and that were apparently retained as receipts for bookkeeping (Aharoni 1981). Oil appears in about half the *ostraca,* showing it to have been a usual part of military rations.

Another important source is the Samaria *ostraca*, administrative documents from a storeroom in the eighth-century-BCE royal palace of the northern kingdom of Israel at Samaria. They are concerned with wine and oil and are apparently bills of lading. Rosen (1986–87) has made an interesting study in which he grouped the recipients of these products in a hierarchical social scale according to the proportion of wine and oil that each received. Three groups emerged. One recipient thought to be the royal household received equal quantities of wine and oil. A second group of three recipients received the products in proportions of 9 wine: 1 oil. A third group of three recipients received no oil.

Very detailed data are available regarding the olive oil production of Late Bronze Age Ugarit. Heltzer (1996) has analysed the taxes of oil paid in kind by the villagers and made a cautious estimate that the 200 villages of Ugarit produced about 5500 metric tonnes of oil per year. He also showed that consignments of oil received by royal servants, probably annually, ranged from 1 to 10 jars, Heltzer suggesting this to be 22–220 litres per year.

Rosen (1986) analysed the data from Ugarit in the same way that he analysed those from Samaria and showed the proportions of oil and wine received grouped the recipients into similar classes, although the proportion of oil was in each case slightly lower. The royal household received 7:1, two recipients received 10:1 and again five received no oil.

Rosen made a similar analysis based on Linear A documents from Bronze Age Hagia Treada in Greece. In this case the proportions were 40:1, 45:1, 55:1 and 60:1.

These limited data suggest that in Iron Age Judaea and Israel and Late Bronze Age Ugarit olive oil was a staple product of importance, although probably not available to the poorest part of the population. In the Aegean in the Late Bronze Age it was apparently produced in much smaller quantities.

The Uses of Oil Lees

The lees are the black fluid left after the oil was separated (Latin *amurca*, Hebrew *mwḥl-my zytym* ['olive water']).

As Pliny (15.33) states, 'Cato prescribed olive lees for all purposes', although in fact this was apparently not the general practice as Varro (1.55.7) complains that its uses are generally not appreciated. He also gives instructions for preparing olive lees (1.64): 'After 15 days the dregs being lighter have risen to the top are blown off, and the fluid turned into other vessels. This operation is repeated at the same intervals twelve times during the next six months...then they boil it in copper vessels until it is reduced to two-thirds its volume.'

The lees were recommended to treat fruit trees, to pour around olives in particular as manure, together with human urine (Columella 2.14.3), to kill noxious weeds (Varro 1.40.7), to smear on vines to keep out insects (Columella 4.24.6; Cato 95), to make an infertile olive bear fruit by wrapping straw around the tree and soaking it with diluted lees (Cato 93) and similarly with figs (Cato 94).

Lees were used to protect grain from insects and mice. The pile of grain was covered with a mixture of lees and chaff on which pure lees were poured (Cato 92). Similarly, dried figs were kept in jars coated in lees (Cato 99), as was oil (Cato 69; 100). Myrtle twigs and berries, and fig branches with leaves and fruit were preserved in lees (Cato 101).

Soaking firewood in lees improved it (Cato 130). Furniture was smeared with boiled-down lees to keep moths from the clothes kept in it, and to prevent decay of the wood and to polish it (Cato 118). Similarly, it was used as grease for belts, shoes and hides (Cato 117).

Lees were sprinkled on cattle food and given to them to drink in order to improve their health (Cato 103; Columella 6.4.4, clearly based on Cato). Sheep were smeared with an ointment of lees, boiled lupins and dregs of wine to treat for scab (Cato 96).

Plaster was made of chopped straw, chalky or red earth and lees (Cato 128), and threshing floors were prepared by adding lees to earth and then tamping it down. The lees kept insects away and prevented the earth from becoming muddy (Cato 91; 129).

3. Olive Oil: Processes of Production

Crushing

In the Bible, oil and wine production are not distinguishable and there is no specific term for installations connected to oil production. There are only two biblical references to the actual production of olive oil. In one (Joel 2.24), both oil and wine flow into the *yqb*, one of the terms used for a winery or wine vat, and the other (Mic. 6.15) speaks specifically of 'treading' olives. Dalman (1928–42: IV, 297) was of the opinion that the term should not be taken literally. There is, however, much evidence that olives were also crushed by treading: the Latin term for a crushing basin, *trapetum,* is derived from the Greek word for treading, and one of the other crushing devices described by Columella (12.52.6) is called *solea*, the sole of a shoe; in Greek there is a term for shoes worn to tread olives (Amouretti 1986: 286; Jüngst and Theilscher 1957: 100), according to the latter, κρουπέζια. Talmudic references refer apparently to treading olives (*t. Ter.* 3.13; *y. Ter.* 3.4, 42b; see Appendix 1) and at Santa Lucia Di Mercurio-Corsica treading olives, albeit enclosed in a sack, has been recorded in recent times.

There is also evidence for other primitive methods for crushing olives. For example, a photograph taken at Jerusalem Karm el-Sheikh shows women crushing olives by rolling a large stone on them.

Archaeological and other evidence shows the many more sophisticated methods of crushing olives. The discussion of these will be deferred to the relevant chapters below.

Olive Oil Extraction and Oil Quality

In the Bible we learn of a fine quality of oil, *šmn ktyt*, and, in the Samaria *ostraca*, *šmn rḥṣ* is mentioned. These are apparently produced by two different processes: the first by pounding with mortar and pestle, the second by the addition of hot water (see Appendix 1).

The section in the Mishnah (*Men.* 8) and related Talmudic literature that discusses the oils to be used in temple ritual deals in detail with the problems of oil quality. In classical literature these subjects are also dealt with.

The quality of oil was affected by four main factors:

1. the quality of the trees and olive groves;
2. at what stage of ripeness the olives were picked;
3. the treatment of the olives between picking and processing;
4. the extraction process itself.

Trees and Olive Groves

In the Mishnah (*Men.* 8) are listed the groves from which oil should not be brought to the temple: (over?)manured groves, irrigated groves, and groves in which other crops were sown between the olives. The implication of this injunction is, however, probably more concerned with ritual uncleanliness than oil quality (see Appendix 1). Incidentally, Columella (5.9.11-12) recommends planting between the olives on alternate years in order to overcome the tendency of the olive only to give fruit every other year.

Stages of Ripeness of Olives

Olives picked early produce a smaller yield of oil of sharper taste and greener colour, known as virgin oil, *huile vierge*, *oleum et olphacium* (Latin: Pliny 12.60.130), green oil (Cato 65; Columella 11.2.83) or summer oil (Columella 12.52.1) (see p. 194). In ancient Egyptian a distinction was made between *bꜣḳ wꜣ*, fresh/green oil, and *bꜣḳ nḏm*, sweet oil, and the biblical *šmn rʿnn* (Ps. 92.11) is perhaps virgin oil (Stager 1985: 175). In Mishnaic Hebrew virgin oil was known as *ʾnpyqnwn* and was one of the oils forbidden in the temple rite (*Men.* 8.3). In the Babylonian Talmud it is explained that it is made of olives not yet a third ripe and that women rub it (in their skin) because it removes the hair and rejuvenates the skin (*Men.* 86a). From Columella (12.52.1-2) we learn that the price for this oil was high but that nevertheless it was usually produced mainly from olives that fell early from the trees. However, leaving picking until late was also regarded as detrimental. The late olives give a yellower oil that lasts for a shorter time.

Treatment between Picking and Processing

In the Mishnah (*Men.* 8) and related texts olives are divided into three categories according to how long they wait before pressing. The best quality are those that are taken directly from the tree to the press; the second quality are those stored on the roof and from there taken to the press; and the third are those stored indoors and then dried on the roof and taken to the press. The problem was that the olives were picked faster than they could be pressed. Cato (3.3-4) and Varro (1.40.2-7) both stress the importance of pressing the olives immediately. Columella goes into a long argument with those who think that olives benefit from storage but, while stressing that olives should be pressed immediately, admits that this is usually impossible and gives detailed instructions for building a well-drained store (12.52.1-3; 12.52.18-20).

Oil Extraction: Pressing and Other Methods

The method used to extract the oil was the main factor affecting quality.

The finest oil was that produced without exerting pressure. One method was after crushing to leave the olive pulp for some time before pressing in order to allow the oil to flow from the pressure of its own weight only. This oil was the 'first oil' for use in the temple defined in the Mishnah (*Men.* 8.4) as that produced by pounding the olives in a mortar and then putting only in a frail—and in a Baraitha in the Babylonian Talmud as oil produced by crushing the olives in a crushing basin and then putting only in frails (*Men.* 86a) (see discussion in Appendix 1). The Latin term for the first oil was apparently *lixivium* (Columella: 12.52.11).

We learn of a second method of producing choice oil from descriptions of pre-industrial processes in Israel and surrounding regions. Hot water is added to crushed olives or the olives are boiled in water and then the oil that floats to the top is skimmed off. This oil is called in Arabic *zet tafaḥ* ('floating oil') or *zet mawi* ('water oil') (Dalman 1928–42: IV, 235-38; Avitsur 1986: 10).

Many ancient sources refer to three qualities of oil. Pliny (15.4.18) says that, as opposed to many qualities of wine, there are only three grades of oil and in Emperor Diocletian's list of official prices three types of oil are given: *olei floris* '(flower [of the] oil'), *olei sequentis* ('following oil'), and *olei cibari* ('popular oil') (Lauffer 1971: 102-103). Pliny (20.6.23) states that *flos* is the first oil to flow out at the beginning of the pressing. In the Mishnaic discussion about the oil for the temple offering, the oil was divided into three grades according to the way the olives were processed. As there were also three grades of olives, this allowed for nine grades of oil in all. Columella (12.52.11) refers to three pressings of each batch of olives, stressing that the oil of each pressing was inferior to its predecessor. This clearly is the explanation for both Pliny's and Diocletian's oil gradings and, although the explanation of the oil gradings of the Mishnah raised much argument reflected in the Mishnah and other Talmudic literature (see Appendix 1), the background to the Mishnaic grading was without doubt also the deteriorating quality of oil in consecutive pressings.

The Greek version of Diocletian's list is not identitical to the Latin, however, and the equivalent of *olei floris* is ὀμφάκινον, *omphakinon*, virgin oil, suggesting that it was not absolutely clear which was the finest oil.

A similar picture emerges from pre-industrial evidence. From Dalman one learns of the slight differences between the methods practised in different villages in the region in the early decades of this century. At el-Kerje the olives were pressed twice, crushed a second time, warmed and pressed a third time. The first oil was called *zet bikr* and went to the owner of the olives, the second, *zet ʾamle*, went to the owner of the press and the third went to make soap. At Nablus the olives were pressed only twice, the second pressing being after soaking the mash in water and after a second crushing. At el-Ṭafile hot water was poured over the first and only pressing (Dalman 1928–42: IV, 245-46). At Miʿilya the olives were pressed only once without addition of water. At Huevar-Huelva, Spain, the olives were pressed twice. Before the second pressing the olive cake was broken up in a cake breaker (T0179). During pressing, hot water was poured on the olive pulp. Brun (1986: 42-43) presents the following model based on practice in France today: three pressings—before the second, hot water was added and before the third the olive cake was re-crushed. This supplied three grades of oil: 'huile de 1ère pression'; 'huile de 2ème pression'; and from the third pressing 'huile de ressence'. The oil that was retrieved from the lees from the three pressings provided a fourth grade, 'huile d'enfer'.

The list of oils forbidden for the temple ritual in Talmudic literature hint at various secondary processes for producing oil. These forbidden oils include olives that were soaked in water, oil extracted from lees and the *šalwq*, seethed oil, presumably retrieved either from poor-quality olives or from the dregs and lees by boiling (see Appendix 1).

To sum up, the written and pre-industrial evidence show that their were six main types of oil: virgin oil, first oil, oils of the first, second and third pressing, and lees oil. In this categorization, factor (a), the quality of the olive groves, and factor (c), the treatment of the olives between picking and pressing, have been ignored.

Oil Separation

After the olives have been crushed and the liquid expressed by the press, the lighter oil rises to float on the watery lees and it must then be separated. The Latin term for the workman who separated the oil was *capulator*, and he both poured the liquid from vessel to vessel and skimmed off the oil. Cato speaks of his using a shell (*conca*) and Columella, about 200 years later, a ladle (*concha ferra*, literally, 'an iron shell') (Pliny 15.6.22; Cato 66; Columella 12.52.8-12). From archaeological and pre-industrial evidence we know of more sophisticated methods of oil separation using special vessels and also by allowing the oil and sometimes the lees also to flow by gravitation (see pp. 174-76). The only hint of such methods in the written sources is Columella's derogatory reference to the 'built up double (twin) container' (*structile gemallar*) (Columella 12.52.10).

Preamble D

History of Previous Research

Pliny the Elder (18.74.317) can be regarded as the first historian of oil and wine presses and his short survey of the development of the press is of great importance. Later studies of ancient agricultural installations, however, were initiated in attempts to understand references to and descriptions of such installations in ancient literature, mainly those found in classical and Hebraic texts. Early interpretations were based primarily on comparison with contemporary appliances known personally to the scholars concerned. On the publication of the first archaeological excavations and surveys, the evidence these provided was used to interpret the written sources further.

The most detailed descriptions of oil and wine presses in classical literature are those of Cato the Elder (18–22) and it was to the interpretation of this text that many studies have been devoted, initially based on contemporary installations only (e.g. Meister 1763). After the first excavations at the Villa Rustica of the Campania in the environs of Pompei at the end of the eighteenth century (La Vega 1783; Ruggiero 1881), interpretations were based on these and thus modified. Later scholars still focused their researches on the interpretation of this text, either alone (Hörle 1929) or in conjunction with other classical texts, primarily those of Vitruvius, Columella, Pliny the Elder and Hero of Alexandria (Drachman 1932; Jüngst and Theilscher 1954, 1957). Nevertheless, room remained for re-interpretation based on more recent archaeological finds (Brun 1986: 236-47). These scholars also attempted to determine the chronological sequence of the development of the press and crushing equipment basing their conclusions partly on the ancient texts and partly on archaeological evidence.

In Hebraic literature the most detailed reference to an agricultural installation is without doubt that to an oil press in the Mishnah (*B. Bat.* 4.5). Already in the Babylonian Talmud the Hebrew terms of the Mishnah were translated into Aramaic and explanations given based apparently on personal knowledge. In the Middle Ages interpretations were added using German and French terms to explain the parts of the press, clearly referring to contemporary oil presses. Scholars and travellers to the Holy Land in the nineteenth century described rock-cut and other installations, and Goldman (1907), in his pioneering treatise on the olive in the Mishnah, referred to classical sources, European archaeological finds and descriptions of finds from Israel. At Gezer, Macalister (1912) excavated an Iron Age oil press and in the excavation report included a detailed survey of the rock-cut installations in the vicinity. Krauss (1910–12), in his studies of the Talmudic realia, also used these and other excavations to explain Mishnaic and Talmudic terms. More recently, publication of installations excavated have been accompanied by renewed attempts at interpreting these terms (Yeivin 1966; Urman 1974; Hirschfeld 1983; Frankel 1981).

Generally, however, from the twenties of this century onwards, with the development of modern archaeology in which a combination of detailed stratigraphy and ceramic typology often enabled the dating of specific assemblages to within a half a century or less, interest in agricultural installations waned, for several reasons. Rock-cut installations did not allow for stratigraphical dating and installations of different periods, in those cases where they could be dated, often varied very little. Possibly the traditional stress on monumental architecture and art also led to neglect of this more mundane subject. As a result, even when oil presses were excavated and published, little research was devoted to the actual press itself, its technology, typology, etc. (e.g. Sebastia–Samaria–02; Salamis-Cyprus–01).

In recent years, however, apparently as a result of greater interest in the economic and technical aspects of history and archaeology and of the greater stress on the survey as an archaeological method, agricultural industries have returned to the focus of attention and a series of regional studies have been devoted entirely or partly to the subject. Examples include North Africa (Camps-Fabrer 1953; Mattingly 1988a for bibliography), Italy (Liverani 1987), the Var region of Provence, France (Brun 1986), northern Syria (Callot 1984), Spain (Fernandez Castro 1983), Cyprus (Hadjisavvas 1988), and Israel (Eitam 1980; Frankel 1984; Ben David 1989). In recent studies greater stress has been laid on the archaeological and pre-industrial installations which are no longer regarded as only aids to understanding the written texts, but have become the main subject of research.

The study of pre-industrial installations has also increasingly gained recognition as a discipline in its own right. Today the traditional pre-industrial installations have largely been replaced by modern industrial equipment, but they were still in use throughout the Mediterranean at the beginning of the century and can still be found functioning in remote regions to this day. Many descriptions and pictorial representations of such installations exist and include not only descriptions by ethnographers but also instructions as to how to make these installations from the period when these were still the most advanced methods available.

France is perhaps the country in which the largest number of wooden presses have survived, in a great variety of forms. As a result the subject has aroused great interest in France, and in French research of the subject of presses as a whole a greater stress has been laid on pre-industrial processes. Already in 1901 Latour published a study on the stages of development of the press based only on pre-industrial presses from the region of Beaune in France. Similar studies followed, some limited to pre-industrial presses (Parain 1979), while others included earlier presses also (Parain 1963; Polge 1967). A particularly rich and varied collection of pre-industrial presses was that amassed by Humbel (1976).

Pre-industrial installations have been studied in other countries as well. Some of the more important papers are those of Paton and Myres (1898) on the presses in the Aegean, of Vincze (1959) on wine presses in Hungary, of Sordinas (1971) on the development of pre-industrial installations in Corfu, of Bel (1917) and Coon (1931) on oil presses in North Africa, of Miguelez Ramos (1989) on those in Ibiza, of Wulff (1966) on those in Persia, and of Casanova (1966, 1968, 1990) on a particularly interesting and unusual group of primitive installations from Corsica. Dalman's (1928–42) monumental work on the ethnology of Palestine was followed by that of Avitzur (1976).

Much important information relating to the history of oil and wine presses and to that of other industries and crafts can be gathered from ancient pictorial depictions. A very rich group of pictures of vintage scenes and wineries is that found in wall paintings from tombs in Egypt. Most of these date to the New Kingdom (Late Bronze Age) but there are earlier examples from the Old and Middle Kingdoms (Early and Middle Bronze Ages) and also later ones from the Ptolomaic (Hellenistic and Roman) Periods. Other important sources are the Athenian vase paintings from the Classical period (in the Levant the Persian period), frescoes from Pompei and Herculaneum and many reliefs mainly from sarcophagi from the Roman period, and many depictions in Late Roman (in the east, Byzantine) mosaic pavements from throughout the Mediterranean world. Many fine representations of presses are also to be found in mediaeval manuscripts, tapestries and altarpieces, these usually associated with the 'mystic press' (see p. 159).

These four interdependent elements, archaeological evidence, ancient pictorial depictions, ancient written sources, and pre-industrial installations, remain the basis of modern research. Pre-industrial installations often retained ancient techniques and are thus invaluable together with ancient depictions in attempting to reconstruct the archaeological remains, of which usually only the stone parts survive. These three then allow for a more accurate interpretation of the written sources, while the latter not only contribute the cultural and historical context for the archaeological finds, but also often provide clear chronological data for the introduction of various techniques.

The ancient depictions, especially those from the Middle Ages, also provided a link between the archaeological finds and the pre-industrial installations, making it possible to follow technological development from ancient to modern times.

The changes in the focus of research can be summed up as follows. The early studies centred on interpreting ancient written sources both classical and Talmudic on the basis of contemporary installations. With the discovery of oil presses and wineries in archaeological excavations, these became the main basis for comparison. In later research the stress moved from the written sources to the archaeological finds and to attempting to understand the chronological technological sequence.

In the study presented in this book these issues will also be discussed, but the main stress will be placed on the regional diversity of these installations.

Chapter 1

Simple Installations (T1)

1A. The Simple Treading/Crushing Installation (Ts111, 121, 131)

1A1. Rock-Cut Simple Treading Installations (T111)

1A1.1. Character and Date. The installation found most frequently in Israel and the surrounding regions is the simple treading installation. It is made up of two parts: a sloping upper surface on which the processing of the fruit—treading, crushing, pressing—was carried out, and the lower collecting vat to which the expressed liquid flowed. The two were connected by channels or bores. The vast majority of these installations were cut in bed-rock, usually in the 'nari', the uniform easily cut rock that forms on the surface of chalk formations throughout the region. One hundred and sixty two such installations (i.e. T111) were recorded in the 1626 sample 10 km square. The total number in the area of the Israel–Palestine Grid probably reached tens of thousands. However, as well as rock-cut examples (i.e. Ts11), built (i.e. Ts12) and free-standing (i.e. Ts13) installations are also found.

Photo 1. Rock-cut simple treading installation/winery (T111): (1626)50/007(02).

The earliest dated installations of this type are apparently Megiddo Inst. 01, dating to the Chalcolithic period or earlier. This was before the introduction of the vine to the region and therefore the Megiddo installations served, presumably, to produce olive oil. However, up to the present, simple treading installations are used to tread grapes (i.e. Balat, Hebron district, Masada), and without doubt these served mainly for producing wine, while other installations evolved to process olive oil.

These installations are clearly the *gt*, *yqb* and *pwrh* of the Bible (see Appendix 1), terms that are usually translated as 'wine press' or 'pressoir'. The term 'press' has here been reserved for true presses (beam presses, screw presses, etc.) and the terms 'treading installation' or 'winery' used for installations consisting of treading floor and collecting vat.

These installations vary in the size and shape of the treading floor and collecting vat, in the manner in which these are connected and in the various devices that are attached to them. Among the earliest dated examples, however, there are some that are carefully built and that conform to exact, albeit varying, geometrical shapes, such as:

Megiddo Inst. 01. Chalcolithic or earlier: round treading floors and collecting vats connected by bores.

Tiᶜinnik (Taanach). Early Bronze Age: rectangular floor and collecting vat connected by two bores.

Afeq, Antipatris. Late Bronze Age: (built) rectangular treading floors, round collecting vats connected by channels.

On the other hand, we have very simple installations that are clearly dated to later periods, such as Burj North (later than Roman) and Yatir Naḥal East and Yatir Naḥal West (both Byzantine).

The vast majority of the installations recorded in surveys (see Chart at the end of List A on the disk) are extremely simple, but cannot be dated by conventional means. The ceramic and other evidence, however, show that settlement density reached its peak during the Roman and Byzantine periods and in fact it would seem that large areas were not cultivated before or after this period. There is, therefore, little doubt that many of these simple installations date from this period.

The simple treading installation does not, therefore, evince clear chronological diversity and it is only rarely possible to date an example by its shape and character alone. Similarly, comparison between simple installations from different parts of the country does not show the existence of regional differences.

By detailed recording in limited areas, differences between simple installations from different periods can sometimes be distinguished (see, for example, section 1C on Iron Age installations). These differences, however, cannot serve as a basis for dating installations in other regions. In the same way, groups of installations found together are often similar to each other but different from those found nearby, showing regional diversity but on a very small scale.

For these reasons all examples of simple installations (T1) have been included only from the sample 10 km square 1626 (see chart at the end of List A on the disk) and the catalogue includes only some examples from other regions in Israel and environs. From other countries all examples recorded appear in the catalogue. Those data will suffice to demonstrate the character, function and geographical distribution of these installations.

1A1.2. Simple Treading Installation (T111): Size and Form. The following conclusions are based on the 163 installations from the sample 10 km square 1626.

Treading floor. Average size: 3.1 m^2, median size: 2.5 m^2. Two peaks can be discerned in the size distribution diagram, chart 2 c. 1.75 m^2 and c. 3.75 m^2. The preference for these two sizes is perhaps connected to the number of people treading the grapes.

Collecting vat. Data for area only. The volume is usually not known in un excavated examples. Average size: 1.24 m^2; median size: 1.00 m^2.

Shape	Round	Square/ rectangular	Trapeze	Oval	Irregular
Treading floor	10.7%	33.9%	12.4%	8.3%	34.7%
Collecting vat	31.4%	46.2%	—	9%	14.8%

The five combinations preferred were

Treading floor	Collecting vat	No. of insts.
Square/rectangular	Square/rectangular	23
Irregular	Square/rectangular	17
Square/rectangular	Round	13
Irregular	Irregular	12
Irregular	Round	12

In a survey carried out in the Taᶜanach region the percentage of installations in which both parts were rectangular was even greater: 85 out of 117 (Ahlström 1978: 20).

Certain minimal conclusions can be drawn from these figures. Greater stress was laid on the form of the vat than of the floor and, in general, straight-sided shapes were preferred. This could be because, while preparing the installation (i.e. ᶜAmqa South), it served also as a quarry. Ready-prepared rectangular building stones were removed while working (see next section). However, rectangular treading floors were universally preferred in built installations as well (Ts1.2, 9), in spite of the apparent advantages of a round floor for treading.

1A1.3. Cutting Simple Rock-Cut Installations. (See also section 6C4) Several uncompleted rock-cut installations were recorded in the archaeological survey of western Galilee which makes it possible to reconstruct how these were cut. Three incomplete rock-cut wineries were recorded. In two, ᶜAmqa South and Yirka East, the collecting vats had been completed and in the third, Shubeika East, the cutting of the collecting vat had not been started. In all three the cutting of the treading floor had been started but not completed and in all three the method used was identical. The outline of the treading floor was marked out by a channel about 30 cm wide. In the wineries of Yirka East and Shubeika East only the channel had been cut. In that of ᶜAmqa South about half of the stone had also been removed. The methods used were similar to those used in quarrying ashlar building stones in the region throughout history (Shiloh and Hurvitz 1975; cf. Olami 1981, sites 15, 29). The stones were extracted almost in their final shape by cutting channels at right angles to each other and then removing the stone by cutting below it. Cutting rock-cut installations was in all probability a dual-purpose operation in which the extraction of ashlar building blocks was an important byproduct. This probably partly explains the preference for square or rectangular treading floors and collecting vats.

Photo 2. Unfinished winery (T03): Yirka East.

Photo 3. Unfinished winery (T03): ᶜAmqa South.

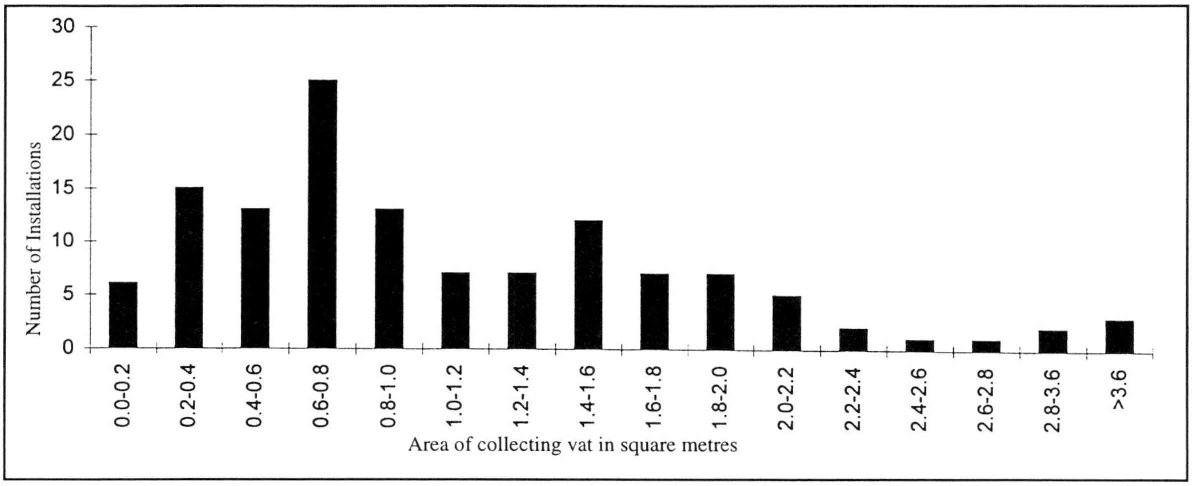

Chart 1. Area of collecting vat in wineries from sample square 16/26 (100 sq. km.)

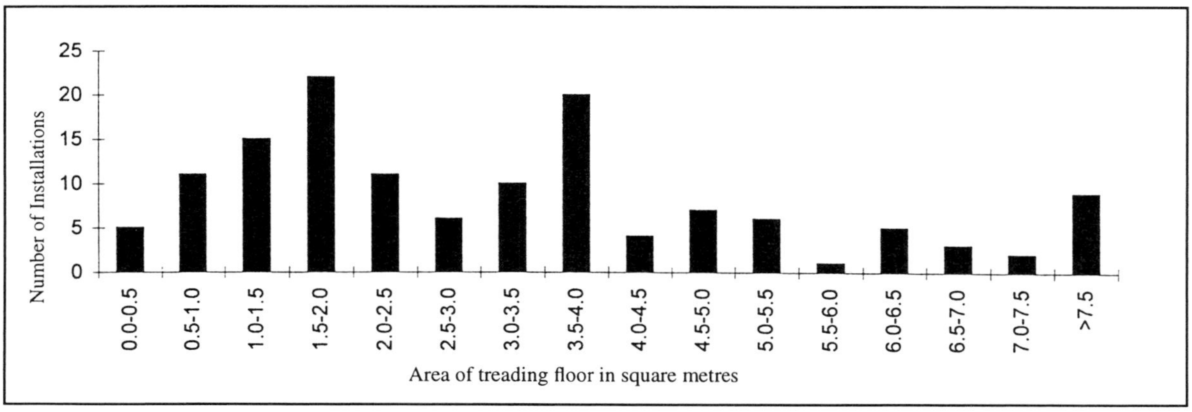

Chart 2. Area of treading floor in wineries from sample square 16/26 (100 sq. km.)

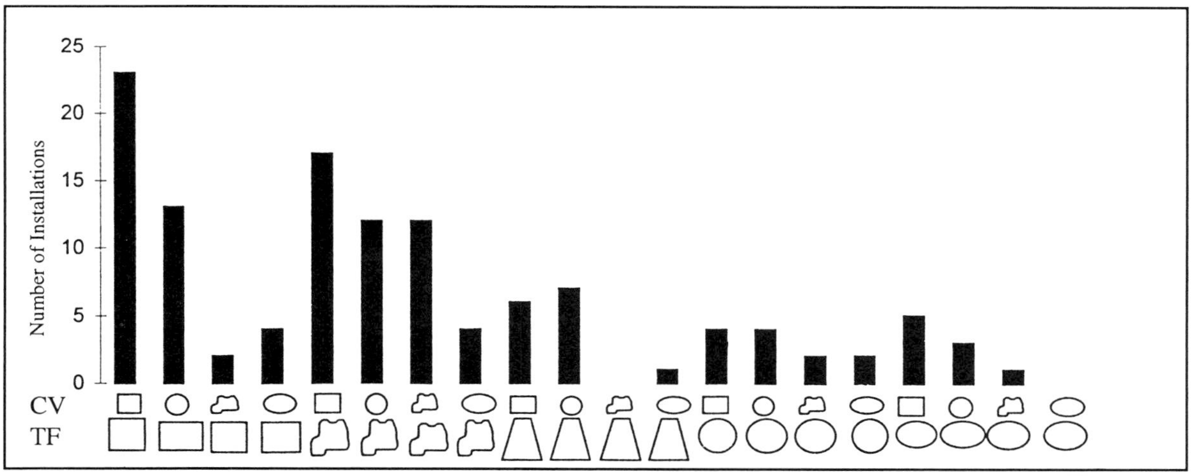

Chart 3. Shape of treading floor and collecting vat in wineries from sample square 16/26

1A2. Built and Free-Standing Treading Installations (Ts121, 131)

Incomparably fewer built and free-standing installations were recorded than rock-cut examples. In the sample 10 km square 1626, one free-standing (Nes ᶜAmin East Inst. 01) and no built installations were recorded. The latter, T121, have been discovered mainly in excavations and include examples from the Middle Bronze Age (i.e. Nahariya) and a pair from the Late Bronze Age (Afeq, Antipatris).

Pre-industrial records (Dalman 1928–42: IV, figs. 97, 98; Frankel and Ayalon 1988: fig. 40) provide evidence for treading grapes in small movable vessels and the Mishnah mentions treading vessels of stone, wood and pottery (*ᶜAbod. Zar.* 5.11) and stipulates that in the sabbatical year treading was allowed in a tub (*ᶜrbh*, *Šeb.* 8.6). The large pottery container from Beer Sheva (T181) with a spout at its base was probably for treading grapes (see section 11B3).

Free-standing and built installations are more likely to have been covered by alluvium than those cut in rock and probably there were many more than the available data would lead one to expect. However, rock-cut installations had many advantages. They were easier to make, were more durable and less prone to the seepage of liquids. Only where no outcrops of suitable stone were available in the vicinity would built or free-standing installations have been used.

Photo 4. Small rock-cut crushing/pressing installation (T112): (1626)74-009.

Photo 5. Pear-shaped small crushing/pressing installation with central groove (T113): Abu Sinan North (see also Photo 7).

1B. Sub-Types of Simple Treading Installation

1B1. Small Crushing/Pressing Installations (Ts112, 113)

In the 10 km square 1626, eight small rock-cut devices (Ts112, 113) were recorded, four attached to larger treading installations and four unconnected to any other installation. These are in effect very small copies of the larger treading installations. Installation 000-1626-74-009, for example, consists of a round rock-cut depression 0.15 m in diameter connected by a channel through a second, 0.07 m in diameter, and to a third, 0.06 m in diameter.

Abu Sinan North Inst. 01 (T113) consists of a pear-shaped groove at the tip of which is a small cup-mark. Another groove bisects the pear and also leads to the cup-mark. This is the only example in square 1626, but eight have been recorded from other sites.

These small installations served without doubt to produce extracts of other fruits and herbs either to add to the wine or to produce perfumes or spices. Columella mentions wines flavoured with wormwood, hyssop, southern wood, thyme, fennel and pennyroyal (12.35), harehound (12.32), squill (12.33) and myrtle (12.38), and Talmudic literature mentions similar additions (see Appendix 2).

1B2. Pairs of Installations

In square 1626 four pairs of simple treading installations were recorded: installations 45-001-01, 02; 53-008-02, 03; 55-001-01, 02; and 63-020-02, 03.

Pairs of wineries from other sites are also known: Afeq-Antipatris, Michal, Inst. 03; Tel Aviv Eretz-Israel Museum Inst. 03; and Qalandiya Inst. 01.

Two adjacent wineries have obvious organizational advantages. Work can be organized on a larger scale and while the must is fermenting in one collecting vat the other installation can be used.

1B3. Small Hole in Centre of Treading Floor (T171)

In four sites from square 1626 and five others small holes were recorded in the centre of the treading floor. These were clearly connected to the pressing of the remaining must from the grape skins and stalks, perhaps by squeezing them in the hole with aid of a stone.

In more sophisticated wineries various other devices that serve the same purpose are located in the same position (Ts81–83, 89).

It has been suggested that this is the *tpwḥ* (apple) of Talmudic literature (Urman 1974: 173 n. 8; see Appendix 1).

1B4. Two Small Cup-Marks on Either Side of the Treading Floor (T172)

In 12 of the simple treading installations (T111) in square 1626 there are two small cup-marks on either side of the centre of the treading floor. The only other similar installation recorded to date was that from Karim, although the latter was slightly different, having a pair of cup-marks on either side instead of one. Half of the installations are concentrated in a limited area of 2 km^2, but the installations are otherwise not identical in character. Two, however, 75-004, and 50-006, are unusual and remarkably similar, consisting of a round treading floor in which a deeper rectangular section was cut, on either side of which were the cup-marks. These two, however, are located 5 km apart.

Wall paintings of vintage scenes from Egyptian tombs provide three analogies that could each explain the purpose of this unusual device. The men treading who appear in the paintings helped to keep their balance by two different contrivances: the first, a horizontal pole above their heads that they grasped (Ts02-2); the second consisting of ropes hanging from such a pole which they also grasped (Ts02-1). In both cases the horizontal pole would have been supported by vertical poles secured in the cup-marks. The third possibility is that the cup-marks held the poles of bag presses (Ts01003, 01004, discussed above, p. 42). There is no evidence of the use of ropes or poles as supports for men treading except from Egyptian wall paintings. Nevertheless, for a small family installation it is the most probable explanation (Ahlström 1978: 33-34).

Photo 6. Winery (T111) with two cup-marks, one on either side of treading floor (T172): (1626)63/010(01).

Photo 7. Winery (T111) with two cup-marks, one on either side of treading floor (T172): Abu Sinan North (see also Photo 5).

Photo 8. Winery (T111) with two cup-marks, one on either side of treading floor (T172): (1626)75-004(01).

1B5. Stone-Cut Tethering Rings (T173)

In the 1626 square approximately 80 rock-cut tethering rings were recorded at 30 sites, 17 of them connected to treading installations (T111). One appears on an Early Bronze Age winery from Tiʿnnik (Taʿanach) and others have been published from other surveys (Olami 1981: 9). Those attached to installations without doubt served to tether the beast of burden that bore the wine jars while the others were probably to tether animals while grazing.

Photo 9. Rock-cut tethering ring (T173): (1626)50/009(01).

1C. Simple Installations from the Iron Age

Archaeological surveys can only be of assistance in dating when a particular type of artefact or installation is found consistently in a significant number of sites that are shown ceramically to be of one period only. The only relevant sites recorded in the 1626 square are some of the Iron Age I, several of which with associated installations (T111) of very similar character. The six installations are (1626)56-001-01; 56-001-02; 56-004-01; 95-004-01;95-004-02 and 95-006-01. In all cases the treading floor is not clearly demarcated and in effect exploits a natural flat surface of bed-rock. The collecting vats are small and either round or rectangular, with rounded corners, their volume being 0.30 m^3, 0.40 m^3, 0.11 m^3, 0.14 m^3 and 0.30 m^3. No far-reaching conclusions can be drawn from these installations except as regards the sites and installations themselves. Without doubt, similar installations were used in other periods, while at other sites of the Iron Age I period the installations were different in character.

Photo 10. Winery at Iron Age site (T111): (1626)95-004(01).

1D. Topographical Distribution of Simple Installations (Ts1): Simple Wine Husbandry

One hundred and sixty-three installations of Ts1 (excluding T173) were recorded in the sample 10 km. square 1626, averaging therefore 1.63 per km^2. In certain parts of the square they appear in far greater density, especially in those areas where *nari* rock is exposed. In square km 63, 18 installations were recorded, in other words one installation for each 5.5 hectares, and in smaller areas (e.g. square km 47) as many as nine installations were recorded in 0.25 km^2, in other words one per 2.5 hectares. In Samaria similar densities have been reported: one per 4.5 hectares and one per 6.0 hectares (Dar 1986: 152).

The concentration of these installations in the areas of chalk formations overlaid with *nari* is presumably primarily because of the suitability of this rock for all types of rock-cut installations. The simple installations are often located at the edge of the outcrops and clearly served the fields in the alluvium nearby. The chalk rock, however, also has the capacity to retain moisture and is therefore particularly suited for unirrigated crops such as vines.

The vast majority of the simple treading installations were without doubt wineries for treading grapes, the first fermentation taking place in the collecting vat. The location in the vineyard, and usually not in the farmstead or village, was characteristic of the vine husbandry of the region and is reflected in the texts from the Hebrew Bible, for example the 'Parable of the Vineyard' in Isaiah 5 and in parallel texts in the New Testament (Mt. 21.33; Mk 12.1). The latter are without doubt based on Isaiah, but nevertheless show that the parable was relevant in later periods as well.

The comparative uniformity of the simple wineries (T111) and their topographical location necessitate a reappraisal of one of the few archaeological studies devoted to wine production in ancient Israel, that of Pritchard (1964) at Jib, Gibeon. While excavating the water system, 56 jar handles with stamp impressions, 40 jar stoppers and a pottery funnel were found (Pritchard 1959). Later two areas of rock cuttings were excavated and Pritchard (1964: 11, 25) suggests that the rock cuttings in square 10M17 include wineries for treading grapes. These, however, are in no way similar to the usual treading installation and it is more than probable that the grapes were trodden in wineries in the vineyards and that the stamped handles and other finds were connected to the storage and maturing of the wine but not to its production.

1E. Oil Production in Simple Installations

The earliest unequivocal evidence for the production of olive oil is from the underwater pottery Neolithic site south of Haifa (c. 5500 BCE) where a large quantity of olive stones were found in a round pit (Galili and Sharvit 1994–95; Kislev 1994–95; Carmi and Segal 1994–95). The bottom of the pit was concave and covered with a layer of large smooth stones, above which were the olive stones, most of which were broken. The olives were probably crushed in another installation and pressed in the pit, the expressed liquid seeping down below the stones. The olives were almost certainly wild olives from the forest on the Carmel.

The evidence from the Chalcolithic period suggests that at this period olives were actually cultivated. In the Chalcolithic sites on the Golan Heights olive pits (Stager 1985: 1) and considerable quantities of olive wood (Liphshitz *et al.* 1991: 2) have been found, as have pottery overflow oil separators. These have a spout at the rim allowing the floating oil to be poured off (T182; Epstein 1978; Epstein 1993; Stager 1990: 97; see section on oil separation pp. 174-76). The installations that were used to extract the oil were apparently the trough-like crushing basins (T136) found at many of these sites (Epstein 1993). It would seem that the olives were crushed in the troughs by pounding. The oil was probably then expressed, also in these troughs, by the direct weight of stones, similar to the method used at the underwater Neolithic site south of Haifa.

It has already been shown that the simple treading/crushing installations (T111) from Megiddo that also date to the Chalcolithic period or earlier pre-date the introduction of the vine to the region, and therefore almost certainly also served to extract olive oil. The olives would have been crushed and pressed on the treading floor and the oil would have flowed to the vat. If the installations from the Golan have been understood correctly, the simple treading/crushing installations (T111) from Megiddo are technologically more sophisticated.

The simple treading/crushing installations (Ts111, 121, 131) from later periods usually served to produce wine. Theoretically, they could have processed both oil and wine, but there is considerable controversy as to whether there were installations that were used for both purposes. In most regions of Israel, olive picking and processing does not begin before the middle of October, whereas the vintage is usually finished by middle of September, leaving time for cleaning and preparation. In Europe the seasons are later, but there is also a similar time gap. Eitam (1980: 26) suggests that the use of one installation for both processes would harm the taste and affect the fermentation process. Others, however, disagree (Boardman 1976: 188; Forbes and Foxwell 1978: 42), and there appears to be no pre-industrial evidence that could shed light on the question.

Usually, however, the simple installations used for oil production can be distinguished from those used for wine, in that the collection vat was usually much smaller. Not only could large quantities of must be expressed by treading in a short time, but from the Talmudic evidence (*t. Ter.* 6.15) we learn that the first fermentation took place in the collecting vat. As a result, even in small simple wineries the volume of collecting vats was usually at least 1 m^3 (see chart 1). Therefore, installations with much smaller collecting vats or vessels can be presumed to be for oil.

A considerable number of simple installations for oil production have been excavated and can therefore be dated. They vary only slightly.

T1223. At Areini an installation consisted of a stone paved floor in the centre of which was a sunken pithos. Olive pits were found and it dates to the Early Bronze Age II period.

T1221. A small built installation in which the treading/crushing surface is a stone slab and the collecting vat a stone basin. A twin installation of this type from the Early Bronze Age II in which olive pits were found was uncovered at Ras Shamra, Ugarit. Many examples were found at sites from Israel from the Late Bronze Age period and the Iron Age.

T1222. In these installations the crushing surface is a stone slab and the collection is in a pottery vessel. The one example is from the Late Bronze Age.

T123. In these installations the treading/crushing surface is plastered. Two examples have been recorded from the Iron Age.

T125. At Yoqneam and Qashish in the north of the country, two identical built installations were found with stone paved floor and steep sides and with pottery collecting vessels, both from the Iron Age I period. In these the olives were apparently trodden, as the steep sides did not allow for any other method of crushing and also helped to alleviate the wastage that this method perhaps involved.

Other types of installations that are also clearly for oil extraction are:

T114. The rock-cut press-bed with small collecting cup-mark.

Photo 11. Rock-cut press-bed with small collecting cup-mark (T114): ʿAmqa East Inst. 03.

Ts1161, 1162. Crushing/pressing installations with small inner collecting vats.

T117. Two connected crushing and collecting cup marks. Each of these three types (Ts1161, 1162, 117) are found only at one site or in one small area—interesting examples of installations typical of a very small region.

T115. The groups of large cup-marks or rock-cut mortars found together at several sites are apparently also for oil production.

Photo 12. Two connected crushing and collecting cup-marks (T117): Bat el-Jebel East Insts. 3 and 4.

T1321. The small free-standing crushing/treading installation.

T1322. The large free-standing crushing/treading installation.

T15. In two installations published—one from the Late Bronze Age, from Beitin 01, and one from the Iron Age, from Bet Shemesh 12—there is no device to allow the oil to flow from crushing/pressing surface to a collecting vat. Olive stones show that these installations were for oil extraction. Possibly the extraction was achieved by the addition of hot water.

T124. The function of these stone-lined rectangular vats is a matter of controversy. Sellin (1904: 76) suggested they were for extracting oil. DeVaux (1951: 428) and Lapp (1964: 30-32) were of the opinion that they were cultic installations and Stager and Wolff (1981: 99-100) returned to Sellin's suggestion. In these installations there is also no arrangement to allow the liquid to flow. The stone objects regarded as steles by those who see these as cultic do not appear to be suitable for crushing the olives, as Stager and Wolff suggest.

1F. Simple Wineries in Other Regions

1F1. Countries South of the Mediterranean

1F1.1. Egypt. A group of very uniform rock-cut wineries have been published from Upper Egypt (Adams 1966; Ts981, 98103). They are from the Roman period, are more complicated than the simple wineries under discussion here, and will be discussed below (pp. 154-55).

In Egyptian tomb paintings of the Pharaonic period there are many representations of vintage scenes. Those from the New Kingdom, especially a group from Thebes, are very elaborate. The floor is raised to some height and covered by a canopy or roof supported by pillars with lotus capitals. As many as seven men tread the grapes, keeping their balance by holding ropes that hang from the roof or canopy (T0211). In two scenes from tombs of the Middle Kingdom at Beni Hassan the collecting vat is not shown and the treaders held on to a horizontal rod supported by two forked poles (T0222), while in a scene from the Old Kingdom, where there is also no collecting vat, the treaders placed their hands on each others' shoulders (T0223). Montet (1913: 118) has suggested that the wineries were in reality not raised above the ground but that their depiction in the paintings in this manner was an artistic convention to enable a more dramatic presentation. To the best of our knowledge no wineries from this period have been discovered in Egypt. In later wineries from Crocodopolis, Theadelfia; Aswan (Simons Kloster) and the 'City of Menes', however, the treading floor is raised. The twin Late Bronze Age wineries from Afeq, Antipatris, that were attached to the residence of an Egyptian governor, are sunken in the ground, but these are probably not Egyptian in character but Canaanite.

Figure 1. Vintage scene: Tomb of Nakht, Thebes (New Kingdom).

1F1.2. North Africa. North African rock-cut installations more complicated than the simple winery (T111) have been published from Algeria (Tigzert 2,3; el-Ma Ougelmine; Azzefoun 1,2) and there described as oil presses, but simpler ones have also been published (Tifrit Na'it el-Hady) from the same regions.

1F1.3. Syria. Many rock-cut or partly rock-cut installations have been published. Most of these are more involved but some are simple (e.g. Deir Mišmiš, Ḡ Siman) and clearly there are many more that have not been published in detail (Butler 1903: 268).

1F1.4. Turkey. Similar installations have been mentioned from Kunya, Turkey, but not described in detail.

1F2. Countries North of the Mediterranean and Mediterranean Islands

In discussion of the production of olive oil in the Greek mainland in Neolithic and Early Bronze Ages, in Crete (Hutchinson 1962: 242; Warren 1972: 25), the Cyclades (Renfrew 1972: 281-87) and Cyprus (Peltenberg 1978: 72-73), the presence of olive stones and wood is referred to but the only device actually mentioned is the pottery underflow oil separator (T181). In this vessel, after the oil has risen above the lees, first the lees and then the oil are allowed to flow out through an outlet or spout placed at the bottom of the separator, as opposed to that at the top of the overflow separator (T182). This method is well documented in the Aegean world in pre-industrial installations (Methana [T0195]) and in classical sites (Nicosia Pasydy Plot, Cyprus; Praesos, Crete). It is probable that most vessels of this type, particularly in the Aegean region, were oil separators. However, vessels of this type can also serve other purposes such as treading small quantities of grapes.

The rock-cut treading installation that is so common in the Levant and apparently also in parts of North Africa and Turkey is not, however, mentioned in these discussions of oil and wine in the ancient Aegean. A small number have been published, a group of three from Petralia, Sicily, one from Athos, Greece, one without collecting vat from Praesos (T118), and one with a raised outlet between treading floor and vat from Monte Cupellazo, Italy (T119). A group of rock-cut installations was published from the Vaucluse region of France (Cavallon district). These also lack collecting vats and were originally all within buildings. The date of all these installations is uncertain. From the discussion in the various publications it was clear, however, that they are unusual in their geographical context.

Examination of the pictorial evidence on Attic vases from the classical period shows that the depictions of vintage scenes are very uniform.[1] They show the grapes being trodden on a wooden platform or table raised perhaps a metre from the ground, and the must flowing into *pithoi* on the ground below (T025). This would suggest that in ancient Greece also the treading of the grapes was carried out in the vineyard, but on mobile treading platforms, the must flowing into vessels that could be removed immediately.

In treading installations found in archaeological sites of the classical periods, the must also flowed into *pithoi* (T983) and, as will be shown below, in later periods as well, the motive of wineries in which must flows into *pithoi* is central to many vintage scenes (Ts023). This method is mentioned in classical literature (Palladius 1.18) and is known in pre-industrial wineries (Stravros, Greece).

Figure 2. Vintage scene: Attic black-figure amphora: Wurzburg 265 (18-902).

The key to understanding this technological difference between treading grapes on permanent treading floors and collection and first fermention in collecting vats in the Levant, and treading and collecting in mobile installations in Greece is to be sought in the climatic differences between various regions of the Mediterranean. Chart 4 shows the differences in summer rainfall among countries to the south of the Mediterranean, where it is negligible, countries to the north of the Mediterranean, where it is considerable, and the islands in the Mediterranean, which fall in between these two.

The danger of rain falling during the vintage was without doubt the reason for collecting the must in *pithoi* and for the rarity of permanent open-air installations. In one of the few published—that from Phaestos—the must was apparently also collected in a *pithos*. It is, however, of interest that there is an injunction in the Tosefta (*B. Bat.* 6.11) that 'if the rain flows from (into?) the wineries and it is a national disaster one need not notify (the buyer of the wine) but if it is not a national disaster one must notify it', showing that, even in ancient Israel, there was a danger of rain during the vintage.

In spite of what has been already stated, the archaeological and pre-industrial evidence shows that in later periods the treading installation (T98) was also widely used in Europe and the Mediterranean islands (see discussion below, pp. 157-58).

Note

1. See Sparkes (1976) for a full discussion of vintage scenes on Attic vases.

Chart 4. Quantities of rainfall in summer months at places in the Mediterranean (millimetres)
(Places marked with * from Bickmore 1958; others from GBHMSO 1964)

Country	Town	June	July	August	September	July and August	Total June to Sept.
Egypt	Cairo*	0.0	0.0	0.0	0.0	0.0	0.0
Egypt	Port Said	0.0	0.0	0.0	0.0	0.0	0.0
Israel	Jerusalem*	0.0	0.0	0.0	0.0	0.0	0.0
Israel	Haifa	1.0	1.0	0.0	0.0	1.0	2.0
Israel	Ramla	0.3	0.1	0.0	2.0	0.1	2.4
Rhodes	Rhodes	0.5	0.0	0.0	2.0	0.0	2.5
Libya	Benghazi	0.3	0.1	0.0	3.0	1.0	3.4
Cyprus	Paphos	4.0	0.0	0.1	4.0	0.1	8.1
Lebanon	Beirut	3.0	0.4	0.4	6.0	0.8	9.8
Greek Island	Samos	9.0	0.1	0.1	4.0	0.1	13.8
Libya	Tarabulus (Tripoli)	1.0	0.4	2.0	14.0	2.4	17.4
Greek Island	Khios	10.0	3.1	0.5	4.0	3.5	17.5
Cyprus	Nicosia	11.0	1.0	2.0	6.0	3.0	20.0
Turkey	Ankara*	13.0	2.0	3.0	5.0	5.0	23.0
Crete	Iraklion	2.0	1.0	3.0	17.0	4.0	23.0
Turkey	Izmir	11.0	1.0	6.0	9.0	7.0	27.0
Greek Island	Milos	5.0	2.0	2.0	22.0	4.0	31.1
Turkey	Adana	16.0	3.0	6.0	19.0	9.0	44.0
Greece	Athens	14.0	6.0	9.0	15.0	15.0	44.0
Malta	Valletta	1.0	0.1	10.0	33.0	10.1	44.1
Tunisia	Tunis	12.0	3.0	4.0	27.0	7.1	46.0
Greece	Patrai	16.0	0.1	13.0	19.0	13.1	48.1
Spain	Malaga	26.0	0.1	2.0	23.0	2.1	51.1
Greek Island	Limnos	7.0	21.1	8.0	21.0	29.0	57.0
Spain	Alicante	17.0	4.0	7.0	38.0	11.0	66.0
Greek Island	Skiros	28.0	8.1	12.0	18.0	20.0	66.0
Sicily	Palermo	14.0	4.0	15.0	52.0	19.0	85.0
Greece	Thessaloniki (Salonica)	42.0	22.0	15.0	34.0	37.0	113.0
Italy	Messina	25.0	10.0	24.0	58.0	24.0	117.0
Majorca	Palma de Mallorca	27.0	5.0	21.0	71.0	26.0	124.0
France	Marseille	26.0	14.0	22.0	68.0	30.0	130.0
Spain	Barcelona	26.0	21.0	40.0	54.0	61.0	141.0
Italy	Napoli	46.0	16.0	19.0	71.0	35.0	152.0
Turkey	Istanbul*	33.0	28.0	43.0	51.0	71.0	155.0
France	Nice	33.0	22.0	32.0	70.0	54.0	157.0
Italy	Livorno	52.0	21.0	26.0	88.0	47.0	187.0

Chapter 2

The Simple Lever and Weights Press (T2)

2A. The Principles of the Lever Press

The essence of technological progress is in the understanding of natural forces and of the laws of physics and finding means to exploit them for the benefit of humankind. Just as controlling fire and the invention of pottery were turning points in the development of material culture, and just as the invention of the wheel and the sail were turning points in the early history of transportation, so learning the effective use of the lever was the first and perhaps the most important technical advance in the exertion of pressure and thus in the production of olive oil and also, but to a lesser extent, in the production of wine.

The simple lever was, almost certainly, the earliest means used by man to achieve mechanical advantage. In a simple lever applying force (applied effort: E) on one end of a lever that presses on a fulcrum (F) at some point along its length will bring great force (effective effort [e]) to bear on the other end, the load point (L). A lever press is a little more sophisticated: the fulcrum, instead of being at some point along the lever, is at its end. This, the fixed end of the press beam, the lever, was inserted, anchored, in a niche in the wall, in a recess cut in a rock face or into some similar contrivance, this acting as fulcrum. The other, the free end of the beam, was the point of applied force or effort.

Pre-industrial presses are known in which the force applied is that of human effort alone (T0110) and others in which weights are placed on the beam (Ts01111, 01112). However, in simple lever presses the weights were generally hung from the beam end (Pre-ind. T01113), while in more sophisticated ones the beam weights were raised by various types of drums and pullies (Pre-ind. Ts01114, 0112, 0113, 0114), or the beam weights were replaced by other devices, the most important of which were winch drums (Pre-ind. T0115; ancient Ts40—5, 6, 9) and screws (Pre-ind. Ts012, ancient Ts40—7, 8).

The material to be pressed, the crushed olives or the grape skins and stalks, were placed under the lever at the pressing point—the load point of the lever. The use of the beam as a lever multiplied the applied force (effort: E) to an effective force (effort: e), to a degree relative to the ratio between the fulcrum to effort distance (D) and the fulcrum to load distance (d). For example, if the pressing point (load point L) was in the centre of the beam in a lever and weights press, the effective force was twice that of the applied weight of the weights: if it was placed at a third of the distance between the anchoring point of the beam (the fulcrum) and its free end (the point of applied force), the effective force would be triple that of the weights. This ratio D over d is the 'mechanical advantage' of the lever. To calculate the exact effective pressure of a press, other ancillary factors would have to be taken into account: the effective weight of the beam itself, which depends on its section along its length, and the specific gravity of the wood, the position at which each weight is hung on the beam and of course their weight, the size of the area, and the volume of material pressed (see, for example, Amouretti *et al.* 1984). However, in the case of ancient presses of which only the stone parts usually survive, most of these factors can only be surmised. Therefore the two factors that will usually be cited in comparing beam presses will be beam length and simple mechanical advantage.

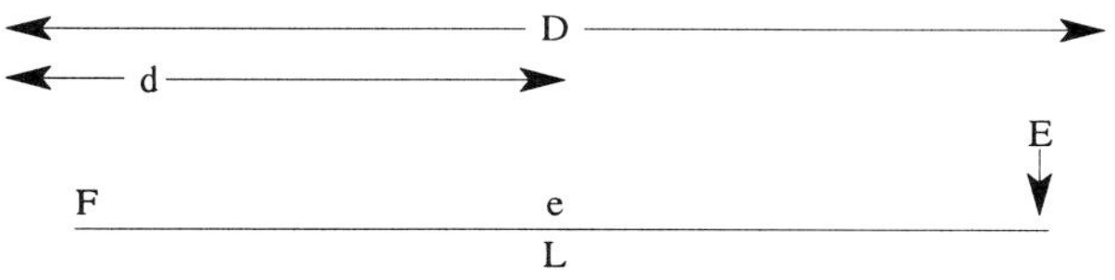

$$e \times d = E \times D$$
$$e = E \times \frac{D}{d}$$

One other characteristic of lever presses is that the pressure is most effective when the beam is horizontal and the pressure vertical, which makes it necessary to readjust the beam during pressing. In some more sophisticated presses various devices were developed to facilitate this readjustment.

The technical solutions found for the three functional points of the beam press—(a) the anchoring point (fulcrum), (b) the pressing point (load point) and (c) the point of applied force (the free end of the beam)—varied greatly regionally and chronologically and will be one of the main subjects discussed in the following pages.

2B. Simple Lever and Weights Press: Israel and Environs

2B1. Date of Introduction

There is clear evidence for the use of beam and weights presses in Syria and Cyprus during the Late Bronze Age and it is possible that such presses were in use in Crete even earlier (see discussion below, p. 64). In Israel, however, this technique was apparently not introduced before Iron Age I. Only one complete lever and weights press has been published from this period, from Dan at the very north of the country, but large numbers are known throughout the country from Iron Age II.

One lever and weights press at Lakhish (Lachish) 02 Cave 4020 was dated to the Early Bronze Age and has since been cited several times including by the present writer (Thompson 1972: 39 n. 4; Frankel 1987: 63; Stager 1990: 105 n. 4). The press was found in a cave and, although no pottery was published from this locus, it was dated by sherds from the fill to the Early Bronze Age. The press, however, is extremely similar to those found in the same region, particularly in caves, dating to the Hellenistic and Roman Periods (e.g. Maresha, T401110002). Only the weights from Lakhish are unusual for this type of press, but not unknown. No other lever and weights presses have been identified in the country that can be dated to earlier than Iron Age I. The presses dated to the Late Bronze Age (Bet Mirsham 01, 02) and many of those from the Early Iron Age (Ts1221, 123, 125) almost certainly functioned without a beam. This all leads to the conclusion that the press from Lakhish was mistakenly dated to the Early Bronze Age on the basis of sherds from fill that does not represent an occupation layer, but was debris washed into the cave from the tell above.

2B2. Typology and Distribution

Very similar simple installations (Ts111, 121, 131, etc.) served not only to produce both wine and oil, but in both cases the various stages of the process were apparently carried out in the same installations. As has been shown, however, some simple installations can be distinguished that are specifically for oil extraction (Ts1221, 1222, 123, 125, 1321, 1322, etc.). With the introduction of lever and weights presses, however, installations that were used for each stage of production were also differentiated. In the case of olive oil, specific devices evolved for crushing, pressing and sometimes also for oil separation. Special chapters are devoted to crushing devices (Chapter 3) and to weights (Chapter 5), and here only the presses will be discussed.

2B2.1. The Anchoring Point (Fulcrum). In simple lever presses in Israel from the Iron Age, in all cases in which the anchoring device has survived it was in the form of a niche into which the beam-end was inserted (Ts211). This is often in the shape of a vertical rectangle, thus allowing the beam-end to be raised and lowered. At Bet Mirsham, in installation 04, the niche measured 0.30 m × 0.70 m. At two sites the niche incorporated specially cut architectural elements (T2112): at Miqne a cap-stone shaped like a reversed 'U' which served to withstand the upward pressure of the beam, and at Gezer, where the niche was cut into a free-standing monolithic standard. The beam lengths of the presses at Bet Mirsham, calculated on the basis of the length of the rooms in which they were located, were 6.0 m, 5.5 m, 6 m, 6 m, 5 m and 5.8 m.

2B2.2. Pressing Point (Load Point): Lateral Collection and Central Collection. The basic principle according to which olives were pressed and olive oil collected is, at all periods and in almost all regions, that already described in relation to the various simple treading and crushing installations (Ts1). The material to be pressed was placed on a press-bed and the liquid flowed into a collecting vat placed beside it—i.e. lateral collection into a lateral vat. Simple presses of this type have been numbered Ts20112 and their press-beds Ts262.

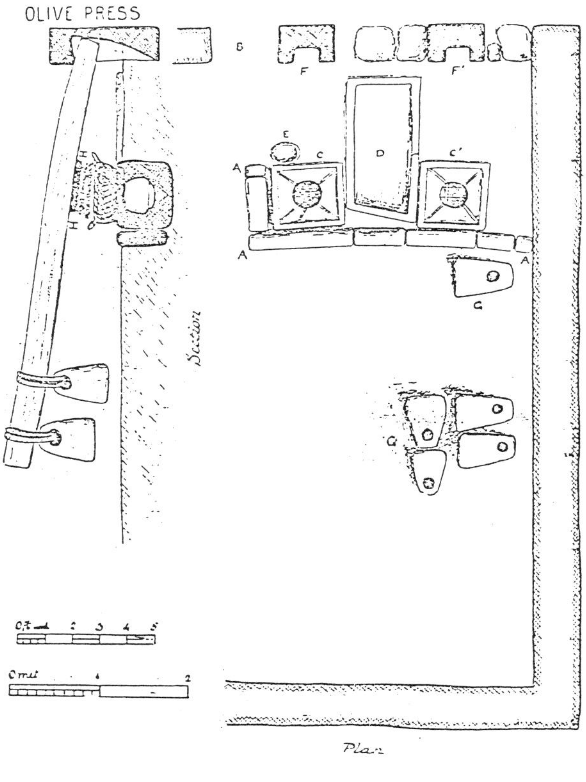

Figure 3. Lever and weights press, square central vat with radial grooves (T2613121): Gezer Inst. 01 (Macalister 1912: 2, fig. 257).

Many presses in Israel, however, operated in a different way: instead of the collection being lateral it was central. The oil collection vat was directly below the pressing point. Instead of two press components, press-bed and lateral vat, there was a single installation, a central collecting vat, the wide rim of which served as a pressing surface. In a pre-industrial example from Ṭafila, planks were placed criss-cross over the opening of the central vat and on these the frails were placed to be pressed. The oil then dripped into the small vat below. A perforated board could have served the same purpose. Simple presses of this type, the Bet Mirsham presses, have been numbered T20113, the central vats Ts261.

Both central vats and press-beds are found both free-standing (Ts26--1) and rock-cut (Ts26--2). The former are more common at the sites in the alluvial coastal plain and the latter at mountain sites where suitable rock outcrops are available in which to cut the installations.

The unusual character of the free-standing central vats of Bet Mirsham, no doubt led Albright (1941–43: 56-62) to explain them as 'dying vats'. Dalman

Photo 13. Free-standing central vats with groove and bore (T26111110): Bet Mirsham Inst. 07 (Albright 1941–43: Pl. 53A).

Photo 14. Rock-cut lever and weights press with central vat (T26112116): Qurnat el-Ḥarmiya.

(1928–42: V, 77-78) was already of the opinion that these were oil presses and he was clearly shown to be correct by Eitam (1979) who based his argument mainly on comparison with rock-cut examples from mountain sites discovered in surveys. In all, around 180 central collecting vats of this group (T261) have been recorded from 34 sites. About 100 of these installations were from one single site, Miqne, but have not as yet been published in detail. All but two of the sites where simple central vats were recorded are south of the Jezreel valley and the main concentration of these sites is in the south. A considerable number of these installations are from stratified contexts and the chronological picture that emerges is as follows:

Bet Mirsham. In Strata C and B—Late Bronze and Iron Age I—no beam presses, only Ts1221, 1222. In Stratum A—Iron Age II—at least nine beam presses with central vats.

Bet Shemesh. Simple installations T1221 were assigned to Stratum II, Iron Age, and beam presses to Stratum IIB–C—Iron Age II. It is not clear, however, on what basis.

Naṣba. Excavators state that the installations are from the Iron Age but continued in use in the Persian period.

Zeror. Persian period.

Mubarak. Hellenistic period.

Megiddo. A central vat was uncovered at this site and although published only in an air photograph, is still *in situ* and can be dated. It is below stables/storerooms of Stratum IV. These are dated to tenth or latest early ninth century BCE. The vat is therefore probably from Stratum V, eleventh or tenth century BCE.

Miqne. seventh century BCE.

Batash. seventh century BCE.

Thus the simple press with central vat can be clearly defined as a southern type, the first evidence for which is in Iron Age I. Its peak was in Iron Age II and it continued in use later.

There are three main sub-types. The most numerous is that with a borehole connecting a groove in the rim with the central vat (Ts2611; 36 examples), the second most numerous has no special device connecting the rim with vat (T2612); 24 samples—ignoring all but one of those from Miqne) and the third has radial grooves (T2613; 7 examples).

The most sophisticated sub-type is that with an auxiliary collecting vat which, as Eitam (1979: 149) suggests, is almost certainly a separating vat, the oil that rose to the top in the central vat then flowing into the auxiliary vat (Ts26112116, 26122106).

Photo 15. Free-standing central vat without connecting device (T2612110): Bet Shemesh Inst. 01 (Grant and Wright 1938: Pl. 20.4).

Because of the limitations of the data, more detailed spacial analysis can be only preliminary, but certain trends can be discerned. The centre of the distribution area of the dominant sub-type, that with connecting bore (T2611) is, according to present available data, to the north-west of Jerusalem, 15–18 WE, 14–18 SN. In this region very few examples of other sub-types were recorded. The second main sub-type without bore (T2612) is found mainly in the coastal plain—west of the 15 N–S coordinate. At Miqne, the biblical Philistine city of Ekron, this type appears in a square not a cylindrical shape, whereas at nearby Batash, and Gezer, it appears in both shapes, square and cylindrical. At Bet Mirsham in the foothills of Judah facing Philistia both these types as well as several others are found. The third main sub-type, that with radial grooves (T2613), is

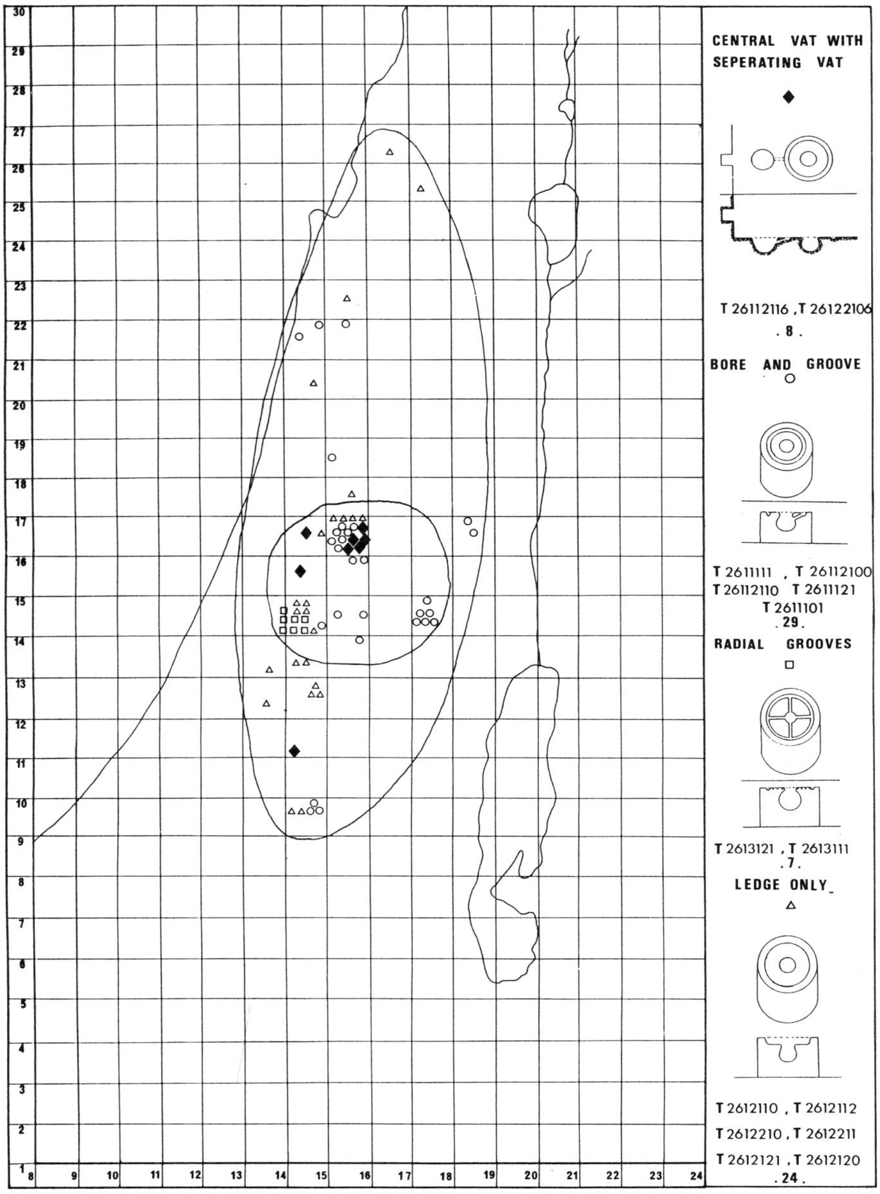

Map 1. Central collection: simple lever and weights press (Iron Age) (T261).

found at one site only, Gezer, but there in two shapes, both cylindrical and square. It is, however, not the only type to be found at Gezer, that is also on the foothills east of Philistia. The second type, without bore (T2612), has been found there in both shapes, but the dominant kind, with bore (T2611), surprisingly has not. Three examples of the dominant form, but square in shape (T2611121), have been found in the Carmel range. A square example of the type without bore (T2612121) has also been found in this region. It should be noted that the variants of the main type, both in form or shape, are located mainly in the periphery of the distribution area to the west and to the north.

Of the 23 simple beam presses recorded with lateral collection, 8 were found south of the Carmel. However, these represent around 5% of the beam presses recorded in that area, whereas the 15 recorded in the north represent around 87% of those recorded in that region. Lateral collection can therefore be defined as the northern type. The earliest and northernmost example of a lever and weights press in Israel is that from the Iron Age I sanctuary at Dan. In this twin press the press-beds are simple slabs of stone and the collection is in sunken *pithoi*. The weights were found *in situ* leaving no doubt as to this being a lever and weights press. By the eighth century BCE the Rosh Zayit press-bed had developed. This was a uniform round and free-standing press-bed with circular groove (T262211) and is to be regarded as characteristic of northern presses. Eight examples were recorded from five sites. At Rosh Zayit there is one example of a free-standing stone vat that apparently served as an overflow oil separator (T186).

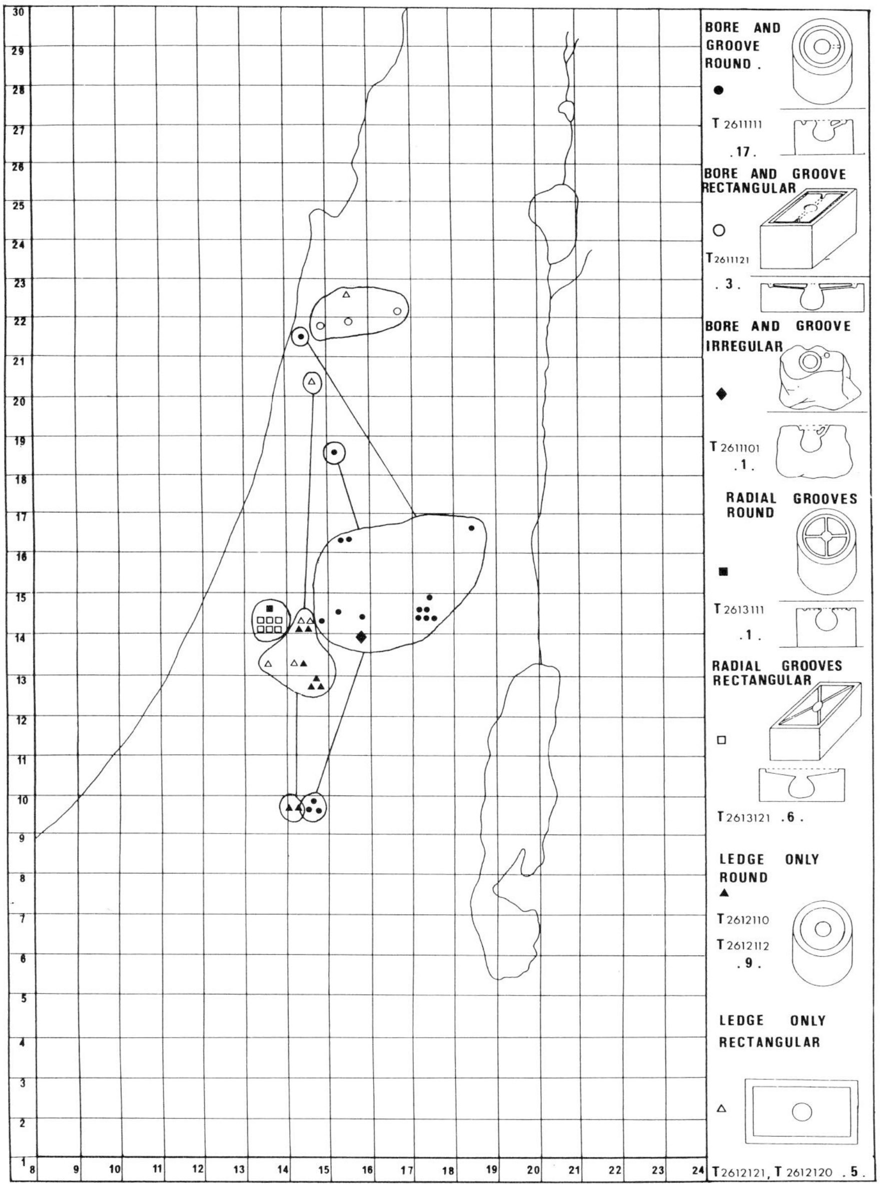

Map 2. Free-standing central vats (Iron Age) (T261-1).

Photo 16. Free-standing round press-bed (T262211): Rosh Zayit Inst. 06.

Photo 17. Free-standing round press-bed (T26221): Shiqmona.

Only two examples of central vats of simple lever presses have been recorded at northern sites. The one at Rosh Zayit replaced a lateral vat at a site where more of the typical northern press-beds were found than at any other. The central vat at ʿAvdon was recorded in a survey and, although it is fashioned more carefully than is usual, and although it was found at a site at which sherds from other periods as well as the Iron Age were found, it is probably to be dated to the Iron Age. Both these presses are to be regarded as a result of southern influence.

Photo 18. Rock-cut lever and weights press with central vat (T26122100?): ʿAvdon Inst. 02.

The lever and weights press was apparently introduced from the north. Those known from Syria and Cyprus have lateral collection and this is the method that continued to be used in the north in the Iron Age and to a large extent also later.

The most interesting question that remains is without doubt the origin of central collection. At Bet Mirsham, in the south, the presses of Iron Age I (Stratum B) are apparently not beam presses and have lateral collection. The beam presses with central collection appear at this site only in Stratum A—Iron Age II. At Megiddo, situated at the northern extremity of the distribution area of this type of press, an unusual type of central vat was found in an Iron Age I context of the eleventh or tenth century BCE. In spite of this paradox, and although as yet no Iron Age I beam presses have been identified in the south of the country, the distribution pattern of the central vat, with a clear centre where the type is uniform and a periphery of variants, strongly suggests a southern origin which will hopefully be clarified in further research (contra Eitam 1987: 27).

Photo 19. Free-standing central vat, square with groove and bore (T26111210): Megiddo Inst. 03.

2C. The Simple Lever and Weights Press in Other Countries

2C1. Syria

Stager (1990: 105 n. 4) has tentatively suggested that a perforated stone found in the northern collecting vat of the Early Bronze Age twin simple installation at Ras Shamra-Ugarit was a beam weight. It is, however, only c. 0.20 m long and too small to be an effective weight. Courtois's (1962: 423) suggestion, not ruled out by Stager, that it is a crusher is to be preferred, especially as there is no evidence for a lever and weights press from so early a context at any other site. However, lever and weights presses from the Late Bronze Age have been found at Ras Shamra-Ugarit in which the collection is lateral, the press-beds being round and oblong in shape (Ts262211, 262212).

2C2. Cyprus

The remains of lever and weights presses from the Late Bronze Age have been published from Maroni, Kalavasos, Myrtou and Apliki in Cyprus (Hadjisavvas 1987, 1988). They have all been dated to the final stages of the period, thirteenth century BCE or later. Here, too, the collection is lateral. Of the press-beds, one is round (T2622111) and three are rectangular, two of these with a spout (Ts2622121, 2622130).

2C3. Crete

From the palace of Knossos Evans (1900–1901: 182-83) published installations from the room south of 'The Corridor of the Gaming Table' (later called 'The Corridor of the Draught Board'), which he defined as press-beds, thus calling the room that 'of the Olive Press'. Later, however, Evans (1921–35: I, 378) revised his opinion calling the device a drain-head and the room 'The area of the Stone Drain-Heads'.

A very similar installation with a similarly long duct has been published from the palace at Vathy Petro from the Late Minoan period close to what is probably an oil separator (T182). It is unlikely that such similar installations should serve different functions at two sites so close chronologically and topographically, but from the photograph alone it is not possible to understand fully the installation from Vathy Petro. A pear-shaped press-bed from the same palace has been explained as part of a lever and weights press (Marinatos and Hirmer 1960: 140). A small round press-bed from the same period has been published from Palaikastro (T262211). Spouted press-beds from the same period (Late Minoan I) have been discovered in excavations at Kommos. Photographs of two-press beds from the palace at Phaestos have been published. Although the publication is not by the excavators, nor from an archaeological journal, it can be presumed they are from the period of the palace. One press-bed is pear-shaped with grooves, the other round with a spout.

As yet, a complete lever and weights press has not been published from Crete or any other area of the Aegean predating the end of the Bronze Age. Neither have press weights or a press niche been reported. However, the square 'press-beds' of Knossos and Vathy Petro have still not been satisfactorily explained and the pear-shaped examples from Kommos, Vathypetro and Praesos were certainly to produce oil.

To conclude, the presses from Cyprus and Syria date from the end of the Late Bronze Age, while the evidence suggests that lever and weights presses appear in Crete already at the end of the Middle Bronze Age.

Chapter 3

Olive-Crushing Devices (T3)

3A. Crushing Devices Associated with Simple Lever and Weights Press T2 (Ts1152, 1153, 126, 133, 134)

There were three simple methods of crushing olives: treading (see p. 46 above); pounding with pestle and mortar; or using a stone roller. With the introduction of the lever press, different types of crushing devices emerged, each apparently suited to one of these techniques.

Round mortars, rock-cut (T1152) or free-standing (T134) were clearly for pounding. Rectangular basins, again rock-cut (T1153) or free-standing (T133), were clearly for use with a roller (T139). It has already been pointed out that the deep vats with flat floors of T125 would appear to be unsuitable for either of these methods and probably served as a treading basin. The crushing basin at Dan is similar and probably functioned in a similar manner.

Apart from Dan, in the other northern sites, Qiri, Shiqmona, Rosh-Zayit, and at Balata slightly further south, the crushing was carried out in mortars. In the excavated sites in the southern coastal plain and in the foothills to the east, Miqne, Batash, Bet Mirsham and Gezer, the crushing basins were rectangular, apparently for use with rollers. Actual rollers were, however, published only from Miqne. Thus, again, a clear pattern of regional diversity emerges: mortars in the north and rectangular basins for use with rollers in the south.

In the southern hill sites the picture is less clear. Most of those installations in which the collection is lateral, all rock-cut, apparently lacked crushing installations. The olives were no doubt crushed on the press-bed itself. Installations in which oil collection was central were usually associated with crushing devices. These, however, were not uniform in character. Those at two sites, Banat Bar and Ḥudash, were round mortars, while those at a third, Qala, were rectangular. The explanation for this situation and its significance must await additional data.

It is of interest that, in the discussions as to how oil was to be prepared for the temple rites in the Mishnaic period, Rabbi Yehuda apparently believed that it should be prepared using the simple methods of the time of Solomon's Temple, and he contended that the olives should be crushed in a mortar (see discussion, Appendix 1).

3B. The Round Rotary Olive Crusher (T3)

3B1. The Round Rotary Olive Crusher in Israel and Environs

The most important development in the crushing of olives was the introduction of the round olive crusher (T3), the operation of which was based on rotary movement. The potential of rotary movement was first exploited in the potter's wheel that was invented in the third millenium BCE (Amiran 1956). This mechanical principle was adapted to olive crushing over two thousand years later, and to the grinding of flour almost certainly later still (Moritz 1958).

One of the advantages of rotary as opposed to reciprocal movement is that it is in one direction only and therefore continuous. In the case of the crushing of olives and the grinding of flour, this enabled the use of animal power and later of that of water, wind and machinery. The fact that rotary movement was first adapted to olive crushing and only later to the grinding of corn is probably to be explained by the fact that wheat was ground by reciprocal movement using a hand quern, while in many cases olives were crushed using mortar and pestle action that was partly rotary in character.

Photo 20. Pre-industrial olive crusher (T01712): Dura el-Qarʿa (Photo: Eli Elgart, Eretz Israel Museum).

Pre-industrial rotary olive crushers (T017) were still in use in many lands until very recently. The last one to function in Galilee at Fassuta was still operating in 1980. From these we learn that they consisted of a round crushing basin in which a round crushing stone (T0171) or stones (T0172) revolved. A horizontal shaft passed through a hole in the centre of the round crushing stone and was usually attached to a vertical pivot which rotated in a socket in the centre of the basin, while the distal end of the shaft usually served as a handle to turn the crushing stone. As a result, the crushing stone rotated both on its own axis and round the crushing basin. Sometimes the crushing stone was located on the shaft between the handle and the central pivot (T01711), sometimes on the far side of the pivot (T01712), and, rarely, the shaft did not serve as the handle and there was a separate handle placed at right-angles to it (T01713) or at another angle to it (T01714).

The date and place of origin of this revolutionary invention has still to be determined. In Israel, mortars and rectangular basins with rollers were in use until the end of the Iron Age. No oil presses from the Persian period have yet been published and the round rotary olive crusher was already in general use in the Hellenistic period.

As opposed to the typology of pre-industrial olive crushers, which can be based on the number of crushing stones and the way they are assembled (Ts017), the classification of ancient crushers (T3) must of necessity be based only on the stone components, specifically on the form of the crushing basin. The three main types are T31, flat crushing surface and sunken socket (41 examples—including sub-type 3102), T32, flat crushing surface and raised socket (86 examples—including sub-type 3202) and T33, concave crushing surface and raised socket (19 examples—including sub-types). The first two types (Ts31, 32) were used with wheel-shaped crushing stones (T38), while the third (T33) operated with convex crushing stones (T382).

Although few examples are stratigraphically dated, all three types are already found in the Hellenistic period.

In spite of the long chronological range of the installation, a clear spacial pattern is evident. Only 9 of

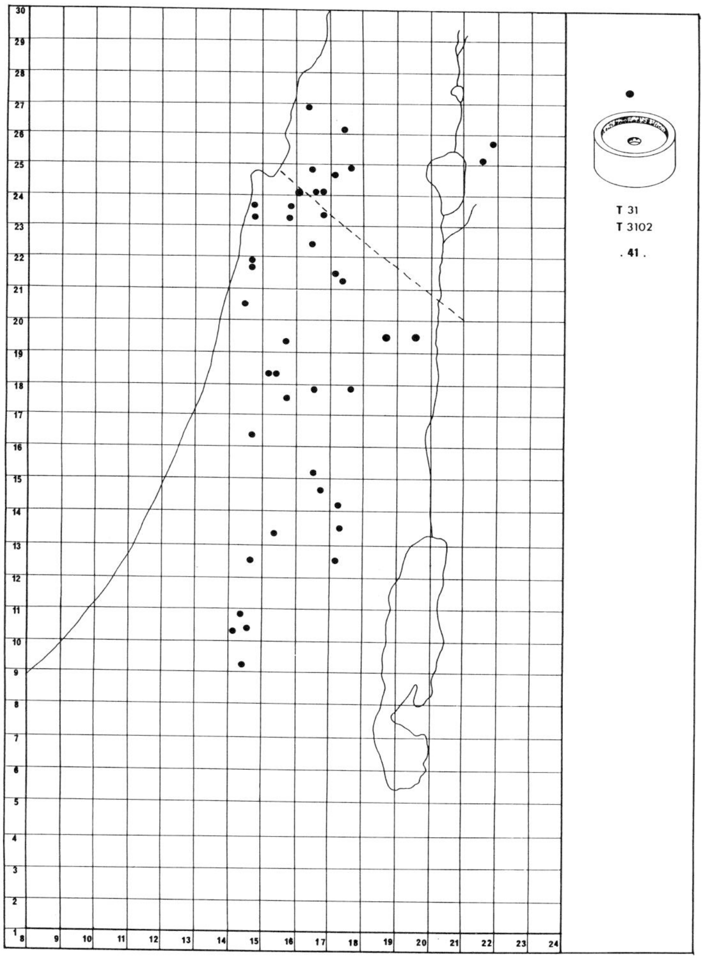

Map 3. Crushing basin, sunken socket (T31).

the 41 examples (21.9%) of those with sunken sockets (T31) are located north of the Jezreel Valley, while only 11 of the 86 (12.5%) of those with raised socket (T32) are located south of it. This picture is, however, almost certainly somewhat distorted by the fact that the survey of the northern areas has been more intensive than that of the south. The true picture would probably show a greater total number of T31, a lower percentage of this type in the north and perhaps a higher percentage of T32 in the south.

Photo 21. Round crushing basin with sunken socket (T31): Ṣur Natan Inst. 02.

T33, that with concave crushing surface, is rare, especially as nine of those recorded are from one site, Maresha, and two from another, Umm el-ᶜAmad. Here too, however, a regional pattern can be discerned. This is apparently the type to be found in the many installations located in the man-made caves in the western foothills of Judaea (the 'Shephala') that date to the

Photo 22. Round crushing basin with sunken socket (T31): Ṣur Natan Inst. 03.

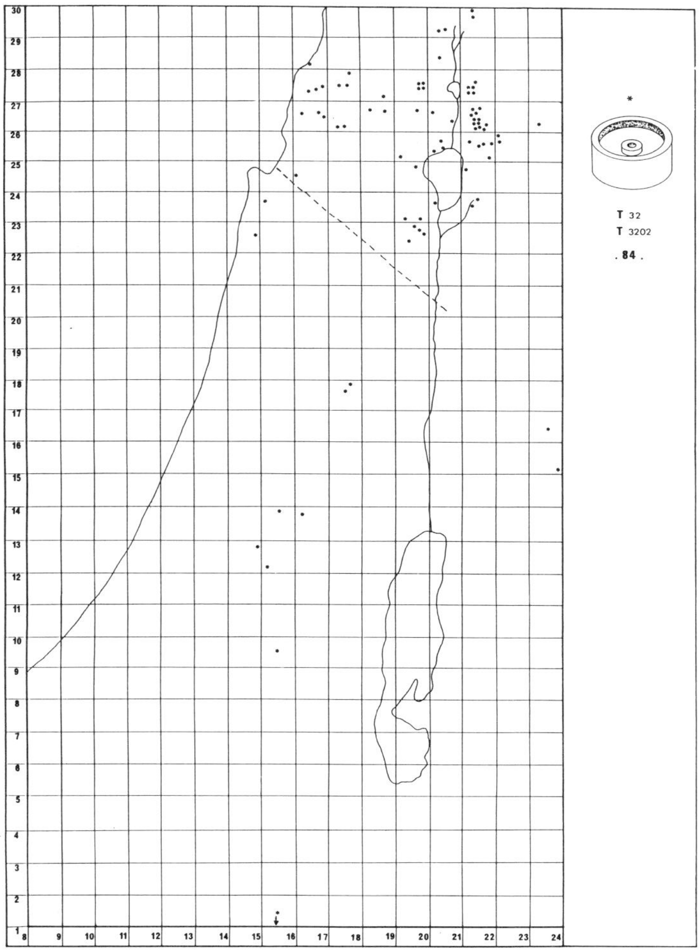

Map 4. Crushing basin, raised socket (T32).

Hellenistic and Roman periods. An interesting detail found at Maresha is the replaceable socket (T361) which perhaps explains the crushing basins with large square sockets (T312, possibly also T313, that with large round sockets).

Another feature of interest is the construction of the crushing surface in segments (Ts3-2, and see also T017002). In some cases this is probably a repair, but in others there is no doubt that it was constructed in this way to allow for replacements.

Ts322, 323 and 332 are types intermediate to the main types, but there are others that are fundamentally different. T341, which is clustered mainly in the square 16–17 WE, 26–28 SN, similarly to T33 has a concave

Photo 23. Round crushing basin with raised socket (T32): Tabgha Inst. 02.

Photo 24. Round crushing basin, concave crushing surface, central protusion, no socket (T341): Zabadi Inst. 02.

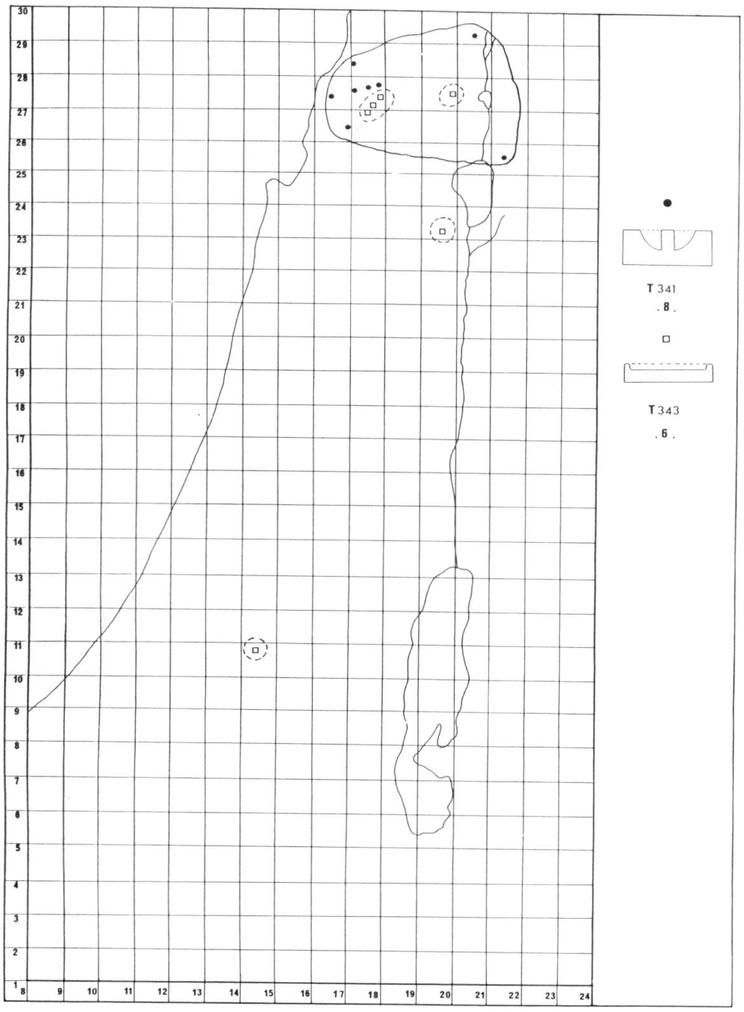

Map 5. Crushing basins without socket (Ts34).

crushing surface but lacks a socket. Two convex crushing stones doubtless revolved in it on one axle, keeping each other in position. In T343 the crushing floor is flat, but both socket and central protusion are lacking. It probably functioned in a similar manner to T341. In T351 from Dinᶜila, the crushing surface is convex but the central protrusion is also lacking. It is not clear whether the small hole in the centre is a socket or not. T352 from Deir is smaller but similar, consisting of a crushing basin with seven grooves on its inner face.

3B2. Crushing Installations in Other Countries

Several crushing methods are known other than that of the rotary crusher with vertical crushing stones. A series of pre-industrial semi-mechanical devices based on the mortar and pestle and the hammer exists, and these are operated by pullies, flexible boards, water or wind power (Ts018).[1] These apparently represent a separate technological development and there is neither archaeological nor written evidence for their use in olive or wine production in the ancient world. Similarly there is no evidence of an early history for the cake breaker (T0179) used in pre-industrial processes in some European countries.

It has been suggested that plain and slotted rollers (Ts861, 862) were used for crushing olives in sophisticated installations in North Syria (Callot 1984: 20-22). It will be shown below, however (see pp. 146-47), that these installations are probably wineries and the rollers probably served to crush grapes not olives.

Two other installations of unusual type certainly did serve to crush olives. The first is the horizontal oil mill (T37). This also is a rotary crusher with a fixed lower component and an upper rotary element. In the centre of the lower fixed part is a conical, or rather bell-shaped protrusion. On this a ring-shaped upper stone, bell-shaped in section, revolved. Whole olives were inserted at the top and olive pulp came out at the bottom. Slightly spiral striations on the crushing surfaces aided the process. A hole in the centre of the top of the bell-shaped portion was for a spindle. On either side of the upper stone were mortices which served both to secure handles and to hold a framework which rode on the spindle and allowed for adjustment of the height of the upper stone (Moritz 1958: 87). These installations have been reported from Volubilis in Morroco, and from the Lower Guadalquivir in Spain. At Volubilis these appear in some installations with the more usual rotary crusher and in others alone. In the former case it has been suggested that they served in the second crushing only (Brun 1986: 79). This mill has very close affinities to the 'Pompeian Donkey Mill' used for grinding grain (Moritz 1958: 74-90). It has been suggested that the conception of the rotary grain mill originated in the western Mediterranean (Moritz 1938: 116). It is therefore possible that in this region the rotary grain mill was adapted to use with olives before the rotary olive crusher (Ts31-33) penetrated this area.

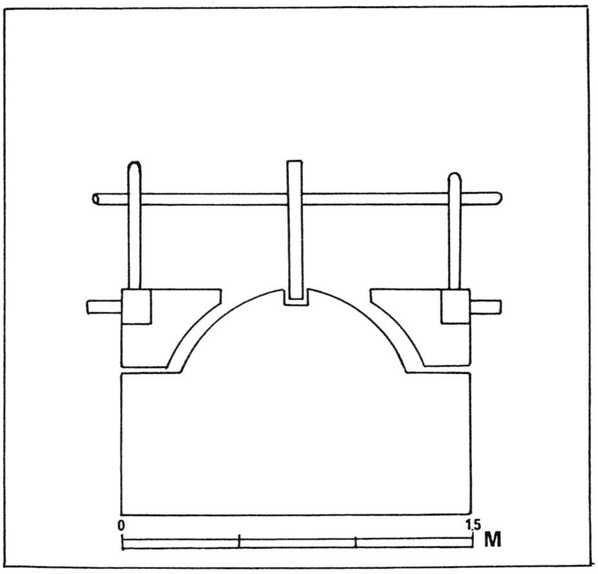

Figure 4. Horizontal oil mill (T37).

The second unusual oil crusher is the horizontal olive hand-mill (T375), a device that has apparently been overlooked in previous discussions on oil producing devices. Again the similarity to the rotary hand-mill for grain is great (Moritz 1958: 103-21), the difference being only the addition of a spout, clear evidence for its use for extracting a liquid. This device

Photo 25. Horizontal oil mill (T37): Volubilis, Morocco (Brun 1986: fig. 24).

has not as yet been reported from a stratified site and cannot therefore be dated. Two examples published are from southern France, and one from Italy.

There are pre-industrial parallels for both these devices. In western Galilee a horizontal oil mill was in use until recently (T01781), as was a horizontal treadmill in Egypt (T01783) and a rotary hand-mill for crushing olives in Tunisia (T01782). These are of interest technically. In this case, however, it is most unlikely that they represent a continuation of ancient traditions.

Apart from these exceptions, the other crushing devices found throughout the Mediterranean are part of same main technical tradition already touched upon in discussing the round rotary olive crushers found in Israel.

Ts31 and 32 are found in the countries of North Africa, including Egypt and those of the Levant and Cyprus. The only possible regional differentiation is that the few examples recorded from Morocco are of T31 and those from Algeria and Tunis T32. In North Africa there are three examples of a bowl-like crushing basin, apparently with central protrusion but lacking a socket (T342). Another variation found in western North Africa is the use of an elongated striated roller as a crushing stone (T389). It has pre-industrial parallels in the same region (T01752). Yet another variation found both in Syria and North Africa is the crushing basin with outlet for fluid (Ts31005, 32005). It is surprising that it is rare in both regions and unrecorded in others.

The distribution pattern of the type with concave crushing surface (T331) is quite different. It is found in

Photo 26. Round crushing basin, concave crushing surface, raised socket, the *trapetum* (T331): from one of the villas at Boscoreale, today at Pompei (Brun 1986: fig. 18).

the Greek mainland, southern Italy and Malta and possibly there is one example from Tunisia. The earliest dated example is that from Olynthus from the early Hellenistic period. The presence of the convex crushing stone (T382) adds Spain and the Greek island of Chios to this group. The one from the site of Pindakos on Chios is the earliest evidence for a rotary olive crusher of any type—and is dated to the Greek classical period. In southern France a variant of convex crushing stone is found more rounded and striated (T383).

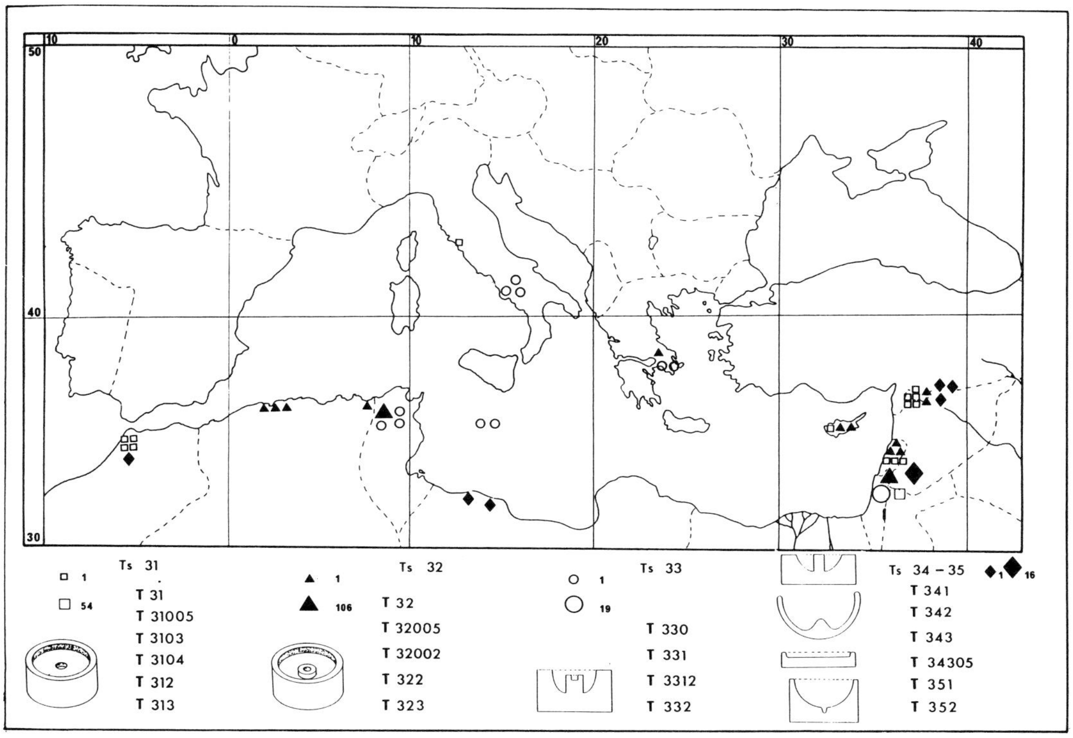

Map 6. Crushing basins, Mediterranean Basin (Ts3).

3B3. Crushing Installations in Ancient Literature and Historical Conclusions

It is usually thought that *ym* and *mml*, the Hebrew terms for two of the components of the Mishnaic oil press (*B. Bat.* 4.5), are the crushing basin and stone, and these terms appear with these connotations in dictionaries of modern Hebrew. I will show below, however, that this interpretation of the text is incorrect and that the Mishnaic terms were actually *rḥym ʿlywnwt* and *rḥym tḥtwnwt*, upper and lower millstones (see Appendix 1).

Cato the Elder, the first of the great Latin agronomists, who wrote in the first half of the second century BCE, gives detailed instructions on how to make and assemble a crushing device that he calls the *trapetum*. He also explains where to buy the crushing stones and gives prices (Cato 20–22). With the discovery in the late eighteenth century of the Campanian Villa Rustica, destroyed in the same eruption of Vesuvius that destroyed Pompei, scholars identified the *trapetum* with the crushing devices found at these sites (Boscoreale-Giuliana; Boscoreale-Pisanella). These were particularly finely made examples of the oil crusher with concave crushing surface (T331) operating with two crushing stones (T382). There is today scholarly consensus on this identification and it is usual to call these devices by their original Latin name, *trapetum* (Drachman 1932: 7-8).

Columella (12.52.6-7) returns to the subject of olive crushers about 200 years later, and explains that '*molae* (mills) are more practicable than a *trapetum* and a *trapetum* more practicable than a *canalis* (channel) and *solea* (sole of shoe). *Molae* are more easily managed since they can be either lowered or raised according to the size of the berries so that the kernel which spoils the flavour of the oil is not broken. Moreover the *trapetum* does more work and with greater ease than the *solea* and *canalis*. There is also a machine that resembles an upright threshing sledge which is called *tudicula*. This performs its function without much trouble except that it frequently gets out of order and if you put a little too many berries into it it becomes clogged'.

Columella has here added three crushing devices: *mola*, *canalis* and *solea*, and *tudicula*, or perhaps four if the *canalis* and *solea* are different devices as the Loeb translation of the *et* as 'or' would imply. The *tudicula* has been convincingly shown by Laporte (1978) to be a bronze object rectangular in shape and eliptical in section at its short end, with rows of knobs on both sides. Examples of these have been found at several North African sites from the Roman period. Different suggestions have been put forward as regards the meaning of *canalis* and *solea*. Hörle based his explanation on the first term and suggested the *canalis* to be the trough of the 'Reciprocal Crusher' (T0177) and therefore that the *solea* is the crusher that rolls backwards and forwards in it, the term deriving from the sole of the shoe figuratively in the same way as the term *trapetum* derived from the Greek term for treading grapes. White (1975: 226-27) and Brun (1986: 69), however, take the term *solea* literally, suggesting that it is actual treading of olives that is referred to, the *canalis* being the treading vat. Brun brings the treading vat from the Villa Rondanini relief as an example.

Brøndsted (1928: 112) has suggested that the *mola olearia* is the rotary crusher with flat crushing surface (Ts31, 32), operating with two wheel-like crushing stones. He bases his thesis on comparison with a relief of what he contends is such a device from Arles and with a pre-industrial installation with one crushing stone from Salone-Environs (T01711). Drachman (1932: 42) supports this view bringing another analogy from the Villa Rondanini relief. Since then it is usual to use the term *mola olearia* for rotary olive crushers Ts31, 32 (White 1975: 228; Brun 1986: 73-78). The literary evidence and the archaeological picture from Italy suggested that the olive crusher with concave crushing surface (T331, the *trapetum*) appeared before the one with flat crushing surface (Ts31, 32, the *mola olearia*).

It would appear, however, that these two olive crushers are too similar to deserve such different names; the latter is not closer to the true *mola*, the grain mill, than the former and perhaps most important, in spite of Drachman's (1932) contentions to the contrary; neither is the latter easier to adjust for height in order to avoid crushing the kernels. There is in fact no evidence for adjustment for height in pre-industrial flat olive crushers although Cato (22.1-2) devotes two sentences to explain how to adjust his *trapetum* for height.

A far better candidate to be the *mola olearia* is the horizontal oil mill (T37) found in North Africa and Spain. It has close affinities to the donkey mill, the true *mola* of this period, and as has been shown above, had a special apparatus to raise the upper stone. The fact that Columella was a native of Cadiz makes it all the more likely that he should recommend a device that probably originated in Spain. If this suggestion is correct, then the Latin term *trapetum* was perhaps not limited to the rotary crusher with concave crushing surface (T331), and it would be preferable also to abandon it as a *terminus technicus* in modern scientific parlance.

One of the questions that has been a subject of study is the number of crushing stones used in the various types of olive crushers. Forbes and Foxhall (1978) have suggested that the concave crushers of Olynthus and Pindakos operated with one crushing stone only, and Kloner and Sagiv (1987) have suggested the same as regards to those from Maresha. In most cases, however, the archaeological data does not suffice to answer this question satisfactorily. It is perhaps significant that the majority of the pre-industrial crushers use one crushing stone. It is very probable that the pre-industrial crusher with two or more crushing stones (T0172) does not represent a continuation of an ancient tradition but is an early pre-industrial innovation.

The point of departure for any discussion of the early history of the olive-crushing process must be to stress that the main types of rotary crusher, both those with flat and those with concave crushing basins, are so similar technically that there can be no doubt that they had a common origin. The questions that then arise are the form of the original prototype, its date and its origin. The concave crusher apparently reached Italy fully developed, probably from Greece. Although the evidence is sparse, the earliest dated component of such a press is a crushing stone from Pindakos in Chios, a Greek island about 10 km from the Anatolian coast. It is perhaps significant as Forbes and Foxhall (1978–79) have pointed out that this crushing stone is slightly less convex than the later examples.

As yet the material from Anatolia has not been published but, in the Levant, Syria, Lebanon, Israel and environs, the flat crusher is numerically predominant and the few examples of the concave crusher are apparently intrusive and not an integral part of the main cultural stream. The picture in North Africa is similar to that in the Levant.

Although no flat crushers as early as the Pindakos stone have been published, there are dated examples from the Hellenistic period and it would appear probable that the more sophisticated concave crusher developed from the simpler flat type. If this is so, it is likely to have occurred in the region where this type is most common and deep-rooted. If the two types already existed side by side at the beginning of the Hellenistic period, some time must be allowed for development and differentiation.

In conclusion, the invention of the rotary olive crusher can tentatively be dated to the second quarter of the first millenium BCE. The region of origin is certainly the eastern Mediterranean, probably the Levant or Anatolia, and the form probably that with flat crushing surface (Ts31, 32).

Note

1. An interesting pre-industrial semi-mechanical mortar and pestle is the Indian 'girani', in which the pestle is rotated by a long beam turned by oxen (Achaya 1993).

Chapter 4

The Improved Lever Press (T4)

No complete oil presses have as yet been discovered that can be dated to the Persian periods. Those from the Hellenistic and Roman period, however, although almost all lever and weights presses, were different from those of earlier periods, not only in size but also in the new solutions found for the technical problems that arose at all three function points of the press, the anchoring and pressing points and the point of applied force. As regards the latter, in later periods presses developed in which methods other than weights were used to apply force to the beam. The main types were the 'lever and drum' and the 'lever and screw' presses, the latter often using screw weights. The various types of beam presses have many characteristics in common and these will be discussed in this chapter, but the beam weights will be discussed in Chapter 5 and screw weights in Chapter 6.

4A. Improved Lever Presses in Israel and Environs

4A1. The Maresha Press: Lever Press, Central Vat, Plain Piers (T40111)

At several sites in the western foothills of Judaea, the 'Shephala', there was a very uniform type of lever press. The beam was anchored in a niche, collection was central and two plain stone piers stood one on either side of the central vat (Ts46111, 46112, 46113, 4617). These plain piers served two purposes. The first was to support the pile of olive mash frails during pressing. The inner faces of the piers are often concave to accommodate the frails (Ts46112, 46113). The second purpose was to keep the beam raised in order to allow work to be carried out under it before and after pressing. Small grooves are often found on the top of the piers. In these the rod was placed to which the beam was tied or on which it rested. In several of these presses an additional small elongated vat was located immediately in front of the central collecting vat (T46113). Apparently, however, liquid did not flow directly from the central vat into this auxiliary vat which must therefore be explained as a work area used before and/or after the pressing.

The earliest presses of this type are lever and

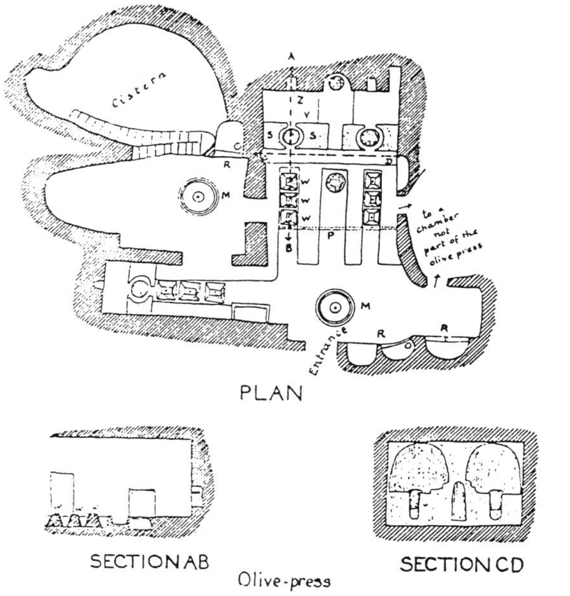

Figure 5. Improved lever and weights press, southern type: Maresha Inst. 09 (Bliss and Macalister 1912: fig. 92).

Photo 27. Improved lever and weights presses (T401110002): Maresha Inst. 15 (Photo: Amos Kloner).

weights presses from Maresha dated to the Hellenistic period, while a lever and screw press of the same type from Bet Loya is from the Byzantine period. The main concentration of the Maresha press (T40111) is in the 'Shephala', where it is the only type of beam press found. Examples of this or very similar types are found both further north (e.g. Gamla with radial grooves [T46121], first century CE; Sumaq, Shush) and further east (Yajuz), while plain piers from unexcavated presses (T4617) have been identified also in eastern Upper Galilee.

The Maresha press is clearly a direct development from the Bet Mirsham press with central vat typical of the southern regions of the country in the Iron Age (T20113). Not only are both presses based on central oil collection, but in both the main focus of attention is on the pressing area. That is the part of the press in which the greatest variation is found in the Iron Age and in which the main change took place in the later press, the addition of plain piers.

4A2. The Zabadi Press: Lever Press, Lateral Collection, Slotted Piers (T40222)

The Zabadi press found in western Galilee is different from the Maresha press in almost every way. Three examples of this press have been excavated at Zabadi, Karkara and Umm el-ᶜAmad. Instead of the niche in which the beam was anchored in the Maresha press, in the Zabadi press it was anchored between two slotted piers (T421). The oil collection is not central as in the Maresha press, but lateral. The press-bed is round and

Photo 28. Improved lever and weights presses (T402221203): Zabadi Inst. 02; from south.

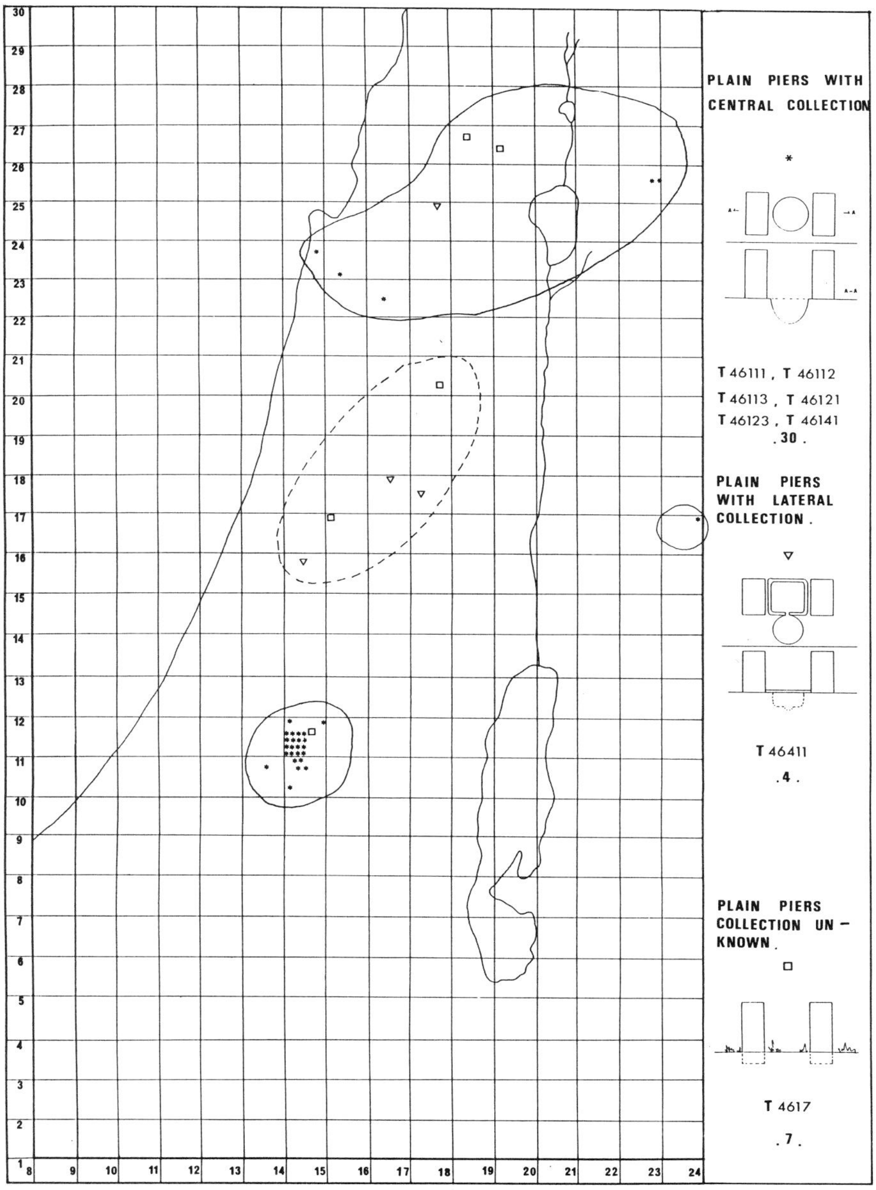

Map 7. Plain piers.

Photo 29. Improved lever and weights presses (T402221203): Zabadi Inst. 02; from north.

Photo 30. Improved lever and weights presses (T402221203): Zabadi Inst. 02; western press from east.

Photo 31. Improved lever weights and screw press (T402221203/8): Karkara Inst. 01.

large, reaching a diameter of 2 m, and there are two collecting vats connected at the rim thus allowing for oil separation by overflow decantation.

The existence of slotted piers in Galilee was, until recently, overlooked and, when this type was recorded, it was confused with types more familiar from other regions (e.g. Matfana-Ḫirbet el-Medfane; Dalman 1928–42: IV, 228; Gichon 1979: 200). The manner in which the slotted piers were operated is learned from pre-industrial lever presses from Europe (e.g. Bosco Tre Case; Chateau De Vinzelles; Champvallon; Loché),

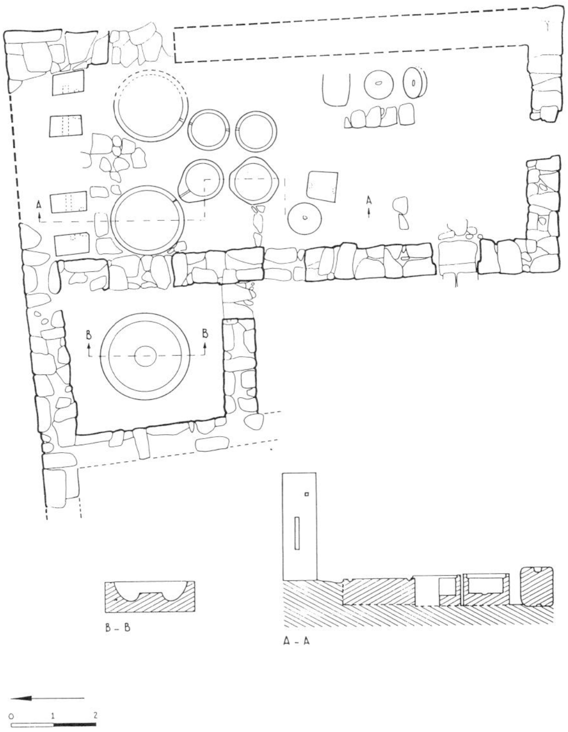

Figure 6. Improved lever and weights press, northern type: Zabadi Inst. 02; plan and section.

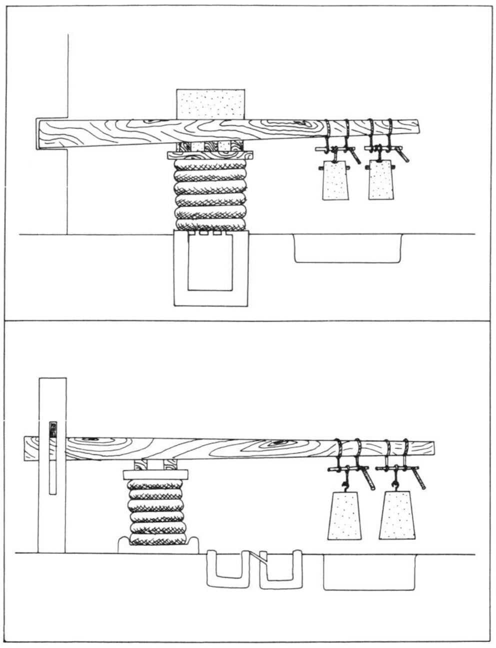

Figure 7. Reconstruction of southern Maresha press (above) and northern Zabadi press (below)—comparison.

although in these the piers are of wood, not stone. The two piers stood a short distance apart (0.70–1.00 m) and a third stone was laid across the space between them,

Photo 32. Slotted piers (T42120): Karkara Inst. 03.

giving them the appearance of a gateway. This 'lintel' adds weight to the press to withstand the forces created by the lever press and sometimes additional stones were even laid on it. On the inner faces of the piers were narrow, vertical rectangular slots, the slots of the two piers being opposite each other. Wooden planks were fitted horizontally with one end in each slot and the beam was held in place between them. Insertion and removal of planks enabled the height of the beam to be adjusted.

The possibility of readjusting the height of the beam is clearly connected to the fact that the olive pulp was pressed several times, producing oil of progressively poorer quality in each pressing (see above p. 45).

Slotted piers appear in several sub-types as a result of there being two groups of auxiliary devices, each serving a different purpose. Those of the first group facilitated the insertion and readjustment of the anchoring planks. One method was for one of the slots to be open, that is cut right through the pier (T4212). At two sites excavated, Karkara, and Zabadi, there were twin lever presses, and in both cases the open slots of both presses faced a central corridor allowing the planks of both to be inserted and adjusted from one position. The second method of inserting the planks was by means of a specially cut insertion channel (T4213). In many cases, however, there was no special device for this purpose and the planks were apparently inserted at an angle and then put into place (T4211). The devices of the second group served to fasten a horizontal beam-rod above the slots. To this rod the beam was apparently attached while the anchoring planks were being readjusted. The first of these devices consisted of a small hole on one pier and a notch on the one opposite (T421-1); the second consisted of a small hole on one pier and an angular groove at the edge of the pier opposite (T421-2); and the third consisted of a hole on one pier and an angular groove at the centre of the top of the pier opposite (T421-3). A fourth category consisted of those piers lacking an auxiliary device for this purpose (T421-0). In these installations the beam was apparently supported by a forked prop or some similar appliance. The combination of the two groups of devices allows theoretically for 12 sub-types (3 × 4).

Photo 33. Slotted piers (T42123): Shubeika Inst. 01.

Photo 34. Slotted piers (T42113): Kafr Yasif Inst. 01.

In spite of the small numbers of examples, certain spacial patterns can be discerned in the distribution of some of these sub-types. Those with no auxiliary device (T42110) and with insertion channel (T42130) are concentrated more to the north, while those with upper hole and channel (Ts42113, 42123) are found farther south. The most sophisticated of the slotted piers are those with open slot, hole and angular channel (T42122). These are usually also those cut most carefully. It is of interest that this type is concentrated in a comparatively small region to the east of the two similar but somewhat simpler types: those with hole and notch (Ts42111, 42121). This distribution pattern suggests that the installation improved as it penetrated eastwards into the less inviting hill country.

Only three presses of the Zabadi type have been excavated, but 71 examples of slotted piers (T421) have been recorded from 56 sites. As will be seen, not all of these were part of presses T40222; two excavated examples were in presses with central oil collection (Quṣeir 01; 03 Stage 1 T40226). However, the slotted

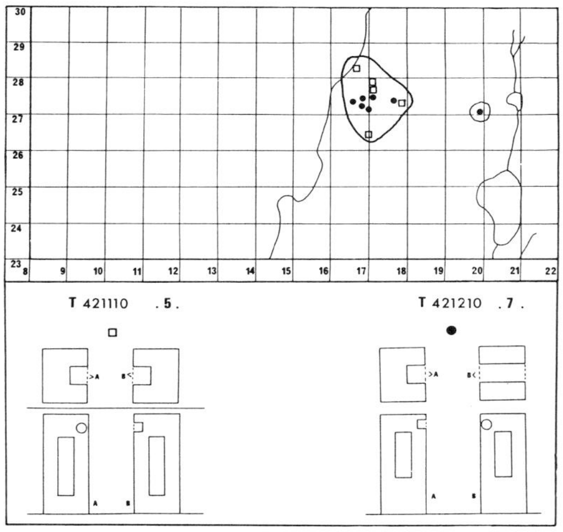

Map 8. Slotted piers with hole and notch.

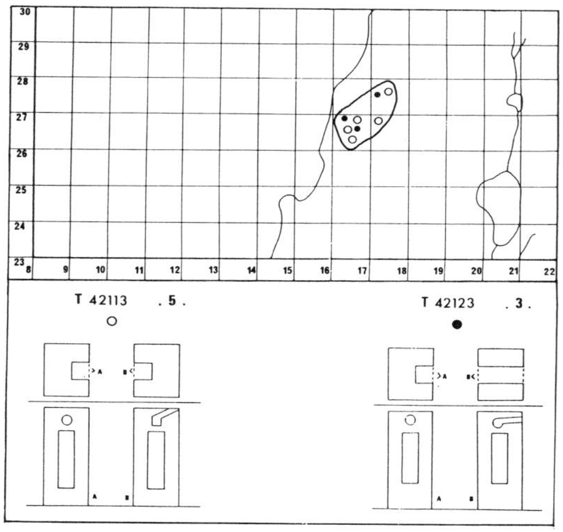

Map 9. Slotted piers with upper hole and notch.

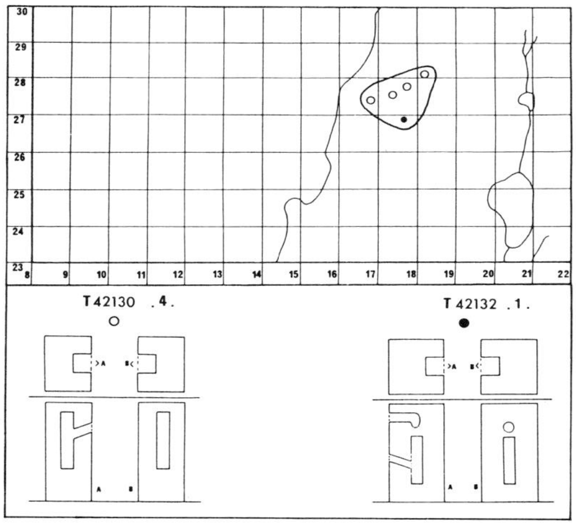

Map 10. Slotted piers with insertion channel.

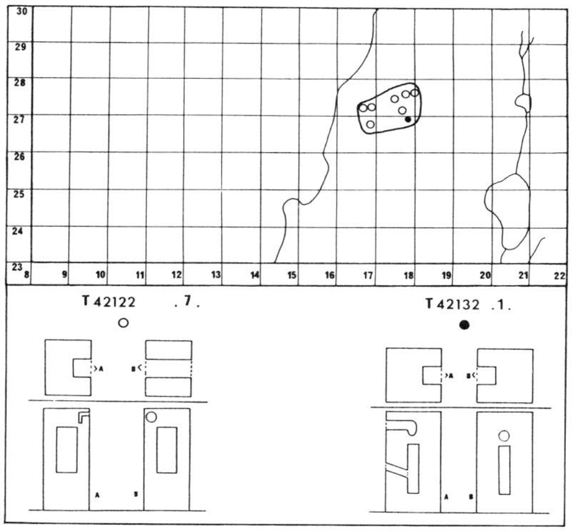

Map 11. Slotted piers with hole and angular channel.

Photo 35. Lever and screw press with central collection and slotted piers: Quṣeir Inst. 03 press B, stage 1 (T402260047); stage 2 (T403160047), complete press.

piers can be regarded as the characteristic element of this press, and its spatial distribution pattern is clearly defined. The northern extreme of the distribution area could not be determined because it is north of the international border, but only three examples are located south of the 26 EW coordinate (Sirin, Usha, Qastra) and only three are east of the 18 NS coordinate, including one of the previous three (Sirin, Marus, Quza). Thus, this type is limited to the region known today as western Galilee. This was ancient Phoenicia and the border of the distribution area of the slotted piers coincides almost exactly with that between Phoenicia to the west and the region to the east known in ancient times as Upper Galilee and later as Palaestina Secunda. The

Zabadi press can therefore be regarded as a Phoenician type. The press shows certain clear affinities to the Rosh Zayit press, typical of the northern part of the country in the Iron Age. The oil collection is lateral and the press-bed is very similar.

The oil press described in the Mishnah (*B. Bat.* 4.5) is of this type. The Mishnah was largely collated in Lower Galilee, which would explain the references to a press of Phoenician type (see Appendix 1).

The Zabadi press has also some important characteristics in common with the press described by Cato (18), the most important being the slotted piers called by him *arbores* (see below, p. 83).

The earliest dated example of this type of press is that from Zabadi from the Roman period. However, a press with a very similar anchoring contrivance, the slotted niche, which consists of one block of stone instead of two piers, is dated to the Hellenistic period (Umm el-ᶜAmad 01, Ts4023, 4221). Three other examples of this type are rock-cut and undated (T4222). This device is very probably the prototype from which the slotted piers developed, although it is likely that the

Photo 36. Slotted niche (T4221): Umm el-ᶜAmed.

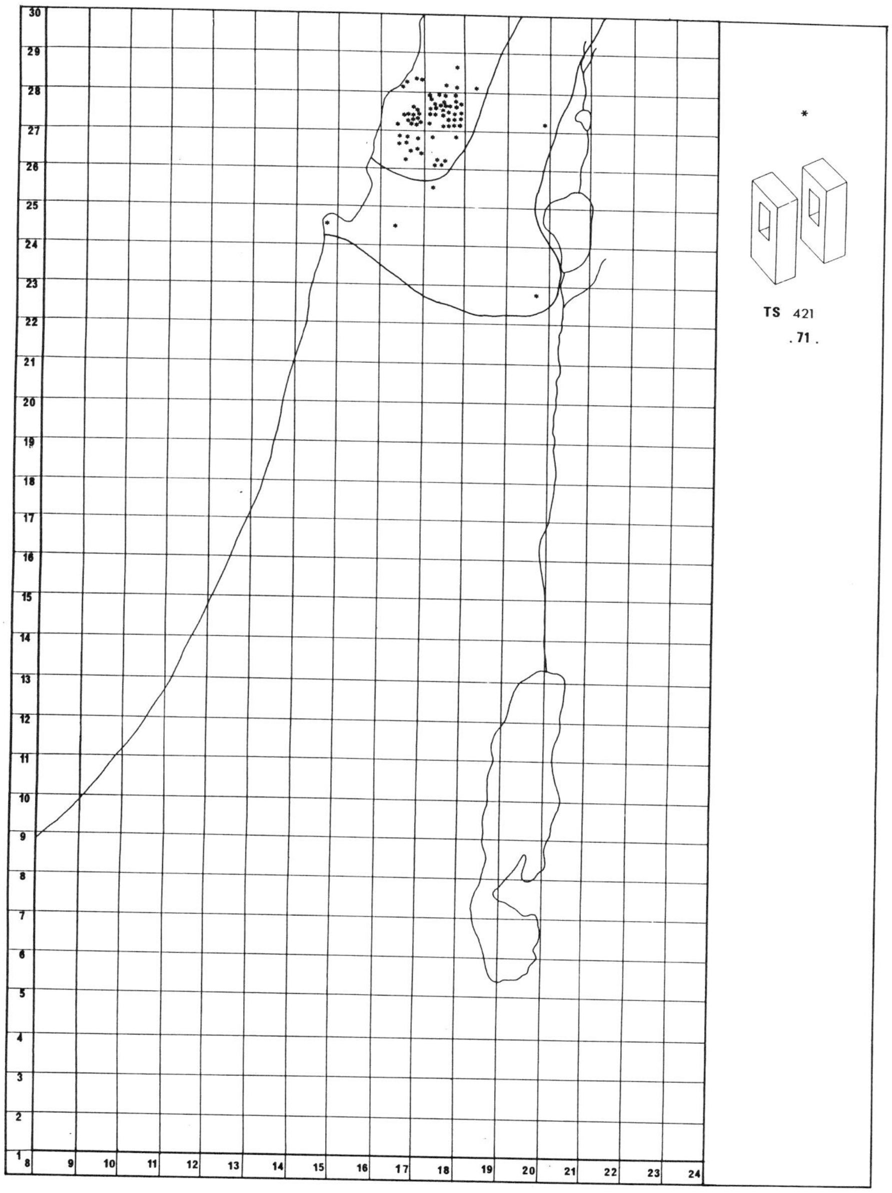

Map 12. Slotted piers (T421).

two types existed side by side. Slotted piers were still in use at the end of the Byzantine period in a lever and screw press at Karkara 01.

4A3. Perforated Piers (T423)

Perforated piers are similar in size and shape to slotted piers (Ts421) and when found *in situ* occupy the same position in the press, showing that they also served the same purpose, that of anchoring the beam of a press. The perforated piers differ from the slotted in that instead of a slot there is a single perforation to secure a rod to which the beam-end was attached. The beam-end was therefore always in one fixed position and could not be raised and lowered as in the case of the slotted piers.

In those cases in which perforated piers were found in association with weights, it was always with screw weights (Ts62; Najjara, Siyar el-Ghanam, Mavo Modiᶜim, Midya, Diniᶜla 01, Quṣeir 03). This suggests strongly that perforated piers were used in Israel only with lever and screw presses (see below, Chapter 6). In the three oil presses from Quṣeir it is possible to follow the transition from one method to the other. In Quṣier press A1 there are two of the simplest type of screw weights—Miᶜilya weights, T62711—together with slotted piers. In press B there is one of the more sophisticated and much larger Bet Ha-ᶜEmeq weights (T62511). In this press in the first stage the piers are slotted, but at a later stage the slots were blocked up and a perforation added (T4232). In press C the same type of weight as in press B is used, but from the start the beam was anchored in a perforated niche (T4123).

The change is probably because the use of the screw allowed for the weights to be much heavier than beam weights. These weights averaged around 1.4

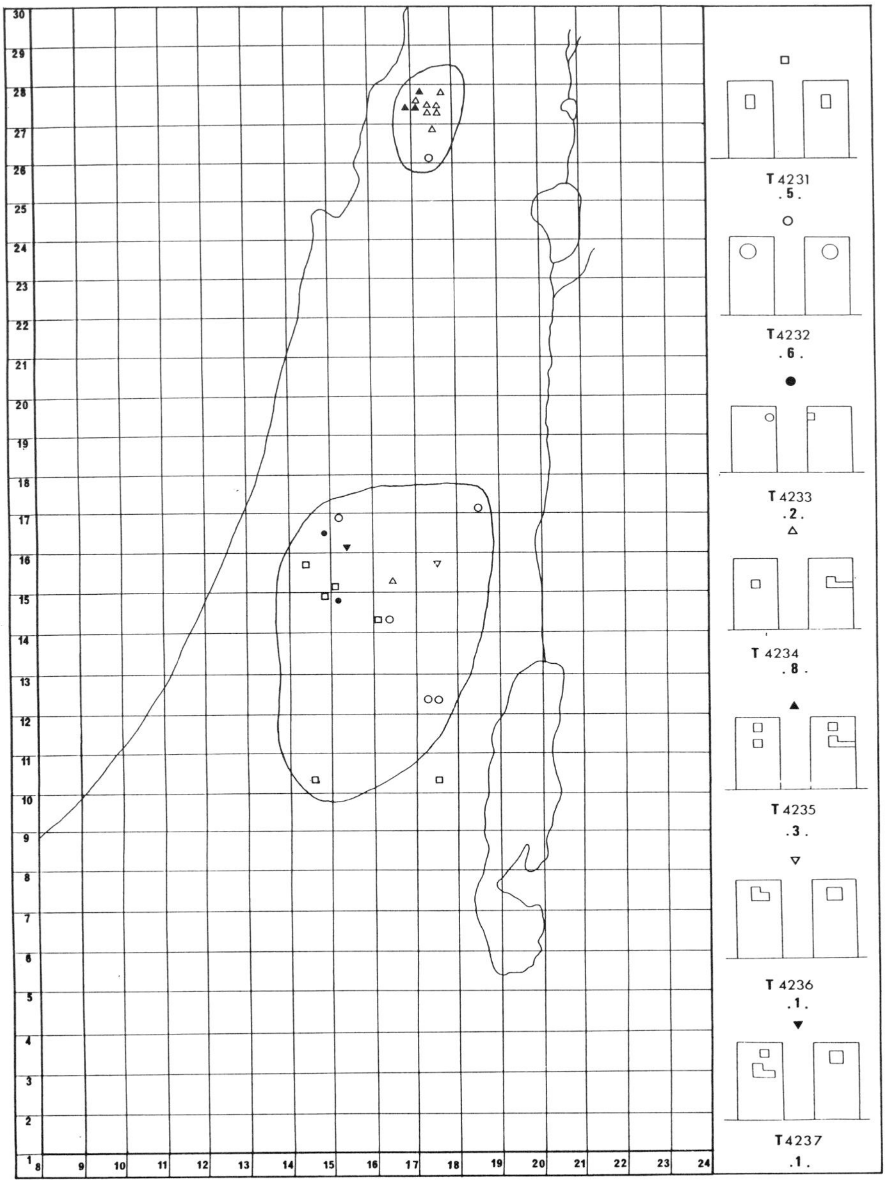

Map 13. Perforated piers.

tonnes but were sometimes much larger—the one from Sebastiya, Samaria, 02 reaching around 3.3 tonnes. The great additional weight resulted perhaps in the second and third pressing being waived. It should be pointed out, however, that in lever and screw presses from other regions and other periods, including pre-industrial presses, this was not the case and second and third pressing was usual. The waiving of the additional pressing in Israel in the Byzantine period was probably a result of the sudden great improvement achieved by the introduction of the new technique. Only later perhaps was a return made to additional pressings.

One other technical point must be raised, however. The slots of the slotted piers were to lower the beam-end in order to keep the beam horizontal as the pile of pressing bags became lower through compression. The same effect could be achieved by adding planks or similar objects between the beam and the pressing board as the pile of pressing bags lowered. This method could have been used both with slotted and perforated piers, but it apparently was not.

Twenty-five examples of perforated piers have been recorded from 21 sites. These were located mainly in two regions, in southern Samaria and in a small region in Upper Galilee. The variation in type of perforated piers, similarly to that of the slotted piers, is mainly because of differences in the way the rod was placed in position, whether by one of the perforations being open (Ts4231, 4232), by the use of an open notch (T4233), an open-angled channel (Ts4234, 4235), or a closed-angled channel (Ts4236, 4237). In some cases an additional perforation is present (Ts4235, 4237). These, however, were almost certainly not an alternative position for the beam, but for a rod to reinforce the main rod in anchoring the beam. The piers in Samaria vary greatly in type whereas those in Upper Galilee are extremely uniform, all using an open angular groove.

It is reasonable to see in the perforated piers of Upper Galilee a direct continuation of the slotted piers found in the same region. The origin of the perforated piers in Samaria is more difficult to determine. It would seem improbable that they are directly connected to those of Upper Galilee.

It should be pointed out that the perforated piers of Israel are different from those of North Africa (Ts424) in several ways. The two most important are that the Israeli piers have one perforation and were used in lever and screw presses, whereas those from North Africa had almost always several perforations and were used in lever and weights presses.

4A4. Guider Mortices (T44)

4A4.1. Two Mortices below Beam (T441). In three presses from Israel, those of Bet Loya 06, Ḥermeshit, and Sebastiya, Samaria, two small mortices were found in the floor of the press on either side of where the beam would have been. These mortices would have held two posts that stood on either side of the beam. Such posts are known from pre-industrial presses (Ts012--5) and can serve two purposes. In some cases these act only to keep the beam in line, for example, Fez, Tezarin Taghzuth, Morocco. More usually, however, these are slotted and serve as intermediate piers. Drachman (1932: 122-27) has described how piers of this type functioned in a press from Bosco Tre Case in Italy. By raising and lowering the free end of the beam and using the cross pieces inserted in the slots of the intermediate piers like the pivot of a see-saw, it is possible to lower and raise the fulcrum/anchoring end of the beam, on condition that the latter is left free. This method is apparently always connected to lever and screw presses of the types in which turning the screw one way will lower the beam and turning it in the opposite direction raises it (see below, Chapter 6, Models 6A–6L [Pre-ind. Ts0121]).

Many similar guider mortices of slightly differing types (Ts441, 4412, 442, 443, 444, 446) are found in North Syria and in Cyprus, all in lever and screw presses. The three presses from Israel where guider mortices were found are also lever and screw presses and are from sites situated not far from each other. In that from Bet Loya the crushing basin (T343) and screw weight (T62441101) are so similar to Syrian examples that there can be almost no doubt that here Syrian influence has been exercised (Frankel, Patrich and Tsafrir 1990). It is therefore possible that the same is true of the other two examples recorded from Hermeshit and Sebastiya, Samaria.

Guider mortices were also found in North African presses, but there it is less certain that they were used in lever and screw presses (see below, pp. 95, 107-108).

4A4.2. Two Mortices below Free End of Beam (T4481). In the presses from Quṣeir A1, B and C, two small mortices were found below the point at which the free end of the beam would have been. These three presses are lever and screw presses equipped with socketless Miᶜilya and Bet Ha-ᶜEmeq screw weights (Ts62711, 62511). In presses of this type it was the nut, placed above the beam, that was turned and not the screw. Turning the nut in the reverse direction would in this case not have raised the beam (see below, p. 112). It must, therefore, be assumed that the two posts that were fixed in the mortices served to hold pulleys and tackle to raise the free end of the beam after pressing.

Mortices placed below the beam-end are known from several presses in North Africa and North Syria but these are usually single and probably served to hold a forked prop that was perhaps used to lift the beam and keep it raised (Ts4482, 4483, 4484). Similar single mortices found below the beam itself in presses from North Africa and Yugoslavia (T447) and central mortices of three found in North Africa and North Syria (Ts445, 446) almost certainly served the same purpose. However, openings in the wall below the beam-end at San Paul Milqi, Malta, and Chersonesos, Crimea (Ts4485, 4486), possibly also held posts for pulleys like those from Quṣeir.

Such devices were clearly needed in beam presses and there is no reason to seek connections between those from different regions or sites except those of particularly unusual character, for example, the octagonal mortice stone (T4482).

4A5. Other Types of Improved Lever Presses from Israel and Environs

The distribution patterns of the other types of lever presses that can be regarded as intermediate, both typologically and geographically, between the two main types, the Maresha and Zabadi presses, are less clearly defined.

4A5.1. Lateral Collection. Of the eight presses with lateral collection and with neither slotted nor plain piers (Ts40112, 40312), four are in Samaria, one in the Carmel region, two in Upper Galilee and one in Golan. Of the three presses with lateral collection and plain piers (T40114), two are from Samaria, including the one from Tirat Yehuda from the Hellenistic period and the third from Lower Galilee.

4A5.1.2. The Qedumim Press-Bed and Vat (T47114). An unusual type of lateral collecting vat is found in six presses from five sites in Samaria. It consists of a round press-bed and square vat with stone cover with a small round opening through which to ladle out the oil.

4A5.2. Central Collection. Of the 17 presses with central collection but with neither plain nor slotted piers (Ts40113, 40116, 4031, 46110, 46120, 46124), one is in Judaea, four in Samaria, three in the Carmel region, three in Lower Galilee, two in Upper Galilee and two in the Golan.

Only two presses have been identified with both slotted piers and central collection, both from Quṣeir, (T40226), an excavated site at the southern extreme of the distribution area of slotted piers (T421). However, several central vats have been recorded from unexcavated sites at the eastern extreme of the distribution area of the slotted piers and these were also probably part of presses that combined slotted piers with central vats (T46124; ʿAlya, Miʿilya, Zawit, Fassuta).

Photo 37. Lever and screw press with central collection and slotted piers: Quseir Inst. 03 press B, stage 1 (T402260047); stage 2 (T403160047), central vat.

4A5.2.2. Central Vats with Radial Grooves (T4612). Numerically, these have been included in the installations with central vats of other categories already discussed. However, they must also be treated as a separate category. Of the 22 recorded, one is in Judaea, four in Samaria, five in the Carmel region, ten in Galilee and three in the Golan.

4A5.2.3. Central Vats for the Production of First Oil. At Quṣeir A2 (02) a central vat with radial grooves was found in a room with neither pressing equipment nor sufficient space for a press. In an adjacent room there was a lever press. This independent central vat was apparently to produce high-quality oil which dripped from the crushed olive pulp without use of a press. Afterwards the remaining oil of poorer quality was extracted in the adjacent press room. It is probable that the central vat at Qedumim 01 (T46124) served a similar purpose. The first oil is that prescribed in the Mishnah for use in the temple rites and is called in Latin *lixivium* (see pp. 47, 194-95).

Photo 38. Central vat for first oil (T46124): Quseir Inst. 02 press A2.

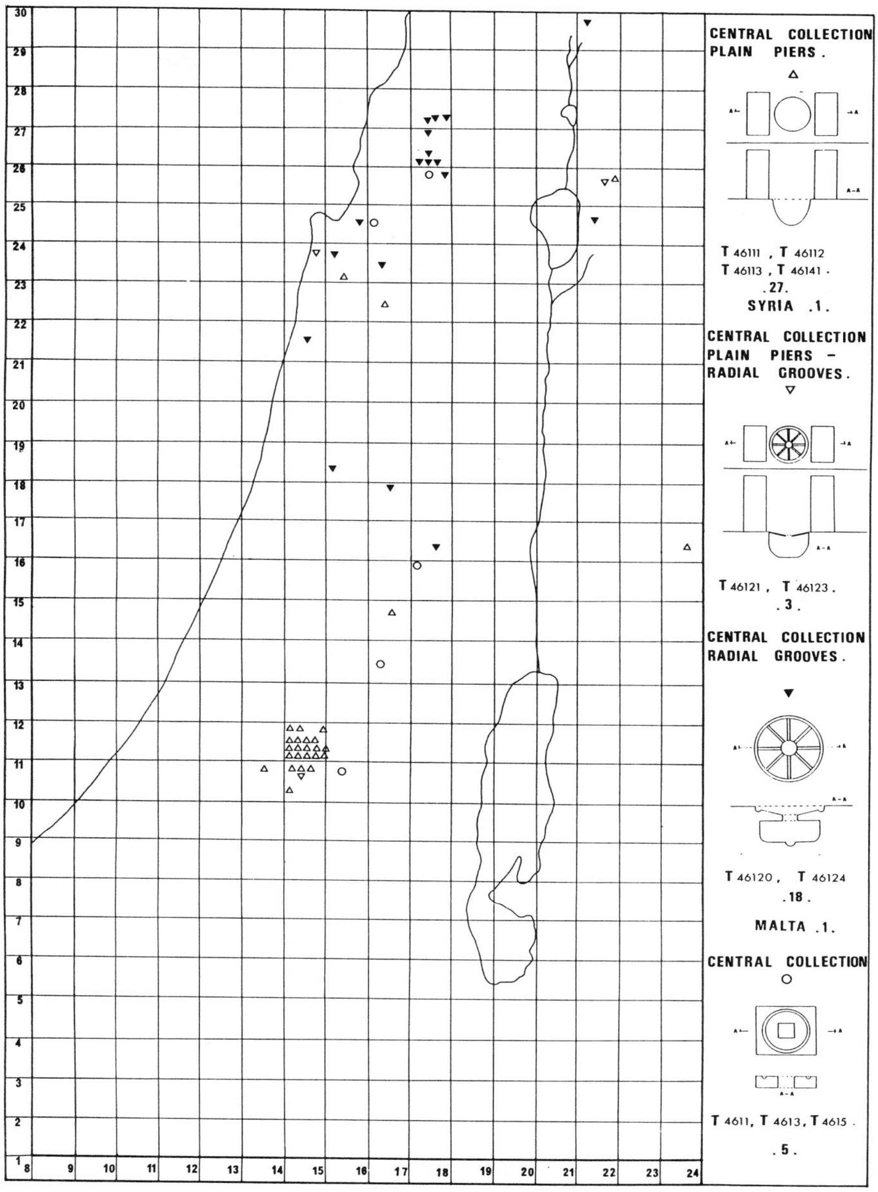

Map 14. Central collection in improved lever presses.

4A5.3. Press-Bed with Radial Grooves (T4624). At Dini[c]la 03 an unusual installation was excavated. It consists of a press-bed with radial and circular grooves and central round opening made up of segments. However, instead of being connected to a central vat, it lay directly on bed-rock and the expressed liquid flowed to twin lateral collecting vats. This peculiar technical hybrid should be regarded as an attempt to adapt a device—radial grooves and central opening—to a technique to which it was not suited: lateral collection (see also the Hamad Press p. 144 below).

4A6. Improved Lever Presses in Israel and Environs: Conclusions

Two clearly defined types of press have been identified: the southern Judaean Maresha press with beam anchored in a niche, central collection and plain piers, and the northern Phoenician Zabadi press with beam anchored in slotted piers, lateral collection and large round press-bed. In the intervening regions the presses are much less uniform. For example, both slotted and plain piers appear with both lateral and central collection. Local variants can also be distinguished: the Qedumim vat and guider mortices below the beam are found only in Samaria, while the central vat with radial grooves is found mainly in Galilee.

In the Sharon and Samaria laterial collection is dominant, whereas on the Carmel and in Lower and eastern Upper Galilee central collection is the main method used. If central collection is regarded as a Judaean characteristic, then these northern regions evince a greater affinity to Judaea than do the intervening regions of Samaria and Sharon. This fits well the historical picture of Samaritan Samaria being located between the Jewish regions of Judaea and Galilee.

4B. Descriptions of Presses in Classical Literature

4B1. Cato the Elder

Cato the Censor, the Roman statesman (234–149 BCE) in his agricultural handbook, *De Agricultura*, describes in great detail how to construct a pressing room consisting of four beam presses (Cato 18) and four olive crushers (Cato 20, 21, 22). He also gives instructions on how to construct a wine press that is very similar to the olive presses (Cato 19) and gives a long list of the equipment needed for olive yard, vineyard and press rooms (Cato 11–13). Excepting minor details, scholars are agreed as to the form of Cato's presses (Hörle 1929; Drachman 1932; Jüngst and Thielscher 1954, 1957; Brun 1988: 236-47), and Cato's technical terms have become those generally used in relevant scientific literature.

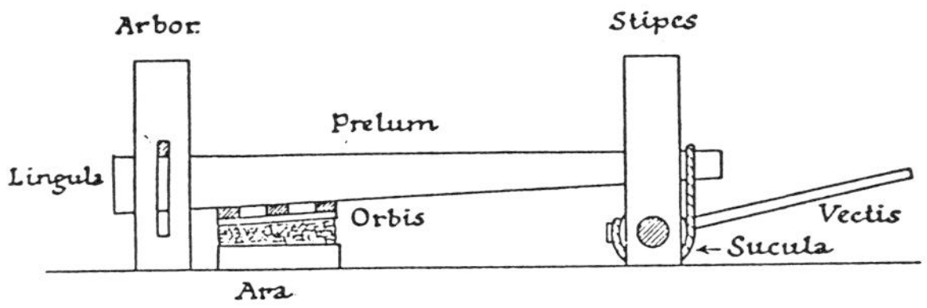

Figure 8. Cato's lever and drum press (Drachman 1932: fig. 12).

The press was a lever and drum press without weights. The fixed end of the press-beam (*prelum*) was narrowed to form a tongue (*lingulum*) which was anchored between two wooden piers (*arbores*) with slots (*foraminae*). The piers were secured in a stone pier base (*forum*? [Brun 1986: 242 n. 34]; *lapis pedicinus*? [Drachman 1932: 105-106]; Brun explains the *lapis pedicinus* as the bottom end of the pier, the tenon that fits into the pier-base mortice). The cross-pieces that were inserted in the slots were perhaps termed *cunei* (Cato 12; Jüngst and Thielscher 1954: 66). On the press-bed (*ara*) was a circular groove (*canalis rotundis*). Above the material to be crushed and below the beam was a pressing board (*orbis olearius*), constructed with 'Phoenician joints' (*punicanus coagmenta*). The free end of the beam was lowered by a drum (*suculum*) that was secured between two wooden piers (*stipites*) and turned by special levers/handspakes (*vectes*).

Beam length	25 ft (7.4 m)
Slotted piers	2 ft × 2 ft × 9 ft (0.59 m × 0.59 m × 2.66 m)
Slots	3.5 ft × 6 fingers (1.05 m × 0.15 m); 1 ft (0.3 m above the ground)
Circular groove of press-bed	Diameter 4.5 ft (1.33 m)
Pressing board	Diameter 4 ft (1.84 m)

There is apparently no mention of a collecting vat.

4B2. Vitruvius

Vitruvius (c. 25 BCE?) in his treatise on architecture, *De Architectura*, devotes a chapter to the farmhouse. There he stipulates:

> ipsum autem torcular si non cocleis torquetur sed vectibus et prelo premitur ne minus longum pedes XL constituatur...
>
> The pressing room itself, if the pressure is exerted by means of levers and a beam and not worked by turning screws, should not be less than forty feet long...
>
> Vitruvius 6.6.3

As the text implies that if a screw press were used the building could be smaller, Drachman (1932: 126) suggests that Vitruvius was referring to a direct screw press (Ts7, Pre-ind. Ts013, 014). Drachman then points out that Pliny the Elder in 77 CE states that the direct screw press (Pliny's press D) had been introduced '22 years ago'. Accordingly, he suggests lowering the date of Vitruvius to the first century CE.

4B3. Pliny the Elder

Gaius Plinius Secundis in his encyclopaedic work *Naturalis Historia* (c. 23–79 CE) presents a short but extremely important historical survey of the history of the press (Plinius 18.74.317). The text is difficult, however. Pliny appears to be describing four presses. Recently, however, Jüngst and Theilscher (1957: 108-109) have suggested the text to be corrupt and by minor changes have restored it to one describing three presses only, the descriptions of B and C being that of one press only—B/C. Pliny's press A reads:

> Antiqui funibus vittisque loreis ea detrahebent et vectibus.
>
> Our forefathers drew them [the press-beams] down by means of ropes, leather thongs and hand-spakes.

This is clearly a description of a lever and drum press similar to that described by Cato. For press B/C, the text as we have it reads:

> Intra c annos inventa Graecanica,* mali rugis per cocleam ambulantibus ab aliis adfixa arbori stella aliis arcas lapidum adtollente secum arbore,* quod maxime probatur

Drachman (1932: 52-53) translated this as follows:

> Within the last hundred years there have come into use presses invented in Greece,* spars with furrows running round them in a spiral, some people putting handles on the spar others making the spar lift up chests of stones with it,* which is very much praised.

Various suggestions have been made as to the form of Pliny's press B (e.g. Brøndsted 1928: 109, fig. 104; Drachman 1932: fig. 14; Roll and Ayalon 1981: 121). However if Jüngst and Thielscher's revision of the text is accepted, these become superfluous. The latter suggest *ab aliis* be revised to *malis* and *aliis* to *a malis*, translating (the Latin text between the asterisks, translation from German):

> In these the thread of a screw is inserted in a nut and a star-shaped turning wheel (*Tummellbaumstern*) is attached to (below) the screw and nut and the press beam screw and nut raise chests of stones.

It had already been proposed that Pliny's press C was a lever and screw press—the second main category of presses (Pre-ind. Ts012; Drachman 1932: 55-56) in this case the type in which the screw raises a box of stones (Pre-ind. Ts012162, 012163). Jüngst and Thielscher's revision shows that the whole of this section (B/C) refers to a press of this type. Their suggestion is to be accepted, not only because none of the proposed reconstructions of press B are satisfactory, but also because the whole passage in its revised form has a logical structure—three sections each opening with a chronological statement and each describing one press. The press D section reads as follows:

> Intra XXII hos annos inventum parvis prelis et minore torculario aedificio breviore malo in medio derecto, tympana inposita vinaceis superne toto pondere urguere et super prela constuere congeriem.
>
> Within the last 22 years people have invented a press with shorter presses and smaller press houses, with a shorter spar straight in the middle bearing down with full weight from above on the lid laid on the grapes and to build a superstructure above the press.

This is a description of a press belonging to the third of the main categories of presses, the direct pressure non-lever presses (Ts7), in this case a single rotating direct-pressure screw press (Pre-ind. Ts0131, 0134).

4B4. Hero of Alexandria

The writings of Hero of Alexandria (middle first century CE) concerned with presses, although written in Greek, have survived only in Arabic translation. In at least two of the manuscripts there are in addition diagrams, some of which were apparently also copied from Greek originals.

There is some controversy as to the dating of Hero of Alexandria. It was thought that he wrote in the first century BCE, because the internal evidence suggested that he was a pupil of Ctesibius who lived in the reign of Ptolemy Eugertes II (Heath 1910–11). However, it has been suggested that the date should be lowered, one of the reasons being the fact that he describes direct-pressure screw presses (presses C and D) which, according to Pliny, were introduced only in the first century CE (Drachman 1932: 125-28). As regards this argument, however, it is far from certain that this type of press was introduced to Rome and to the Levant at the same time. Hero is, however, certainly not to be dated to later than the first century CE.

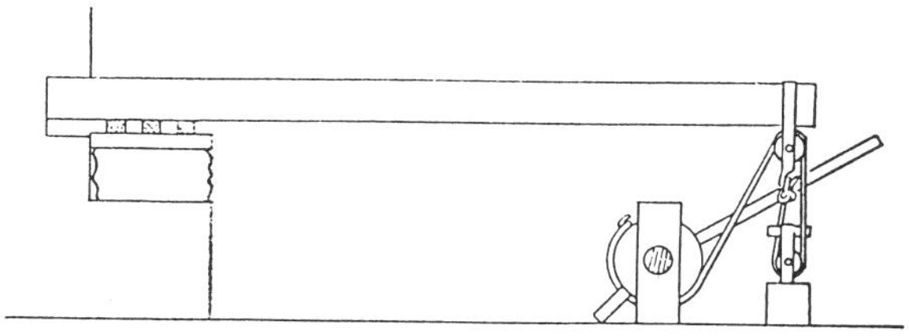

Figure 9. Hero of Alexandria's press A: the lever, drum and weight press (Drachman 1932: fig. 20).

Hero Press A. (Hero, *Mechanica* 3.13-14; Drachman 1932: 63-67, fig. 20; 1963: 110-15). This is a lever weights and drum press. The beam is anchored in a niche in the wall. A rope attached to a drum secured to the ground passed through a pulley at the end of the beam, then through a pulley attached to the weight and finally is apparently attached to the beam-end. Turning the drum raised the weight, which was then tied to the beam using special attachments found on both the weight and the beam. The substances to be crushed, grape skins and stalks, were enclosed by a rope wound around them in the form of a cylinder. In the text the collecting vessel is not mentioned, but on the diagram on the manuscript from Leiden a jar is shown. The beam was 25 cubits long (11.56 m) and the stone weight weighed 20 talents (524 kg).

The comparable pre-industrial installations did not operate in exactly the same manner. In T0113 from Isfahan, Persia, there is a drum attached to the ground and a pulley on the beam but not on the weight, and during pressing the weight was not tied to the beam but hung freely on the pulley and drum. In T0114, found in several countries, there is no pulley on either beam or weight, but after the weight and beam are raised by the drum the weight remains attached to the beam as in Hero's press.

Hero Press B (Hero, *Mechanica* 3.15; Drachman 1932: 70, 72, figs. 23, 24; 1961: 115-22; Jüngst and Thielscher 1957: 105-107). This is a lever and screw press. There is no mention of the way the beam was anchored in this press or of the form of the press-bed or collecting vats, but as the description is a continuation of that of press A, the stated intention being only to suggest a preferable way to lower the beam because of the dangers of the hanging weight, it can be presumed that the other parts of the press were as in press A. The top end of the screw was hinged to the beam-end by a complicated block, the description of which is not clear. The lower end was screwed into a wooden nut of the same length as the screw. This was made by cutting a rectangular block in half lengthways, cutting a female tread in the exposed cut and gluing the halves together. The lower end of this nut was attached to a screw weight in such a manner as to allow it to rotate but not to work free of the block. There was a socket in the centre of the upper surfaces of the weight and the lower end of the wooden nut was attached to the weight by metal wedges (Drachman 1963: 121-22). Crossed holes in the bottom of the nut served to hold hand-spakes. Turning these caused the nut to turn and swallow the screw and to lower the beam.

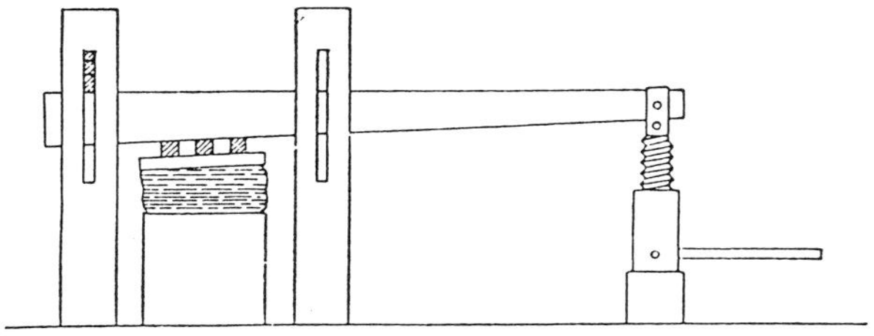

Figure 10. Model 6L: Hero of Alexandria's press B (Drachman 1932: fig. 23).

There are apparently no pre-industrial presses comparable to Hero press B. The weight, however, would be suitable for use with presses of T01216 and the Dinᶜila screw weight T622 would fit Hero's description (see below, p. 113). As is apparently often the case, Hero probably suggested an improved version of an existing press that was probably never produced.

This press, however, solved one of the problems inherent in the usual models of lever and screw presses: the tensions caused by the end of the beam moving in a circle and the screw being straight (see discussion, p. 107). It is also of great importance because it showed that Hero knew of the existence of a screw weight—thus providing a *terminus ante quem* for the introduction of this device.

Hero Press C (Hero, *Mechanica* 3.19; Drachman 1932: 57, 73-76, figs. 25-26; 1963: 126-33, figs. 50a-50d). This is a double-screw direct-pressure press in which the screws are attached to the screw base in such a way as to allow them to rotate without working free. This is achieved in a similar manner to that in which the screw nut was attached to the weight in press B and the same Arabic word, *ḍb*, is used for the wedges that were apparently inserted into dovetail channels to hold the lower end of the screw. Two female screw threads were cut in the pressing board and turning the screws in these lowered it and brought pressure to bear. This press has been reconstructed in different ways:

1. With a frame and handles at the upper end; Nix's reconstruction (Hero, *Mechanica*, fig. 60; Drachman 1963: fig. 50d) (Model 7D1 below, fig. 32).
2. Without a frame and with handles at the upper end; Drachman's first reconstruction (Drachman 1932: figs. 25, 26) (Model 7D2 below, fig. 33).
3. Without a frame and with handles at the lower end; Carra de Vaux's reconstruction and Drachman's second reconstruction (Drachman 1963: figs. 50a, 50c). On the British Museum manuscript there is a diagram illustrating this press with handles at the upper end. Drachman, however, decided to ignore it in his second reconstruction because he considered it so unlike the description in the text. Several pre-industrial presses work on this principle (T0144). However, it would appear to be very unlikely that these continue an ancient tradition (Model 7D3 below, fig. 34).

Hero Press D (Hero, *Mechanica* 3.20; Drachman 1932: 57, 76, fig. 27; 1963: 133-35). This is a single rotating screw press within a frame. There is no evidence in the text as to the position of the handle, but in the diagram in the Leiden manuscript it is at the top end, as it is in Pre-ind. T0134.

4C. Lever Presses in Other Countries

4C1. The Eastern Mediterranean

4C1.1. North Syria. The impressive presses of northern Syria have been known to Western scholars for over a hundred years (de Vogue 1865–77; Butler 1903: 8, 37, 268; Tchalenko 1953: 360-72), the most recent and fullest publication being that of Callot (1984). Callot's aim was to present the great variety of installations found in this region and to attempt to explain all the possible ways in which these could have functioned—often one installation is reconstructed in several ways. As a result, only a small part of the wide survey on which this work was obviously based was actually published. Therefore any quantitative assessment based on the published material must inevitably be partly distorted. Callot presents all the installations he publishes as oil presses. However, two basically different types of installation can be distinguished. Although these have certain components in common, primarily the beam niches and press weights, the pressing area and collecting vats are completely different. In the first type there are a round press-bed and one (T404-213) or twin (T404-214) square collecting vats. In the second type there is no press-bed but instead a

Figure 11. Winery from Behyo, Syria, Inst. 50 with typical monolithic beam niche with dovetail base (T41410001): note plain rollers (T861) (De Vogue 1865–77: Pl. 118a).

large pressing platform and a bell-shaped collecting vat (T404-298).

As will be shown below, in Israel and its immediate environs large pressing platforms/treading floors are found only in wineries and oil presses are equipped with clearly recognizable press-beds. In other regions, however, it is not always easy to distinguish wineries from oil presses. In France, Italy and parts of North Africa, installations are found that are almost certainly oil presses, but lack press-beds and are associated with small pressing platforms (T467) or large ones (T466).

In North Syria, however, there is a clear differentiation between wineries and oil presses. The installations of the first type with press-bed and square collecting vats (Kafr Nabo 01, 02, 11, 12; Taqle 01; Sergible 01; el-Kjeir 03; Sarfud 07; Banaqfur 50) are all associated with olive crushers (T3), and are clearly oil presses. Of the other type, those with large pressing platforms (T466), none is connected to olive crushers, but they are often associated with stone rollers (Ts862, 862), which in Israel, as I will show below (pp. 146-47), are connected to wineries. Two of the installations under discussion lack lever presses (Deir Mismis 04, Behyo 11) and are clearly wineries and not oil-separating plants as has been suggested (Callot 1984: Pl. 98) and there is little doubt that the other installations with large pressing platforms and bell-shaped vats (T404-298) are also wineries.

Of the seven oil presses, six have twin square collecting vats (T47121) and these without doubt functioned as overflow oil separators similarly to the twin round collecting vats of Galilee (T4722).

If the bell-shaped collecting vats (Ts476) are of wineries, then their suggested function as separating installations (Callot 1984: Pll. 92, 93, 95, 96) must be revised. The fact that the mouth of the vat is higher than the opening through which the must flowed was because the treading floor was lower than the working level from which the must, probably after first fermentation, was taken out. The drawing of vat T4762, the only one that does not allow for free flow of the must to the vat, is based on conjectural reconstruction of an unexposed part of the installation and is probably incorrect. The lateral basins (T476--1) are small intermediate vats to collect sediment. The cup-marks round the mouth of the vats are similar to those on the edge of treading floors in the Syrian wineries (T805). In spite of the great difference in size, these presumably served similar functions to the compartments found on the edge of treading floors in Israel (T801), which were probably connected to the addition of various substances to the must.

As in most ancient presses, the most impressive part of the Syrian presses and that which appears in the greatest variety of forms is the anchoring device, the characteristic beam niche (Ts413, open niche; 414, closed niche; 415, pier niche). In spite of the clear weighting of the sample published toward exceptional types, in order to attempt to define what is the characteristic niche, some quantitative analysis will be attempted. Of the 68 examples, 50 are true closed niches, 15 open niches and only 3 pier niches, all three from one site, two of them being rollers in secondary use. Twelve of the niches are made of monolithic blocks, and 12 (6 included in previous 12) incorporated dovetail mortices in their construction. The most striking characteristic of these niches is the great variety of grooves for horizontal and vertical boards and posts to enclose the beam-end, quite a different system from the cross-pieces of the slotted piers found in Israel (T421). Fifty-five of the 68 niches had such grooves, the two most common types being two vertical grooves only (26 examples), and one horizontal and two vertical (9 examples).

The vast majority of the presses are lever and screw presses and the very characteristic screw weights (T62) will be discussed below. Devices that, as already shown in discussion of Israeli presses are there connected to lever and screw presses, are the guider mortices (Ts441). These appear in Syria in large numbers and a variety of forms (22 examples in six sub-types). In Israel only three examples were recorded and it is probable that those found there are a result of influence from Syria.

Five unique beam weights were published (Ts5352, 535202, 5362) from the region and will be discussed below. They are from three sites and the two that were found *in situ* are from oil presses, too small a number on which to base far-reaching conclusions.

The oil and must collection in all the installations in North Syria is lateral except in one case, that from el-Bara where the collection is central. There is a crusader inscription on this installation, but it could be a later addition. The questions that arise are clearly whether this is a lone example of this type and what, if any, are the connections to similar installations in Israel.

A device appearing in the catalogue at only one site in Syria, Dehas 02, is the auxiliary press-bed (T4626), an additional press-bed connected to the main bed but not aligned with the beam. The pressing frails were probably placed here before and/or after the pressing, allowing the oil to flow into the collecting vat at these stages of the work also. Similar installations were found at single sites in Cyprus and Yugoslavia. There is no reason to seek a connection between these different examples. It is perhaps surprising that this obviously useful device does not appear more often.

4C1.2. Greece, Crete, Delos, Kalymnos. Although knowledge of oil production in ancient Greece is not great, nevertheless certain characteristics can perhaps be delineated. Three presses from three very different periods are remarkably similar. A press from Halies of the classical period, one from the Laureatic Olympus from the sixth century CE and one from Athens—the Agora from a century later—all have pear-shaped press-beds (T463) and single round comparatively small collecting vats. Pear-shaped press-beds similar to these have been found at other sites in the Greek mainland and in the islands. The sub-type with splaying spout (T4632 found at Athens, the Agora, and at the Hellenistic press from Praesos, Crete) is of particularly pleasing shape as is the one with double press-bed found on the island of Kalymnos. This press-bed has perhaps roots in earlier periods as evidenced by the Cretan Bronze Age press-beds (Ts2622140, 2622151, 2672161). Although the dividing line between round and pear-shaped is not always easy to define, nevertheless the latter is clearly an Aegean type. Therefore the presence of the pear-shaped press-bed (T463) in other contexts is to be regarded as evidence of direct or indirect Greek influence.

Photo 39. Pear-shaped press-bed with splaying spout (T4632): Laureatic Olympus, Greece (Frantz 1986: Pl. 76e).

At none of the three sites mentioned above is there clear evidence of the manner in which the beam was anchored, suggesting that it was in a niche in the wall, especially as that is the case in the oil press from the Hellenistic period at Praesos where there are two niches one above the other to allow for adjustment of the height of the beam. At Olynthos 03 two small depressions in a plaster floor were explained as bases for press piers (T4317). There is, however, as yet no evidence for press-beams being anchored in this manner at other sites.

At Halies and in a survey on the Methana Cape beam weights were found of type T5712, at the Laureatic Olympus a screw weight of type T62132, and at Pindakos, Chios, beam weights of type T5110; these will be discussed below.

4C1.3. Cyprus. The few beam presses published from Cyprus are not uniform, as comparison between the presses from Pachna Sykes, Kouklia Styllarka, Salamis, Mari Kopetra and Kouris Valley shows. The press-beds, for example, include both round (T46210) and spouted examples (T4631) and in one case pressing was apparently carried out on a pressing platform (T467). This variability is probably the result of the play of different cultural influences, as is so often seen in Cypriot material culture and finds expression in other aspects of presses: for example, beam weights (see pp. 100, 101, 103, 104).

There is, however, one type of installation that is apparently unique to Cyprus and the function of which is still in dispute, and that is the Cypriot slotted pier

Photo 40. Cypriot slotted pier (T427) (Hadjisavvas 1992: fig. 191).

(T427). Cesnola (1877: 188-89) already mentioned these monoliths and the magic qualities attributed to them. Guillemard (1888a, b) was, however, the first to describe them in detail, followed shortly after by his colleague Hogarth (1889: 46-52). According to Guillemard, the monoliths are on average 1.2 m wide, 0.8 m thick and 2.5–4.0 m in height. The slots average 0.23 m in width and 1 m in length. Some of the slots do not go right through the stone and in several cases the floor of the slot is at a slope. Hogarth (1889: 47-48) brings a list of 50 examples. Guillemard and Hogarth were both of the opinion that the monoliths were connected to oil presses. Hadjisavvas (1988: 115 n. 4; 1991) has recently returned to the subject. He has excavated several such installations at Yerovasa, Pachna Ayios Stefanos, Pachna Sykes and Kouklia Styllarka, and has proved these to be oil presses (see map in Hadjisavvas 1991: fig. 1). He compared the Cypriot monoliths to the Galilean slotted piers (T421), calling them *arbores*. They are, however, usually found singly and, according to Hogarth (1889: 48), only in one case do they stand with the slots opposite each other. Therefore both Guillemard (1888a: 475; 1888b) and Hogarth (1889: 50) suggested that these served as beam niches, the beam-end being actually inserted in the slot. There are pre-industrial presses in which the beam is anchored in this way (e.g. Saint Lauren de La Plaine-Layon and Cheillé; see Photo 91). The Cypriot monolith is then functionally more similar to the Syrian beam niches (Ts413) than to the Galilean slotted piers.

As noted above (p. 90) at one site in Cyprus, Mari Kopetra, an auxiliary press-bed (T4626) was found. In this case it was between two press-beds. The pressing frails were apparently placed on it before and after pressing. Similar installations were found at single sites in Syria and Yugoslavia.

4C1.4. The Crimea. A large number of wineries have been excavated in the Crimea dating from the Hellenistic to the late Roman period (see pp. 155-56 below). Associated with them are many lever and weights presses. In no case is there clear evidence as to how the beam was anchored and it is therefore to be presumed that it was in a niche in the wall. The press-beds vary, but a pattern can be discerned. The round (T46210), spouted and pear-shaped (T4631) and square press-bed with spout (T4642) and one example of a press-bed consisting of blocks of stones (T4651) are mainly from the earlier installations, whereas the largest group—the square press-bed (T4641)—are predominantly later. This is apparently a local development and a similar development in the plan and character of the wineries can also be discerned. The weights associated with these presses (Ts55320, 55321, 55326) are of great significance and are discussed below (p. 104).

An interesting device found in the press from Chersonesos is an opening in the wall at the free end of the beam (T4486), probably to secure equipment to raise and support the beam. Similar devices were found at Quṣeir in Israel and in presses from North Africa and Malta.

4C2. The Northern and Western Mediterranean

4C2.1. Italy. In the late eighteenth century several oil presses and wineries were uncovered in the excavations carried out at the Campanian Villa Rustica situated below Vesuvius and destroyed in the same eruptions that destroyed Pompei, of which the final one was in 68 CE. The plans, sections and descriptions of these presses were published over a hundred years later by Ruggiero (1881), but even before then, and certainly for long after, these were regarded as the typical Roman oil and wine presses and as such were the main archaeological evidence on which scholars based their attempts at reconstructing the presses mentioned by classical writers, in particular that described by Cato. This was until recently the main focus of research into ancient agricultural installations, and as late as 1932 Drachman, in his authoritative work on the subject, mentions no other Roman presses from Italy.

However, Rossiter (1978) has shown that, although the Campanian presses are very uniform, they are not the only type of Roman press to be found in Italy. He called the Campanian presses 'platform presses' and a second type 'circular bed' presses.

The Campanian 'Platform Presses' (T4091-9–9) were lever and drum presses and, while varying somewhat in the character of the collecting vats, were uniform in the three other main parts, the beam-anchoring device, the pressing area and the drum attachments.

What remains of the beam-anchoring device is two shafts connected at the bottom (T439). One shaft, the narrower, was to secure the standard in which the press-beam was anchored. This was presumably of wood with a slot in which to insert the beam, similar in form to the Cypriot slotted piers. The second shaft was wider and served as a manhole to enable someone to reach the bottom of the anchoring standard for servicing. The anchoring shafts measured c. 0.35 × 0.35 m, the manholes 0.55 × 0.55 m, and the depth of both types of shaft was 1.8–2.5 m.

There was no press-bed and the anchoring standard was secured on a slightly raised area within a large pressing platform (T466). The areas of these varied examples were 13.6 m^2, 16 m^2, 18 m^2, 20 m^2, 24 m^2, 24 m^2 and 27 m^2.

At the free end of the beam outside the pressing platform were additional shafts to secure two drum piers, both connected to the same third wider shaft that also served as a manhole for servicing (T459). In several cases two presses were placed on either side of a central working area in which the drums of both presses were operated, creating the central court plan (T492).

A lever and drum press is depicted in a wall painting from the house of the Vettii at Pompei (04) in which the beam is clearly anchored in a single standard. Several different types of pre-industrial lever and drum presses are known. Those raising weights (Ts0113, 0114) have been discussed in relation to Hero's first press and are clearly not related to the Italian Roman presses. The type found in Corsica in which the drum is connected to the beam and not to the ground (T0116) is also not of the same type as the Campanian presses. A press known as the 'Casse-cou' press (T0115; Bouzon 1970, see Photo 91), however, several examples of which are found in France, is very similar to the Campanian 'platform press,' although lacking both anchoring piers and manholes, and also without large pressing platforms.

The Campanian 'platform press' was used both for wine and oil but, in spite of the fact that both oil and wine plants were equipped with large pressing platforms and lacked press-beds, they can be distinguished (Rossiter 1981). In the oil presses there is always a round crushing basin, and the method of collection was also different. In the wineries the must often flowed to an adjacent room in pipes, while the oil flowed directly to collecting vats. An interesting collecting vat clearly connected to oil is T4725, found in two presses at Boscoreale. This is a round collecting vat divided into two compartments, almost certainly a device for oil separation by overflow decantation. This device could be the *gemellar* disapproved of by Columella (12.52.10). These two presses were certainly in existence in Columella's lifetime.

These Campanian platform presses (Ts.4091----9) are clearly defined both geographically and chronologically: all 11 presses included in the accompanying catalogue are in the same region—degree square 1440—and pre-date the destruction of Pompei.

The two press elements characteristic of Rossiter's second type the 'circular bed' press, are the 'Tivoli pier base' with two square mortices (T4311) and the round press-bed (T4621), the press type that incorporates both, being T4051-1. Twelve such presses from Italy appear in the catalogue, one of the two from Sette Finestre, a winery. The collection method varies, but a dominant characteristic is discernible. In 10 of the 11 oil presses the collecting vats are in a separate room, in four installations one such vat (T40519130?, T405191305) and in six two connected vats (T40519140?, T40519141?). Only in the oil press at Sette Finestre are there twin square collecting vats connected directly to the press-bed, similar to those from North Syria (T47121).

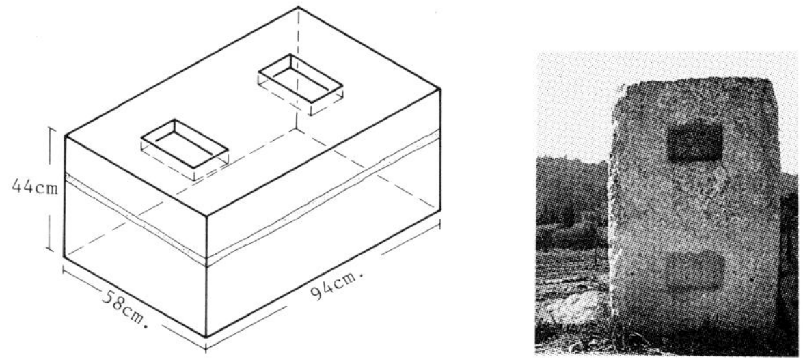

Photo 41. The Tivoli pier base (T43111): Sanary Saint Ternide (Brun site 91), France (Brun 1986: fig. 179).

Of the 12 presses, 7 are in the region north of Rome but the others are situated throughout Italy, examination of the distribution pattern of the Tivoli pier base (T4311) making this even more clear. The dated examples date from the first to the fourth century CE, although from the archaeological evidence alone it is not clear if any predate the 'platform' presses.

The manner in which these presses operated is far from obvious. The 'Tivoli pier base' is clearly to secure two wooden press piers—probably slotted piers similar to the stone Phoenician types (T421, Cato's *arbores*). The frails of olive pulp were pressed on the round press-bed. The question that remains is, however, what force was exerted on the beam? One, that from Capena-Monte Canino, is a lever and drum press, as is shown by the presence of two drum pier bases (T453). In the others, however, there are neither weights nor evidence for the presence of a drum. There is evidence for the existence of weights in Italy that include one dated example from Francolise Posto. These were not found in association with the presses under discussion and therefore cannot be of assistance in the present enquiry. They will be discussed below (p. 105).

Theoretically, these presses could have been lever and weights presses in which the beam was weighted down by stones in boxes, suspended on ropes, thus leaving no archaeological evidence. This, however, is unlikely for several reasons. Pre-industrial presses of such a type are not known, and it would seem improbable that a lever and drum press would be universally replaced by a more primitive type. The only suggested evidence for a lever and weights press in Roman Italy is the relief from Rome-Villa Rondanini (Drachman 1932: 69). In the relief the suggested weight is, however, a simple block of stone and there is no obvious connection between the rope that is clearly discernible and the weight, and therefore as evidence this relief should be treated with caution.

The presses under discussion, Rossiter's 'circular bed press' (T4051-1?) are much more probably lever and screw presses in which the screws raise a box of stones, thus also leaving no evidence. There is ample pre-industrial evidence for such a press (Ts01262, 01263) and Pliny the Elder (above, pp. 86-87) specifically states that such a press was introduced to Italy from Greece 'during the last hundred years', a dating that fits the archaeological evidence extremely well.

The question that remains is the historical relationship between the two types of press. Is the 'platform press' a regional type, an earlier type, or perhaps both? Where did the 'circular bed press' originate and, specifically, where did the Tivoli pier base T4311 originate and when?

The two characteristic components of the 'circular bed press', the 'Tivoli pier base' and the round press-bed are exactly those described by Cato in the second century BCE. Cato was born at Tusculum in Latium and grew up near Reate in the Sabine country. This and the distribution pattern of the circular bed press which shows a concentration in this region strongly suggests that originally this was a northern type and the 'platform presses' a southern type. Later, apparently, the 'circular bed press' type replaced the 'platform press' in the south as well. In the early stages both types were lever and drum presses. It was then that Cato described the northern press, which was also Pliny's press A. Later the northern press developed to become a lever and screw press which raised a box of stones, Pliny's press B/C. The southern 'platform press' was no longer in use at that time so that all examples known are lever and drum presses.

4C2.2. Yugoslavia. In Yugoslavia two different types of presses can be distinguished, each from a distinct well-defined region. Both groups are connected to the Roman presses of Italy, both incorporating the two main components of the circular bed press, the 'Tivoli pier base' with two mortices (T4311), and the press-bed in all cases but one round. In the north-west group, as in most of the Italian examples, the collection vats are in a separate room and there is no evidence for the source of applied force, suggesting these too to be lever and screw presses raising boxes of stones (T40519130?, T40519630?). In this group we meet for the first time a number of industrial plants, those in which there are three presses or more in one structure (T491). Four were recorded in this region, one at Barbariga with 12 presses and a very complex system of 13 collecting vats. There is also one example of what is apparently an auxiliary press-bed (T4626) from Pola similar to the single examples of this device recorded from Syria and Cyprus.

The presses of the second group, the most important of which is Salone Kapljuc, are from southern Yugoslavia. These also have 'Tivoli pier bases' and round press-beds, but they are lever and drum presses, the collecting vats are square and under the beam, and they are single presses (Ts405121306, T405125316). The explanation for the difference between the two groups is perhaps that the technical change to a lever and screw press did not penetrate to this more peripheral region and that there the earlier techniques continued in use.

4C2.3. France and Germany. France is without doubt the richest source of information on pre-industrial presses, dearly cherished remnants of the rich wine-producing tradition from throughout the country. Olives, however, can only be grown in the regions in the extreme south of the country and even there once or twice in a century severe frost kills the trees.

Most of the information on ancient installations derives from the southern region, largely from the detailed survey of the Var region in Provence carried out by Brun (1986).

As in Yugoslavia, in France also the connections to the Roman presses of Italy are marked. However, the differences between the installations in France and those in Italy are considerable and without doubt reflect other influences.

Only five complete presses appear in the catalogue. However, the data from the survey confirm the conclusions that can be drawn from this limited number.

The most marked similarity to the Roman presses from Italy are the 'Tivoli pier bases' (T4311), of which 26 examples are recorded in the survey. This pier base appears in four of the five complete presses. These four presses are also similar to the Italian presses in that the collecting vats are in a separate room, but all four lack the typical circular press-bed and instead there is a large pressing platform (T4661). This picture is also confirmed by the survey where only seven press-beds were recorded.

Two of the four complete presses were lever and drum presses (T405199305, T405199405) and two were lever and weights presses (T405199404). This is clearly the main difference between the Italian and French presses. As we have seen in Italy, no lever and weights presses were recorded. Here again, the picture is confirmed by the survey: 38 'Semana beam weights' (T55121), were recorded, a type not found in Italy, which I suggest below to be an element that originated in Greece. In Provence no presses lacking both weights and drum pier bases were found, the type that in Italy and Yugoslavia I have regarded as lever and screw weights raising a box of stones. However, many screw weights (T62) were recorded in the survey, although none of these have as yet been published *in situ*. However, it is almost certain that in France the lever press passed directly from that using a beam weight to that using a screw weight, often a beam weight adapted to the purpose. The question is only when the transition took place.

One press was published from Ponteves, a survey site that was very different from the four large built presses mentioned hitherto. It is a small rock-cut press with circular press-bed and dovetail mortice, probably for a wooden anchoring standard (mortice T4162; press T404421104), a type not otherwise recorded in France but known from North Africa.

Only six other press-beds were recorded in the survey, four of which were pear-shaped (T4631), and of the latter two were from pre-Roman sites. In Provence, therefore, press-beds are apparently a pre-Roman press component and almost certainly a result of Greek influence from the Greek/Phocaean colony at Marseilles, as are probably the Semana beam weights as well.

The differences between the Italian and the French presses are therefore to be explained by the fact that there was a technical tradition present, presumably of Greek influence, before the Roman conquest. The early beam weight was retained, while the press-bed was abandoned. The Roman 'influence' is seen in the lever and drum press, the 'Tivoli pier base' and possibly in the large pressing platforms, although the 'Tivoli pier press' is a north Italian element and the large pressing platform is an element from southern Italy. In Germany, beam presses were found associated with wineries. As the only significant element reported are the weights, the discussion will be deferred to the relevant chapter.

4C2.4. Spain. Only two complete presses from Spain are in the catalogue. The one from Sentroma Tiana in the north-east of the country conforms to all the characteristics of the Roman 'circular bed presses' from Italy (T40519130?), including the round press-bed that was lacking from the examples from Provence.

The press from Manguarra y San Jose in the south-west (T40619630?) has elements such as the pier base with four mortices (T4336) and the square press-bed (T4641) characteristic of North Africa.

Also worthy of note is the presence of pear-shaped press-beds (T4631) in two different regions in the south-east and north of the country.

A considerable number of beam and screw weights were recorded, including the Semana weight (T55121) already mentioned in connection to Provence. These will be discussed below.

4C3. The Southern Mediterranean: North Africa and the Islands

The North African installations have certain characteristics in common both in the components that are present and in those that are lacking. There is, however, also very marked regional diversity within the region itself.

The characteristic element found throughout North Africa is the Semana beam weight (T55121), already mentioned in relation to France and Spain. There is also a tendency to use a square or rectangular press-bed (Ts4641, 4642). Twenty-five of the 35 beds of these types appearing in the catalogue are from North Africa, and of the 44 press-beds recorded in North Africa 25 (56.8%) are square or rectangular. A third characteristic that is of significance when comparing the North African presses to those from southern Europe is the absence of the Tivoli pier base (T43111). Another characteristic of North Africa is the presence of industrial presses type T491. Fifteen were recorded, one with 21 presses and another with 16–20 presses.

4C3.1. Libya–Tunis: Perforated Pier Press (Ts4022, 4032, 424). In western Libya (Tripolitania), Tunis and eastern Algeria a type of press is found, the main characteristic of which is the imposing perforated piers (T424). These are all lever and weights presses, the beam weight being the Semana weight (T55121).

The piers are similar to and function in the same way as the slotted and perforated piers from Israel (Ts421, 423), two piers together with a 'lintel' serving as the fulcrum of the lever press. As in the case of the piers from Israel and of the Syrian niches, much attention has been paid to the aesthetic aspect of these

Photo 42. North African press with perforated piers (T424311000): Bir Sgaoun, Algeria (Brun 1986: fig. 48).

installations which reach monumental character and proportions. This led Cowper (1896, 1897), the first to publish these installations in detail and whose publication remains the main source for data on the subject to this day, to see in them cultic remains.

About half of the piers are monolithic, but those constructed of two stones or more are usually arranged symmetrically. Stones often project from the vertical line of the pier. This may be to add weight, but they are also symmetrical and add to the appearance of the piers.

There are three types of cuttings. The first are perforations on the inner facing sides of the piers, two to four in number, sometimes open, sometimes closed, and sometimes joined by a channel. Their purpose is clearly to insert cross-pieces in order to secure the fixed end of the beam, the number of perforations showing the number of times the olive mash was pressed, in each consecutive pressing the beam being lower than in that preceding it. The second type of cutting consists of pairs of notches cut in the inner corners of the piers. The pairs number from one to four. In those cases when the piers are made of several stones, care was taken to cut the notches at the joints between the stones. The third type of cutting is a pair of holes or notches in the lintel. These two latter types of cuttings without doubt served to secure rods to which the beam was attached while adjustments were made to the cross-pieces, in a similar manner to that in which the notches, holes and angular channels functioned in the Galilean slotted piers (Ts421-1, 2, 3).

The piers appear in a great variety of combinations of these three devices. Twenty-seven of the 39 pairs of piers that could be clearly defined had two perforations and 14 of these had no additional device, six had three perforations and only two had four. Five had two perforations, one pair of notches and upper holes, four had two perforations and one pair of notches. Other combinations appear in smaller numbers. Four pairs of piers had no perforations at all, but only notches and upper holes, in one case four pairs of notches and upper holes. In these installations the rods, probably of iron, placed in the notches and upper holes presumably secured the piers during pressing. The great variety in the sub-types of these piers is reminiscent of that of the Syrian pier niches and together with the aesthetic aspects shows the wealth and quality of this rural society.

In the catalogue these piers group spatially into two clusters. Forty installations are in a south-east cluster, degree squares 1331, 1332 and 1432, and 11 in degree squares 0835, 0935 and 1038?, some 500 km to the north-west. Presumably this gap is because of a lacuna in the data and similar presses are to be expected in the intervening regions. However, certain differences can be discerned between the two clusters. While four of those in the north-west cluster are monolithic with two perforations only, the most common sub-type in both regions (T42422100), all three examples of piers with connecting channel (Ts424--9) are from this region as are all three examples of slotted piers (Ts424821, 424921). As the installations were categorized almost entirely on the basis of photographs, there is the possibility that those piers defined as slotted were actually perforated with a connecting channel, although Saladin's reconstruction of the press at Henchir Choud et-Battal suggests otherwise.

The two types of press-beds associated with these piers are the one consisting of a circle within a square (T4622) which appears in both clusters and the square/rectangular free-standing press-bed with spout (T4642) that appears in the south-east cluster only. Two of the latter have been recorded in Italy and two from the Crimea.

The press at el-Amud in the south-east cluster was excavated and no press-bed was found, only a plaster pressing platform and a collection vat in a separate room.

At Henchir Choud et-Battal in the north-west cluster the collecting vat is square and constructed of a stone panel with grooved corner stones (T47112), a type recorded at four other sites all in degree squares 0835 and 0836, a good example of a regional device. At this site beam-guider posts (T441) appear, however, as the publication is in the form of a reconstruction there is the possibility that they were not actually found.

4C.3.2. Algeria

4C3.2.1. Dovetail Beam Niche Press (Ts4044, 416). In eastern Algeria a type of lever press is found in which the beam was anchored in a dovetailed niche. As opposed to the Syrian beam niche, however, which was of stone, this was of wood (Christofle 1930b) and what usually remains is only the dovetail mortice in which the niche was secured (Ts4161, 4162, 4163). In two cases, part (Christofle 1930b: 31) or all (T4164) of the upper part of the hollow in which the wooden niche was fitted has been found. In the latter case the upper part of the niche was in the shape of a cross, the arms adding extra strength. The distribution area of this type is also clearly defined: and is located to the west of the distribution area of the perforated and slotted piers, apparently without overlapping: degree square 0236, 1 example; degree square 0436, 7 examples; degree square 0536, 4 examples; degree square 0636, 2 examples, degree square 0836, 1 example; degree square 0937, 1 example.

Apart from the Syrian beam niche mortices, the only examples of similar devices in the catalogue are one from France, Ponteves, clearly a press, and one from Italy, Mariffi, which is, however, possibly an anchoring device for another type of installation.

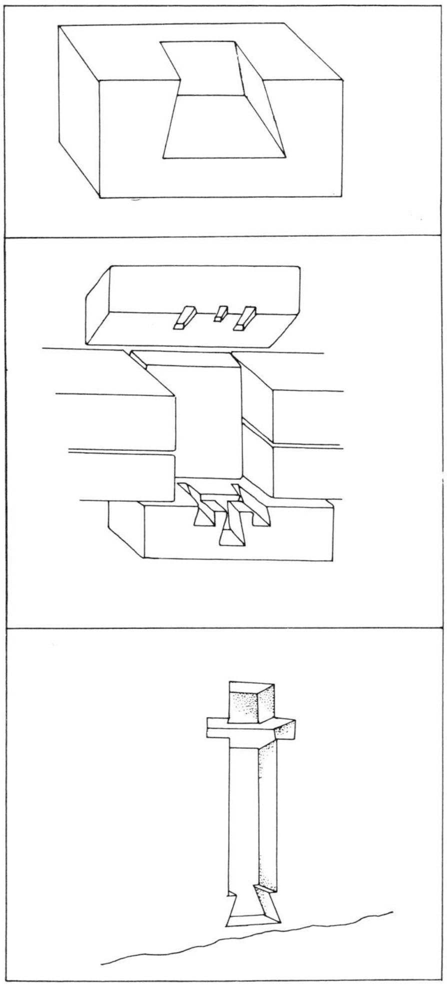

Figure 12. North African dovetail beam niche (Ts416).

Most of the complete presses in the catalogue (Ts4044) are small rock-cut presses, including one lever and drum press, the only one in the catalogue from North Africa (T404421305). One complete large press that appears is from Madaure, a site with eight or more presses. The example in the catalogue (T404421564), however, is a composite reconstruction made by Christofle (1930b), incorporating elements from several presses. There is a collecting vat with panels and grooved corners (T47112), already mentioned above. Two other elements incorporated in Christofle's composite reconstruction of the Madaure press are beam-guider mortices (T447), and an additional single mortice at the free end of the beam (T4487). The latter no doubt served to secure the bottom end of a pole to support the beam when not in use or to hold tackle or a pulley for the same purpose. Similar devices are found at other sites in the region. At Thuberbo Majus these were of two different types (T4484 and T4482), the latter octagonal in shape and found also at Oued Athmenia several hundred kilometres to the west.

4C3.2.2. The Madaure Oil Separator (T47132) (see Fig. 47, p. 175). At Madaure an extremely sophisticated type of oil separator was found, combining the two techniques that have already been mentioned: overflow and underflow decantation. In the Madaure separator the collecting vat is divided into two unequal receptacles that are connected by a hole at the bottom. The expressed liquid from the press flows into the larger half and the hole is blocked. After the oil has begun to float on the watery lees and only lees remain at the bottom, the hole is unblocked and the lees enter the smaller receptacle and there rise to the level of the liquid in the larger receptacle. There are outlets close to the rim in both receptacles, the lees flowing out into a channel through one and the oil flows through the other to a second vat and then through another outlet near the rim into a third vat. Thus the oil is separated from the lees without aid of human hand and only pure oil reaches the third vat. There are hints that a similar installation was found at Ras el-Hammam in Libya but it has not been published in detail.

It is of great interest that this technical breakthrough was apparently not forgotten but did not spread afar. Pre-industrial installations working on the same principle have been recorded from several places in the western Mediterranean from Elche (T0191) and Huevar (T0192) in Spain, from the Spanish island Ibiza (T0193) and from Fez, Morocco (T0194).

4C3.2.3. T-Shaped Pier Mortice Press (Ts4071, 435). In the region of Mauretanian Caesarea yet another type of press has been recorded in a detailed survey carried out by Leveau (1984). It consists of a T-shaped pier mortice (T435), a small plastered pressing platform (T467) and a Semana beam weight (T55121) (complete press T407128004). The pressing platform was built with considerable care of three layers—one of stones, one of rough plaster and one of fine plaster (e.g. Thalefsa East, Leveau site 20) and ranged in size from c. 1.4 m × 2 m to 3 m × 3 m or a little more (e.g. Sidi Rihane, Leveau site 84). These installations apparently lacked both press-bed and collecting vat. Possibly the former was of wood and the latter of pottery. The number of crushing basins recorded in the survey is also very small and perhaps the olives were also crushed on the pressing platform in some way.

Thanks to Leveau's detailed survey, 18 T-shaped pier mortices (Ts435) and 17 pressing platforms (T467) were recorded all in one degree square: 0236. It is at the extreme western end of the distribution area of the dovetail beam niche press (T416). Leveau specifically

Photo 43. Examples of the T-shaped pier base (T435) (Leveau 1984: fig. 232).

states that the one example of the latter that he recorded (Bekkouch, Leveau site 180) is the only one he found. It is of course probable that the T-shaped pier mortice press is distributed over a wider area.

4C3.2.4. The Taourienne Pier Base; The Double Dovetail Mortice Pier Base (T4315). At Taourienne (Leveau site 197) a stone block was found which is possibly an unusual pier base. It consists of what could be two connected closed dovetail mortices. Single dovetail mortices are found in many types of installations, particularly in Galilee (see below, pp. 165-66 and T041 for all installations and types equipped with this device).

Two connected closed dovetail mortices do not appear in the catalogue from other sites. However, Meister (1763: fig. 2; Drachman 1932: fig. 3) reconstructs Cato's pier base, the *forum/lapis pedicinus* in exactly this form, suggesting that he knew of such a device.

4C3.2.5. Single-Socket Pier Base T436. It has been suggested that in the press at Aquae Sirenses, Algeria, the beam was secured to a single post held between a socket on the ground found in the excavation and a similar socket in the ceiling (Ts436, 408126304). No similar press has been found, neither ancient nor pre-industrial, and it is very probable that the excavation data were misinterpreted.

4C3.3. Morocco: The Four-Mortice Pier-Base Press (Ts4061, 433). The characteristic component of the presses of western North Africa is the four-mortice pier base (T433; complete press T4061). Apart from the

Photo 44. Four-mortice pier base (T433): Volubilis, Morocco (Photo: Shlomo Grotkirk).

four Moroccan sites in the catalogue where these elements were recorded, it appears also at one site in Spain, Manguarray San Josë, and one from central Algeria, Oued Athmenia. These sparse data do not enable clear conclusions to be drawn as to the centre of the distribution area or the origin of this device. The similarity to the Tivoli pier base with two mortices (T43111), however, suggests that it originated in Spain. The four-mortice pier base appears in a series of sub-types (Ts4332–4337). Of each, only one example appears in the catalogue and some could well be a result of repairs or use of faulty stones. However, T4336, in which the mortices appear on two different stones, and T4337, in which there are two long mortices width-ways, could well be significant sub-types.

The site from which the majority of presses of this type come is Volubilis, where over 50 oil presses have been recorded. In the other components of these presses there is considerable variability. Of those in the catalogue, the press-beds of ten are square (T4641) and of only four are round (T4621). In over half of the presses the collecting vat is in a separate room. Volubilis is also rich both in number and variety of guider mortices (Ts441, 442, 445, 447). T445 is of particular interest, consisting of two mortices to secure two poles to guide the beam and between them a third socket apparently to secure a third pole to support the beam when pressing was not in progress. A somewhat similar arrangement appears in a pre-industrial press from Isfahan, Iran (Persia). However, there it is in a lever drum and weights press and there the two outer mortices are for drum piers.

Morocco is the only part of North Africa where other weights in addition to the rectangular Semana beam weights (T55121) are found. A screw weight is known from Rabat but apparently not from an archaeological context. At several excavated sites there are cylindrical Volubilis weights of T55111 and at one site the Cotta weight (T56911). It is agreed that these weights were introduced at a later stage than the rectangular weights, and a suggested date for the transition is 150–70 CE (Akerraz and Lenoir 1981–82: 97). A question that is not easy to answer, however, is whether these are beam or screw weights. This and other aspects of these weights will be discussed below.

4C3.4. Lever Presses in North Africa: Conclusions. The spatial distribution of the four main types of North African presses conforms to a remarkable degree to the ancient political divisions of the region. Perforated and possibly also slotted piers are found in the Carthaginian region, later the Roman province of Africa, the dovetail beam niche press in Numidia, the T-shaped pier mortice press in Caesarean Mauritania and the four-mortice pier base press in Tinjanian Mauretania. This regional diversity demonstrates in a striking manner that the political divisions conformed to deep-rooted cultural differences.

4C3.5. Malta. The two sites from Malta in the catalogue, San Paul Milqi and Nadur, show some similarities to those from Italy and others to those from North Africa, and are also of interest in their own right.

At both sites there are press-beds consisting of a circle within a square (T4622), a type found in North Africa and at San Paul Milqi there are beam weights (T55221), similar but not identical to the Semana weight of North Africa, while at both sites the olive crusher is of the classical Italian type—the so-called *trapetum* (T331).

In two of the presses at San Paul Milqi there are unique openings in the wall at the free end of the beam (Ts4485, 4486) which, similarly to other devices situated at this point in the press (Ts448), presumably served to secure tackle to raise the beam-end. In this case, however, they could have been to guide the beam. In one of the presses at this site two beam weights were fastened one above the other (T5622), thus making them undoubtedly too heavy to raise and turning them into a drum base.

At Nadur there are two square collecting vats next to each other (T47121), an arrangement common only in Syria. One example has also been found at Sette Finestre in Italy. However, this is perhaps such a simple arrangement that examples so far apart do not necessarily indicate any connection.

At San Pauli Milqi a round flat stone was discovered with central hole and radial grooves very similar to the cover of central collecting vats (T46124) in Israel. It was not found in the oil press but in its immediate vicinity. It was perhaps part of an installation to extract the first oil before pressing as at Quṣeir 02(A2) in western Galilee. In this case it seems unlikely that there is no connection between the two installations, in spite of the great distance between them.

Chapter 5

Beam Weights (T5)

5A. Simple Beam Weights in Israel and Other Countries: Unworked Field Stones (Ts5110), Weights with Horizontal Bore (Ts5112, 5113), Weight with Vertical Bore (Ts521)

The Late Bronze Age weights from Ras Shamra, (Ugarit) Syria are apparently simple unworked field stones with a natural bore (T5110), but that from Maroni, Cyprus, is bell-shaped with a horizontal bore at the upper end (T5113) and appears to have been fashioned to that form.

The majority of Iron Age beam weights from Israel are of these two types, perforated field stones (T5110) and fashioned weights with horizontal bore (Ts5112, 5113). There is, however, also a third type, called by Albright (1941–43: 62 n. 8) 'doughnut-shaped', a flattened stone with vertical bore, round (T5211) or square (T5212). This type was probably attached to the

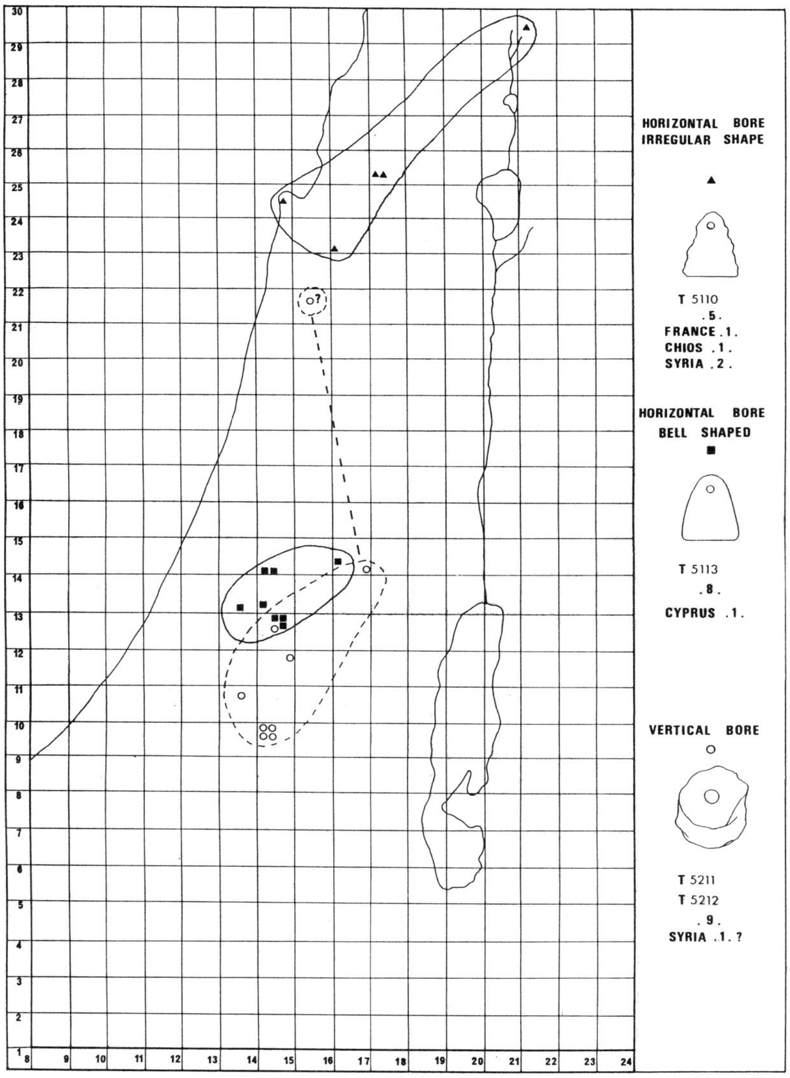

Map 15. Simple beam weights (Ts51, 52).

Photo 45. Bell-shaped weights with horizontal bore—Iron Age (T5113): Bet Shemesh Inst. 13 (Grant and Wright 1938: Pl. 21.4).

Photo 46. Doughnut-shaped weights—Iron Age (T5212): Bet Shemesh Inst. 14 (Grant 1934: Pl.18).

beam by a rope being inserted through the bore and then to a stick placed horizontally under the weight.

At Bet Mirsham only 'doughnut-shaped' weights with vertical bore were found (T5211) and at Gezer only bell-shaped with horizontal bore (T5113). Albright was of the opinion that Gezer was no longer occupied in Iron Age II and therefore suggested that the weight with horizontal bore (T5113) was an earlier type and the one with vertical bore (T5211) a later type, the transition having taken place around 800 BCE.

The recent excavations at Gezer have shown that the site was occupied in Iron Age II. Bell-shaped weights with horizontal bore have also been found in Iron Age II strata both at Miqne-Ekron and Batash-Timna, and there is in fact no evidence as yet for the presence of weights with horizontal or vertical bore in Iron Age I levels.

However, while the weight with horizontal bore (T5113) is found at sites both in the coastal plain and in the hills, the weight with vertical bore, Albright's doughnut-shaped weight (Ts521), including some examples from later periods (e.g. Adderet, Qalandiya), is found at hill sites only. The difference, therefore, between the weights with vertical and those with horizontal bores was not chronological as Albright suggested, but regional. The doughnut-shaped weight with vertical bore was a local variant used in the Judaean hill country.

In all four Galilean Iron Age oil presses in which weights were found—Shiqmona, Yoqneam, Rosh Zayit and Dan—the weights were unworked perforated field stones (T5110), again a regional characteristic.

Similar simple weights have been published from two pre-Roman sites in southern France, Entremot and Ollioules and a late Roman site at Pindakos, Chios.

5B. Complex Weights in Israel, the Southern Levant and Cyprus: The Reversed-T Weight (T53), Weight with Hook (T54), Large Weight with Horizontal Bore (T512)

In lever and weights presses of Hellenistic and later periods in Israel and the surrounding regions, we find three main types of weights. The first is type T512, a larger version of the Iron Age weight with horizontal bore (T511). Of the few weights found in Upper Galilee the majority are of this type (7 out of 10). This weight is also found in small numbers in other regions, in a Hellenistic press at Haifa Rommema, at Barom in the Jordan Valley, at three sites in the Golan, at one in Samaria, and at two sites in Cyprus, one Hellenistic and one Byzantine.

The second type is that in which a ring or hook is fixed into lead in a hole in the top of the weight (T541). Examples were found in two excavated oil presses in western Galilee at Zabadi from the Roman period, and at Karkara, from the Byzantine period. It is not clear whether the weight from Siyar el-Ghanam is of this type because the publication leaves some doubt as to its size.

Photo 47. Weight with hook (T5411): Karkara Inst. 01 (see also Photo 34).

The third and main type of beam weight in Israel and surrounding regions in Hellenistic and later periods is the reversed-T weight, the bore having the shape of a reversed 'T' (Ts531, 532). The manner in which this weight functioned is demonstrated by a pre-industrial example from Ṭafila (Dalman 1928–42: IV, fig. 56). The rope was inserted through the vertical section of the reversed-T bore to be attached to a rod that was inserted in the horizontal bore of the T.

Photo 48. Reversed-T weight (T5310): Kurqush Inst. 03.

Sixty-five examples of this weight were recorded in the region of the Israel–Palestine Grid, three in Cyprus and two in Lebanon. In the Golan there is a characteristic sub-type tapering in shape and with a large upper opening (Ts5323), the difference being at least partly as a result of the basalt rock of which these weights were fashioned there.

Photo 49. Reversed-T weight with large upper opening (T5323): Khawka.

Two other sub-types of this weight must be mentioned. T533—with W-shaped bore—cannot be distinguished from the usual type, except by very close examination. The bore connecting the two lateral holes is not horizontal and as a result no rod could have been inserted through the weight. Presumably it was suspended by tying ropes into the two U-shaped bores that form the W.

The second type (T534), was found at one site only—Weradim. In it there are two crossing horizontal bores. The screw press base from this small site not far from the west coast of the Sea of Galilee is also exceptional (T73223).

The reversed-T weight (Ts531, 532) is well attested for in the Hellenistic period, but it is possible that it originated earlier, having been recorded at Govit, a site at which no pottery later than the Persian period was found.

The reversed-T weight could be a development from the doughnut weight with vertical bore (T521). As the stone became larger, it became difficult to bore through it from top to bottom so that, instead of placing the stick to which the rope was tied under the weight, it was inserted through a horizontal bore near the top. Until more is known of distribution, dating and history of both types of weight throughout the Levant, however, judgment on the subject must be reserved.

The fact that the reversed-T weight is found both to the north and south of Upper Galilee, but is rare in this region where the more primitive type continued in use, must be explained by the fact that this is a comparatively mountainous region to which innovation would penetrate slowly.

It is of interest that each of the three main types of beam weights found in the regions in ancient times are attested for in pre-industrial presses in the region: the type with horizontal bore at Bet Guvrin (T5122), and ᶜAjlun (T5123); that with reversed-T bore at Ṭafila (T5311) and that with a hook or ring at Kfar Hay (T5411). All four are from pre-industrial presses (T01114) in which the weight is raised in the same manner, by a drum which is either hung below the beam or laid across a fork at the end of the beam. Hand-spakes that were inserted in holes in the drum were used to rotate the drum and thus raise the weight. If the hand-spake was released, the weight would cause the drum to turn in the opposite direction and the hand-spake would then make impact with the beam and act as a brake. In those cases that the drum was laid across a fork at the beam-end, a rod was specially placed across the fork to arrest the hand-spakes. This is without doubt also the manner in which the weights were raised in ancient presses in this region. The drum is the *glgl* of the Mishnaic oil press (*B. Bat.* 4.5) and probably 'the seat...that they fixed to the press beam' also mentioned in the Mishnah (*Kel.* 20.3) was for the man who turned the drum when it was laid across the forked beam.

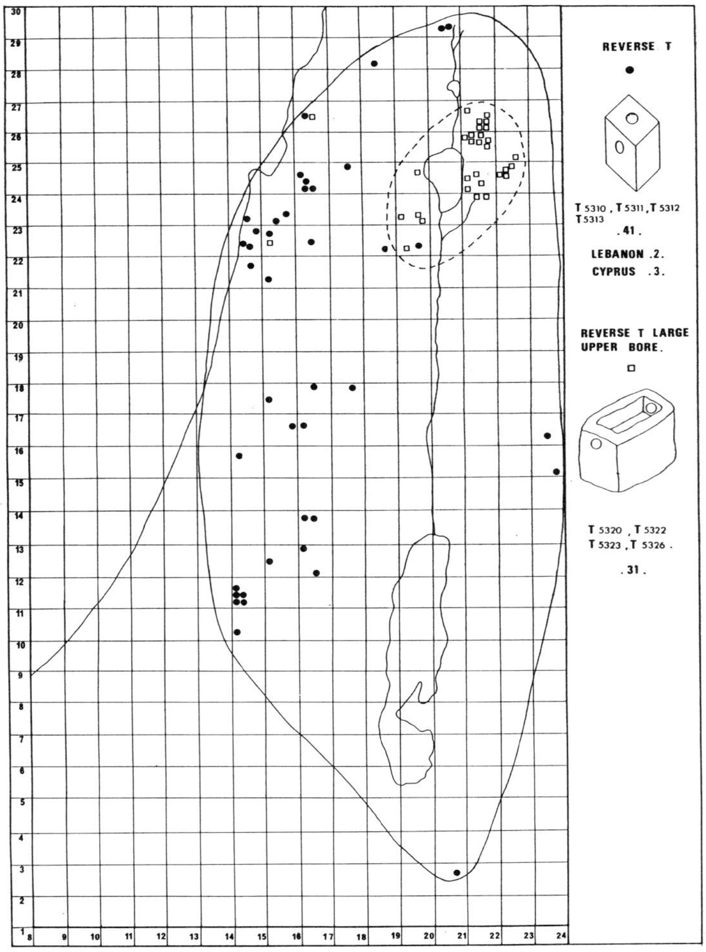

Map 16. Weights with reversed-T bore (T53).

Photo 50. Pre-industrial lever and weights press (T01114) raising the weight: Bet Guvrin (Photo: American Colony, Jerusalem).

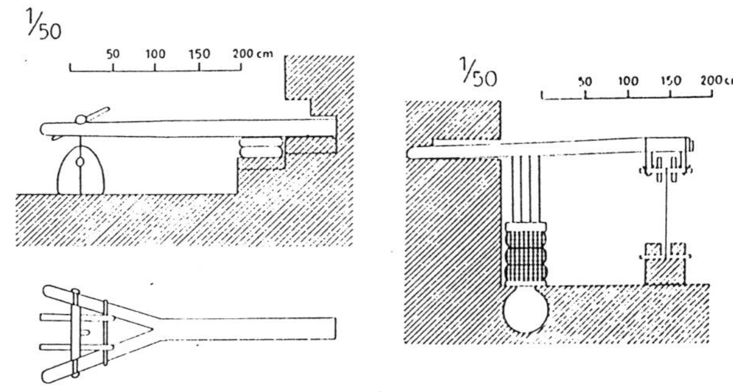

Figure 13. Pre-industrial lever and weights presses from ʿAjlun (left) and el-Ṭafile (right) (Dalman 1928–42: IV, figs. 55, 56).

5C. The Semana Weight, Rectangular Beam, Weights with External Mortices (Ts55, 56)

The most common weight in the ancient world, the beam weight with external mortices (Ts55, 56), is not found in Israel at all and apparently also not in Lebanon.[1]

In the catalogue 182 examples appear in 26 sub-types from nine countries and four islands, and there is no doubt that the catalogue is incomplete on all three counts. The most common type is the Semana weight (T55121), rectangular with dovetail closed mortices on the short sides and with a groove joining them on the top of the weight. Seventy-seven examples were recorded in North Africa, 38 in France and 1 in Cyprus. The way this weight functioned has been a matter of controversy. Paton and Myres (1898: 216), while discussing a similar weight from Amorgos (T5562), raised two possibilities: that the latter was either a beam weight with tackle attached or a screw weight. Drachman (1932: 97, fig. 32) in discussing the more common Semana weight (T55121) suggested that it was a screw weight; Christofle (1930b) had, however, already put forward the idea that this was a beam weight. Gadukevych (1958: fig. 76) in the 1950s still reconstructed presses from Crimea, equipped with weights of this type, as lever and screw presses, and Hadjisavvas (1990) did the same for a press from Cyprus even more recently. Christofle's suggestion is confirmed, however, by North African pre-industrial analogies (T0112) in Algeria and Tunis and is to be accepted (Oates 1952–53: 3; Mattingly 1988: fig. 5).

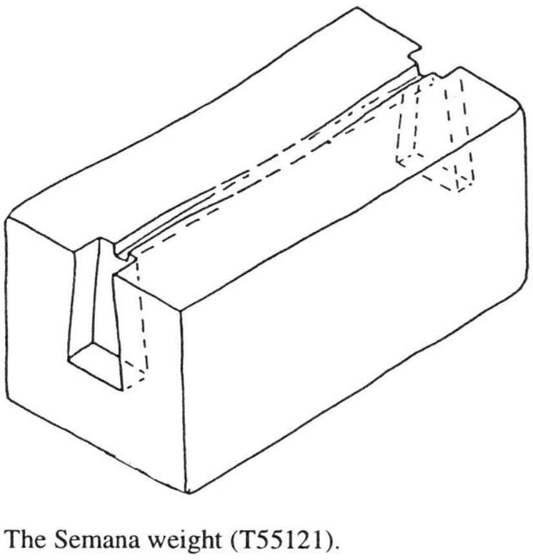

Figure 14. The Semana weight (T55121).

Wooden boards, fastened in the mortices of the weight and held together by a rope or metal rod inserted in the groove, secured a drum. A rope was tied to the free end of the beam and attached to the drum, which when turned raised the weight. Thus, while in all the Levantine presses the drum was attached to the beam, here it was attached to the weight. If the weight were not lifted, this would become a lever and drum press, which at least in one case is what happened (T562). There are several elements that distinguish the different sub-types of this weight. The first is shape. Cylindrical variants are found in Morocco and Italy (see discussion below, p. 105) but the vast majority are rectangular. In Algeria there are types in which the narrow section of the weight was trapezoidal, resulting in the sides of the weight being parallel to the mortice (T55171) or rounded (T55181), two aesthetic embellishments. The second main element in these weights that varies is the shape of the mortice which can be closed and dove-tailed reaching only part way down the weight as is usually the case (Ts551) or open, reaching the bottom of the stone—in this case either dovetailed (T552) or straight (T553).

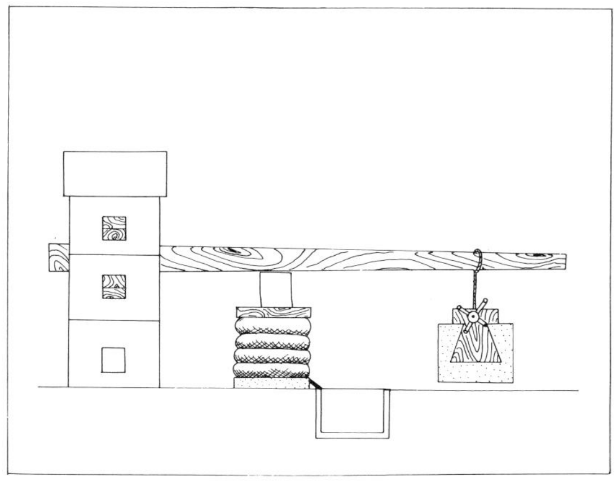

Figure 15. Libyan press with perforated piers (Ts424) and Semana weight (T55121)—reconstruction.

The third main element is the groove on top of the stone—being present, as is usual, or absent. Where there is no upper groove, often various small ancillary mortices are added on the top of the weight and joined to the external mortices on the sides of the weight (Ts556, 557, 558, 561). Tenons inserted in these mortices replaced the rod that was inserted in the groove to keep the side-boards in place. The main question that arises relating to the Semana weight and its many variants found in places so far apart is its origin and history. The cylindrical weights in North Africa were apparently not introduced until the second century CE and their absence from Italian Roman presses of the first and second century suggests that there the situation was similar. The discussion of these weights will be deferred, therefore, to a later stage and this section will focus on the rectangular weights.

The absence of the rectangular weight from Italy clearly indicates it not to be Roman in origin. There is in fact evidence that this weight pre-dated the Roman presence in the western Mediterranean. In France at Entremot one such weight is pre-Roman (Goudineau 1984) and in North Africa at Gammarth such a weight is dated to the second century BCE, and at Constantine to the fourth century BCE. However, as in many other cases in this study, the chronological evidence is insufficient even to attempt the construction of a model to explain the origin and diffusion of this type of weight, which must of necessity be based primarily on typology and spatial distribution.

It is very probable that the prototype of the weights with lateral mortices (Ts55, 56) is the simplest type: without groove, and with open mortices (Ts5522, 5532). Four of the seven weights published from Delos are of this type, one was from Chersonesos in the Crimea from the Hellenistic period, one from Kouklia Styllarka, Cyprus, and one from Piesport, Germany.

Two other variants are close to this suggested prototype. One has an open mortice but also an upper groove (Ts55221, 55224, 55321, 55324, 55326). Of these, three were recorded from the Crimea, one from Delos, six from France, including the early example from Entremot, three from Spain and four from Malta, the only types found in the latter two countries.

The other variant that is close to the suggested prototype is that in which, instead of a groove, there are small ancillary mortices (Ts5562, 5572, 5582, 5612). Of these, one is from Delos, one from Amorgos, and the others from Madaure in Algeria and Oued el-Htab in Tunisia. In both the Greek islands and North Africa there are examples with open and others with closed mortices.

It should be noted that the standard Semana weight found in France and North Africa (T55121) does not appear in the eastern Mediterranean, except for one example from Cyprus. Three weights from the Crimea and from Delos have a groove on top, but the mortices in these weights are open (T55224, 55321, 55326).

The fact that the standard Semana weight is almost unknown in the eastern Mediterranean, whereas the suggested prototype and related sub-types are found in several different countries in that region, strongly suggests that this group of weights originated in the east.

Very tentatively, therefore, a model can be proposed by which this weight originated and developed in the Aegean world and was diffused by Greek colonists in two waves, in each case in a slightly different form—the weight with upper groove and open mortices to Malta, Spain and France, and the weight with ancillary mortices only to North Africa. The standard Semana weight, however, developed in the west.

Two main objections could be raised against this model. The first is that the earliest dated example is from North Africa and that nearly all those on which this argument is based are from a much later date (Amourretti 1986: 175). In answer, it must be stressed that this model and others of a similar nature are based on the assumption that the presence of a dominant type in a region indicates a deep-rooted tradition with a long history.

The second possible objection is that the distribution map of types T55, 56 does not conform to the pattern of Greek settlement as reflected in historical documents. As yet no examples have apparently been found in the Greek mainland. Southern Italy lacks this type of weight and, on the other hand, the parts of North Africa where the largest number of examples have been recorded was an area of Punic and not Greek settlement. In answer, it can be suggested that this weight originated only in certain parts of the Greek world. Delos and Amorgos are both in the Cyclades, and it is probable that these weights were in use also in the Anatolian mainland, a region apparently rich in a variety of screw weights, some of which are very similar to the rectangular weights with lateral dovetail mortices

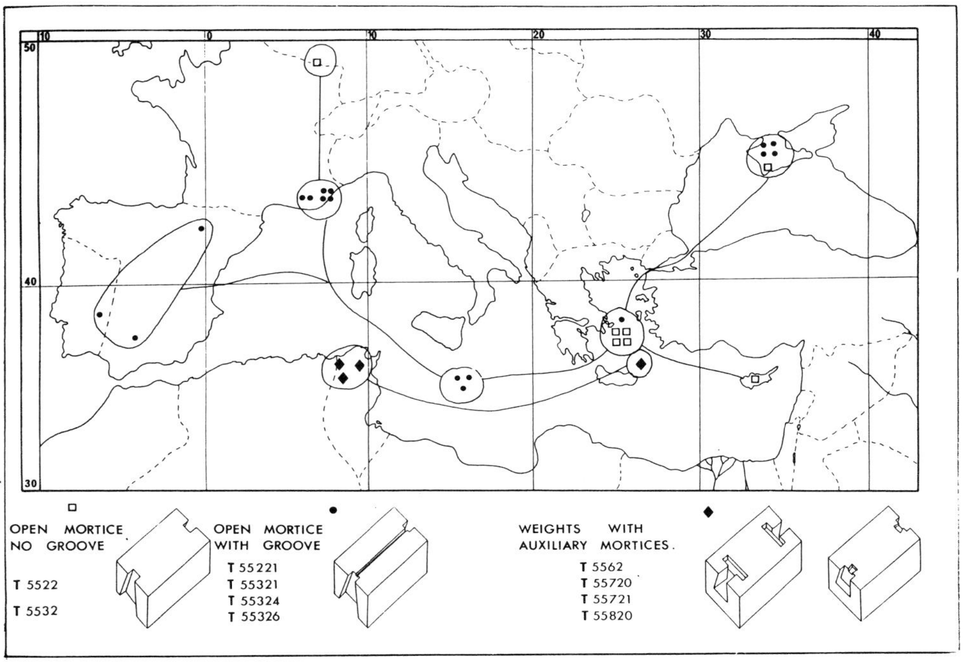

Map 17. Semana weight—main sub-types (T55).

under discussion (e.g. Pontus 01–07 [see fig. 23]). It is known that the Greek settlers of Marseilles, for example, came from Phocaea in Anatolia. It should also be stressed that Greek influence was not limited only to the area of their colonies. Greek grave goods were found in the earliest Punic graves at Carthage.

5D. Cylindrical Weights with External Mortices

5D1. Italy: The Francolise Posto Weight (Ts55110, 55190, 5541)

Three cylindrical weights from Italy with external mortices, but lacking a central socket, appear in the catalogue. Two are from unstratified and undated contexts and the third from Francolise Posto was uncovered *in situ* in an oil press, the excavation of which had not been terminated. The latter is dated to between the middle of the first and the middle of the second century CE.

If the presumption proposed above, that in Italy oil presses passed from the stage of lever and drum to lever and screw presses weighted by boxes of stones, is accepted, then these weights that are clearly from a later stage must represent the beginning of the development of the screw weight in Italy. As such, they will be discussed again in the next chapter.

5D2. Morocco: The Volubilis and Cotta Weights (Ts55111, 55412, 56911)

In Morocco, as noted above, during the second century CE rectangular Semana weights were replaced by almost identical cylindrical Volubilis weights (T55111), which are very similar to the Italian weights discussed immediately above, even including one with four mortices. The difference is only that they have a groove or grooves on the top like the normal North African Semana weight (T55121). The Cotta weight (T56911) is an interesting variant with zig-zag mortices.

In this case also the question that arises is whether these weights are beam weights or screw weights. The fact that these weights are so similar to the Francolise Posto weight and were introduced at the same time makes it very probable that they are also screw weights. In addition, at Volubilis these weights are associated with guider mortices, a device that is usually found only in lever and screw presses. It is true that two lever and weights presses have been reported from North Africa with guider mortices (Madaure; Henchir Choud et Battal). These are, however, both published as reconstructions, thus raising doubts as to the reliability of the data.

A more conventional screw weight has been published, apparently not from a datable context, from Rabat, and in addition it is worthy of note that, as opposed to the pre-industrial presses of eastern Algeria and Tunisia which are lever and weights presses, those from the Fez region in Morocco are lever and screw presses (Fez, Tezarin Taghzut, Morocco [T01218050]). This situation, which probably reflects the situation in the regions in earlier periods, makes it even more likely that the Volubilis and Cotta weights are screw weights.

5E. Beam Weights with Two Vertical Bores (Ts57)

5E1. The Methana Weight, Greece (T5712)

Hamish Forbes and Lyn Foxhall in their current survey of the Methana, Cape, have recorded several examples of a type of weight that they were kind enough to describe to me. It is rectangular in shape, T-shaped in section lengthways, with two small rectangular bores in the arms of the T (T5712). The weight from Haleis in the Argolid, not far distant to the west, is apparently of the same type. This weight has certain characteristics in common with the Semana weight (T551-552). Both are rectangular and both have attaching devices on their short side. It is very probable, therefore, that the Methana weight worked on the same principle as the Semana weight. Two wooden components in the shape of a reversed 'T' were probably inserted through the holes from below and held a drum that raised and lowered the weight. Until more is known of the distribution pattern and chronology of both types, however, it is difficult to determine whether this is the prototype from which the Semana weight developed, or whether they developed and existed side by side.

5E2. The el-Beida Weight, Libya (T5732)

An unusual rectangular weight was discovered at el-Beida in Libya. It had two square vertical bores and cuts connected the bottom of the bores to the edge of the weight. The weight was large: 2.25 × 1.00 m, height 0.60 m. A drum was probably attached to the weight and the size of the weight makes it possible that it was a drum base of a lever and drum press rather than a beam weight. In spite of a certain similarity, it is doubtful if this weight is directly connected to the Methana weight.

5F. The Taqle Weight with Reversed-T Bore and External Mortices, Syria (Ts5352, 535202, 5362)

Three variants of the Taqle weight were reported from North Syria. These are rectangular weights with a bore in the shape of a reversed 'T' and external mortices. In one the mortice is rectangular (T5352), in another it is rectangular and there is an additional bore across the weight (T535202), and in the third the mortice is dovetailed (T5362). The first two are from oil presses, the third not *in situ*.

These weights are clearly a combination of the Levantine weight with reversed-T bore (T53) and the rectangular weight with external mortices, the Semana weight (T5512).

The location of these presses in North Syria is an additional argument for placing the origin of the Semana weight (Ts551-552) in Anatolia.[2]

5G. Beam Weights: Conclusions

Simple beam weights with horizontal bore are found in many regions and there is no reason to seek connections between them. It is of significance, however, that there are regions such as North Africa in which they have not yet been found, and others such as Italy where both simple and complex beam weights are absent. This phenomenon suggests that olive-pressing techniques were introduced into these regions after they had developed to an advanced stage.

The simple weight with vertical bore—the doughnut weight—has been reported from southern Israel only, suggesting that it developed and was used exclusively in this limited area. The later hooked weight was also apparently limited to the Levant.

In complex beam weights two technical traditions can be discerned: the reversed-T weight, which is limited to the Levant and can be called the Levantine or Phoenician weight; and the Semana weight and its many sub-types which apparently originated in the Aegean world and became the main weight in North Africa, southern France and Spain. The Syrian Taqle weights represent perhaps the meeting point of the two technical cultures.

Notes

1. See, however, the note on the 'Ashqelon weight', below note 2.
2. After the text of this book was completed and the catalogues closed, an additional type of weight was recorded at two sites in southern Israel, in an oil press from the Byzantine period excavated by Y. Israel 5 km north of ancient Ashqelon and at a site near Kibbutz Erez. This, the 'Ashqelon weight' (T551711), is square in shape, higher than it is broad and tapers towards the top. It is equipped with two lateral dovetail mortices, an upper grove and horizontal bore connecting the two mortices. While the shape of the Ashqelon weight and the horizontal bore are reminiscent of local beam weights, the mortices and groove and the manner in which the weight was operated derive from the Semana weight, and it was clearly introduced from one of the regions where this weight was the dominant type. At the same site where the oil press was found there were kilns where the Gaza amphora was produced. The trade relations with other countries doubtless led to technical exchange.

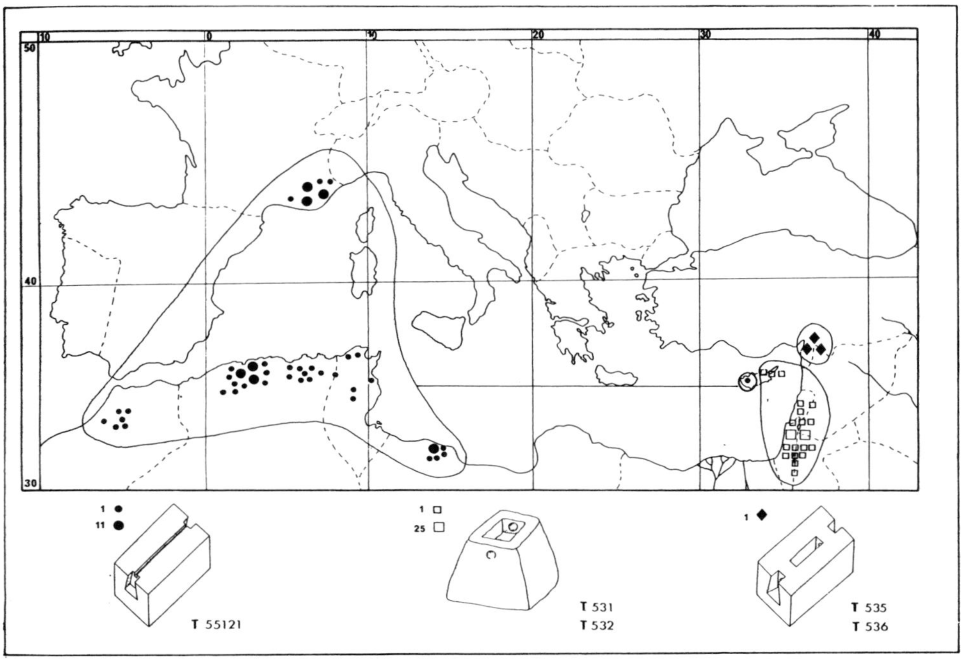

Map 18. Developed beam weights—main types: Mediterranean Basin.

Chapter 6

The Lever and Screw Press (T6)

6A. The Introduction of the Screw Press

6A1. Principles and Chronology

The introduction of the use of the screw in oil and wine presses was the next important technical turning point in the history of the press after the introduction of the lever. Hero of Alexandria (*Mechanica* 3.15) stresses the greater safety that the screw presses provided, but there is no doubt that the main benefit is in the mechanical advantage achieved, which is the multiple of the circumference of the end of the screw/nut hand-spake and the distance between the screw threads.

There were two basically different types of screw press: the lever and screw press (T6) and the direct-pressure single- or double-screw presses (Ts7, 8.1–8.3), in all cases with variants in which the screws rotated and others in which the screw was fixed and the nut turned.

As is specifically stated by Pliny the Elder (18.74-317; above, pp. 86-87) and implied by Hero of Alexandra in the order in which he presents his presses, the first type to be introduced was the lever and screw press. As is often the case in the history of technology, the new invention was first applied to an already-existing device and only later was its full potential realized. However, just as there were regions such as Italy that never went through the stage of the lever and weights press, but apparently started directly with the lever and drum press, so, as will be shown below, there were regions in which the stage of the lever and screw press was omitted and which passed directly from lever and weights presses to direct-screw presses.

Determining the chronology of the introduction of the lever and weights press is dependent on archaeological evidence only and, as a result, remains a matter of dispute, and awaits further data before a clear picture can be drawn that includes all the lands of the Mediterranean.

The chronology of the introduction of the screw press can be determined, however, to a great degree of accuracy with the aid of the written sources. Cato in the second century BCE clearly did not know of the screw press. Vitruvius (6.6.3) mentions a screw in pressing, but there is some doubt as to the date of Vitruvius (Drachman 1932: 126; above, p. 82). There is no doubt, however, that Pliny the Elder lived and wrote in the second half of the first century CE and he speaks of the introduction of the lever and screw press 'during the last hundred years' and of the direct-screw press during the 'last twenty-two years' (Pliny: 18.74.317; above, pp. 86-87). Thus the lever and screw press was probably first introduced into Italy during the first century BCE and the direct-screw press during the first century CE.

6A2. Technology: Lever and Screw, Intermediate Piers (Ts441-446)

In the lever and screw press the free end of the beam is lowered with the aid of a screw. This was achieved in a variety of ways (Models 6A–6M). There are certain characteristics that all or some of these models had in common.

The first of these is that, as the end of the beam moves in an arc of a circle and the screw is straight, operating the press can result in mechanical tension causing the screw to stick. In Hero of Alexandria's lever and screw press this problem was solved by a special hinging device (Model 6L below; above, pp. 87-88), but apparently the difficulty was usually overcome by the nut being attached above the beam in a way that allowed it to move a little while the screw was slightly flexible and well oiled.

One other characteristic that all the models except 6C, 6F and 6M had in common, was that, by turning the screw in one direction, the beam was lowered and, by turning it in the reverse direction, it could be raised. In many cases a pair of intermediary slotted or perforated piers were placed on either side of the centre of the beam. A cross-piece that was inserted in the slot or perforation of these piers under the beam functioned like the pivot of a see-saw. If the anchoring end of the beam was left unattached, it was possible to raise it by lowering the free end of the beam, and also the reverse was the case and it was possible to lower it by raising the free end of the beam (Drachman 1932: 122-24, fig. 41; Violet 1938: 165).

Intermediary piers of this type are found in many pre-industrial presses. However, there are cases in which there are intermediary piers that do not function in this way but serve only to guide the beam and prevent it from moving sideways. The intermediary piers in the Moroccan lever and screw presses are of this type (Tezarin Taghzuth; Fez). In pre-industrial presses it is usually possible to distinguish between these two types of intermediary pier, although this has not been done in the catalogue where both are numbered T012---5. There is, however, clearly no way to determine which of the two types of intermediary pier were affixed in guider mortices found in ancient presses (Ts441-446).

Slotted intermediary piers of the first type can only function in lever and screw presses. Guider piers, however, could also be part of lever and weight or lever and drum presses. In Israel and Syria in ancient times guider sockets were apparently found in lever and screw presses only. In North Africa, however, there is perhaps evidence for their appearing in lever and weights presses. The evidence for the presence of guider mortices in lever and weights presses in North Africa, however, comes from two presses, both of which were published as reconstructions only, thus raising some doubt as to the reliability of the data (Madaure; Henchir Choud et Battal).

6B. Models of the Lever and Screw Press Based on Pre-industrial Presses, Classical Written Sources and Ancient Artistic Representations: Models 6A–6M

Model 6A: Lever and Screw Press—Rotating Screw Anchored at Lower End to Frame of Press (Pre-Ind. Ts012111): The 'Grand Point' Press

In this press the nut is attached above the beam. The lower end of the screw is made fast to the lower frame of the press so that it could rotate but not work free. Turning the screw in the nut lowered the beam. The lower end of the screw was spherical with a narrow neck between it and the screw proper. The spherical screw-end was inserted through a hole in the lower frame of the press and kept in position by two boards, both with semicircular indentation that were fitted round the neck. These boards were fixed either under the press frame or into a slit cut into it. This type of press is recorded in an eleventh-century manuscript from Echternach, Luxemburg, and is common in the Bourgogne region of France but is attested for also in Champagne, southern Germany and Hungary. It is known in French as 'Pressoir à Grand Point' (Violet 1938).

Photo 51. Model 6A: the grand point press: Chateau de Vinzelles, France (Humbel 1976: Pl. XX).

Model 6B: Lever and Screw Press—Rotating Screw Pivoted in Upper Frame of Press and Ground (Pre-Ind. T012112)—'The Pivoted Screw Press'

In this press the female thread is in the beam-end which rides up and down on a rotating screw that remains in a fixed position by being pivoted at both ends. It was apparently a suggestion for a new type by Rozier (1776: 10-12, Pl. 3). It should be pointed out that Rozier calls this press 'Pressoir à Martin', a name usually given to Model 6G, the 'Languedoc press'. This is also the manner in which Drachman (1932: 54, fig. 14) reconstructed Pliny the Elder's questionable second press (see pp. 86-87 above). It is far from certain that such a press ever actually functioned. The mechanical tension between the beam-end moving in an arc and the straight screw would, in this model, be particularly problematic.

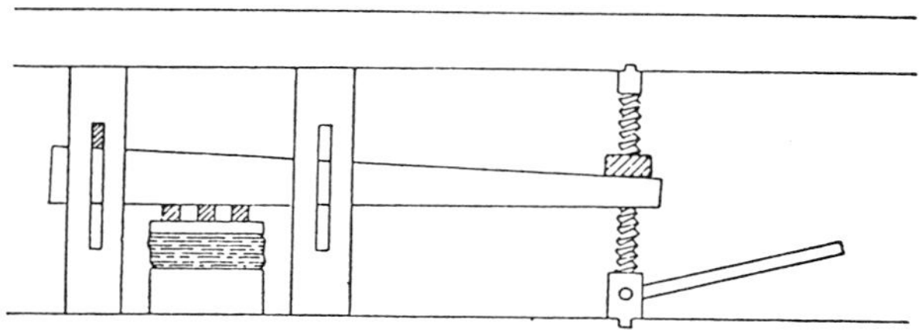

Figure 16. Model 6B: the pivoted screw press (Drachman 1932: fig. 14).

Model 6C: Lever and Screw Press—Rotating Screw Screwed into Lower Press Frame, Tied to Beam-End (Pre-Ind. T012113)—'The Tied Screw Lever Press'

In this press the female thread is in the lower press frame, and the screw is screwed into it from above. The top end of the screw is tied to the press-beam. There is only one example of this press in the catalogue.

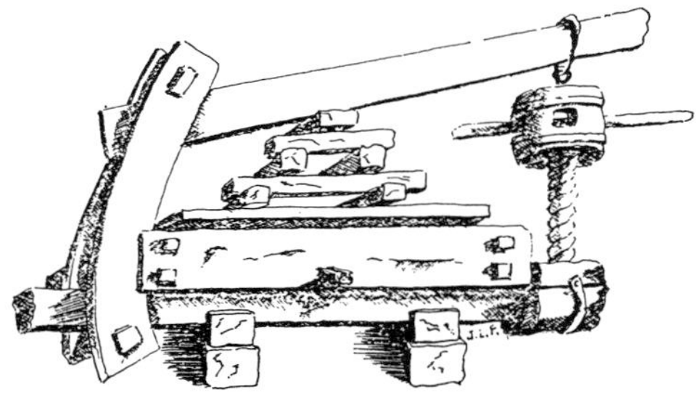

Figure 17. Model 6C: the tied screw lever press (Humbel 1976: fig. 15).

Model 6D: Lever and Screw Press—Rotating Screw Anchored between Two Posts Fixed in the Ground (Pre-Ind. T012120)—'The Taissons Press'

This press functions similarly to Model 6A. The screw-end is also anchored in a similar manner. The two boards with semicircular indentation are in this case inserted into cuts in the anchoring posts, and sometimes fastened by tenons or pins. This press is found in Bourgogne, and in other regions of France, and possibly also in Hungary. It is known in French as 'Pressoir à Taissons'.

Photo 52. Model 6D: the Taissons press: 'Maison Rustique 1749' (Humbel 1976: Pl. IX).

Model 6E: Lever and Screw Press—Rotating Screw Anchored in Fixed Stone Block (Pre-Ind. Ts012122)—'The Fixed Block Press'

This press functions in a similar manner to Models 6A, the 'Grand Point', and 6D, the 'Taissons press'. In none of the examples in the catalogue, however, is the exact manner in which the screw is attached to the block quite clear. From a photograph or diagram alone it is not always possible to distinguish this press from Models 6H or 6K in which the screw is attached to a weight. In fact it is probable that scholars describing the presses sometimes also confused them. The presses at Tezarin Taghzut and Fez in Morocco are almost certainly presses of the same type and they have here both been numbered T0121805. There is no mention of a weight in the description of the first, however. It is also probable that in some cases the way in which the press was operated changed and, if originally the weight was lifted, later the screw was turned without actually lifting the weight, which then became a block fixed in the ground. There could therefore well be some mistakes in typology in the catalogue regarding these types of presses.

Photo 53. Model 6E: The fixed-block press: Zürich Switzerland (Humbel 1976: Pl. VIII).

Model 6F: Lever and Screw Press—Press-Bed above Beam, Rotating Screw Pivoted on Ground Raises Beam and Press-Bed (Pre-Ind. T012130)—'The Raised Press-Bed Press'

The female screw-thread is in the beam-end. Turning a screw, the bottom end of which is pivoted on a block placed on the ground, raises both beam and press-bed.

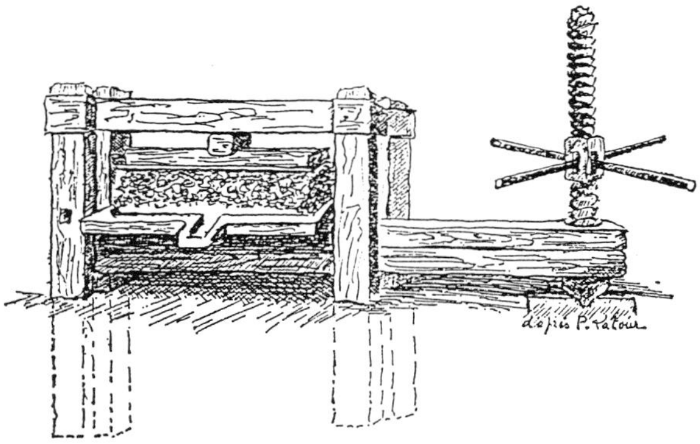

Figure 18. Model 6F: the raised press-bed press (Humbel 1976: fig. 11b).

Model 6G: Lever and Screw Press—Screw Fixed to Weight, Screw and Weight Rotating Together (Pre-Ind. Ts01214, 01215)—'Languedoc Press'

In this press the nut is fixed above the beam and the bottom end of the screw is fixed to a weight. A block that goes right through a square hole in the weight is fixed to it by a horizontal pin that goes right through both weight and block. A metal pivot is attached to the bottom of the block that protrudes below the weight. Turning the screw together with the weight will first lower the beam and after it can descend no further because of the resistance of the substance being pressed; additional turns of the screw will raise the weight from the ground. This press appears in a fifteenth-century manuscript illustration, is called in Diderot's Encyclopaedia 'Pressoir dit a Grand Banc' from Languedoc and Provence. It is also called 'Pressoir à Martin' and pre-industrial examples can still be seen in Provence. The additional weight in T01215 is apparently not of technical significance, but is an easy way of making the weight heavier in cases such as this where it is perforated.

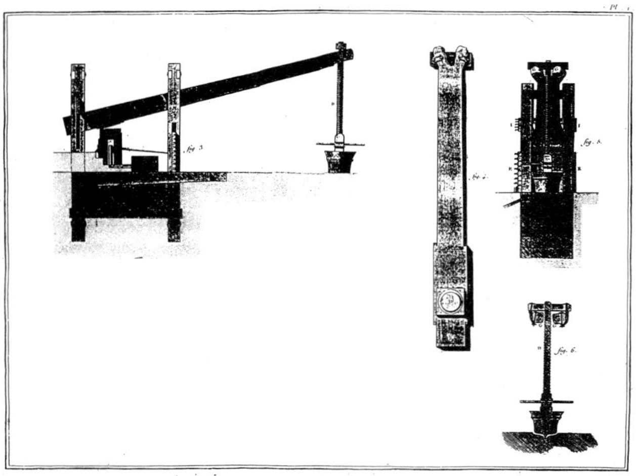

Figure 19. Model 6G: the Languedoc press (D'Alembert and Diderot 1762: Pl. 1).

Model 6H: Lever and Screw Press—Rotating Screw Anchored to and Lifts Weight that Rotates Separately (Pre-Ind. T012161)—'Arginunta Press'

In this press the nut is fixed above the beam-end and the bottom end of the screw is attached to a weight in ways similar to those in which the screw is attached to the press frame in Model 6A, 'the Grand Point', and to the fixed posts in Model 6D, 'the Taissons press'. In the presses from Arginunta, Lania and ᶜEin Hod the screw-end was held by two boards with semicircular indentations as in the presses mentioned above, and in the latter two these boards were held together by the insertion of long round pins. The boards were then attached to mortices in the weights by additional tenons.

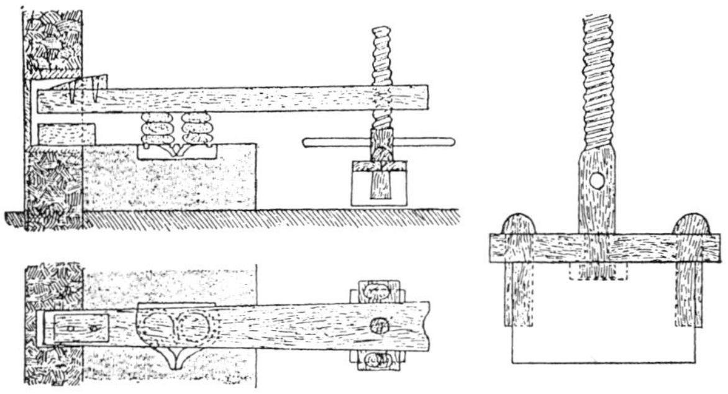

Figure 20. Model 6H: the Arginunta press (Paton and Myers 1898: fig. 1).

The screw turned in the weight without the possibility of working free. By turning the screw in the nut, above the beam, it lowered the beam and, as in the Languedoc press, Model 6G, when the beam could descend no further, turning the screw raised the weight off the ground. This press is recorded in manuscript illustrations from Spain in which the screw weight is shown in section, showing the round screw-end in the weight. Pre-industrial examples are found in Israel, Portugal, Spain, Cyprus, Kalymnos, Italy and Hungary. It is apparently rare in France. This model is easily confused with the 'fixed-block press', Model 6E (see above).

The various types of very characteristic screw weights of the Arginunta press constitute the main archaeological evidence for the existence of lever and screw presses in the whole Mediterranean Basin.

Model 6J: Lever and Screw Press—Rotating Screw Anchored to and Lifts Box of Stones or Board with Stones Placed on it (Pre-Ind. Ts012162, 012163)—The 'Box Press'

This model works in exactly the same manner as Model H, the Arguninta press, only loose stones replace the weight of the latter. The first type, that in which a box of stones is lifted, is found in France and known there as 'Pressoir à Cage'; the second, in which the stones are on a board, is found in Rumania, Switzerland and Hungary. This is Pliny the Elder's press B/C (see above pp. 86-87).

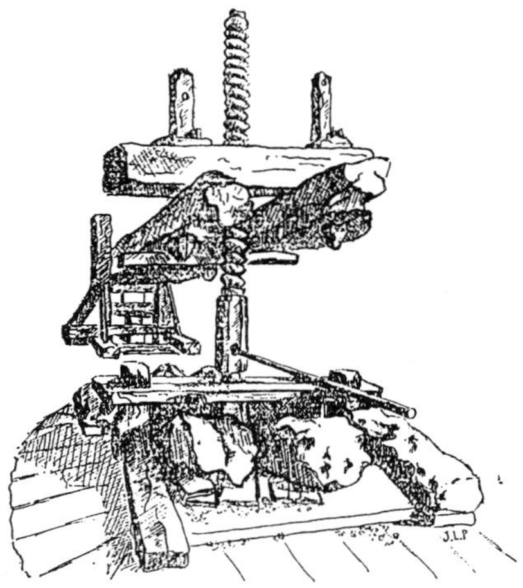

Figure 21. Model 6J: box press—variant that lifts stones placed on board (Humbel 1976: fig. 17).

Photo 54. Model 6J: the box press: 'Maison Rustique 1749' (Humbel 1976: Pl. X).

Model 6K: Lever and Screw Press—Screw Rotates in Perforated Weight or Weights (Pre-Ind. Ts01217, 01218)—'Perforated Weight Press'

This model, which is found mainly in Spain and Morocco, functions in the same manner as Models 6H, the Arginunta, and 6J, the box press. The screw rotates in the perforated weight and lifts it. The perforation makes it possible to increase the weight by adding a second stone (T01217). As in Model 6H there is a possibility of confusion with Model 6E, 'the fixed-block press'.

Photo 55. Model 6K: the perforated-weight press—Elche, Spain (Photo: Carter Litchfield).

Model 6L: Lever and Screw Press—Screw Hinged to Beam, Rotating Nut Anchored to Weight (Hero's Press B; see Fig. 10, p. 88)

In this press the rotating nut swallows the screw, thus lowering the beam and finally raising the weight. This is Hero of Alexandria's second press (*Mechanica* 3.15, see above). No pre-industrial presses of this type have been recorded. It is obviously closely related to the Arginunta press, Model 6H, and is presumably an improved version of it that was suggested but was probably never made. The screw weight of this press could also serve in the Arginunta press.

Model 6M: Lever and Screw Press—Screw Fixed to Lower Press Frame and Nut Rotated (Pre-Ind. T01221)—'The Fenis Press'

In this press the screw is fixed to the lower press frame and the nut above the beam is turned. There are only two examples in the catalogue from Italy and Hungary.

Most of the models here described would leave no archaeological remains, Model H, the Arginunta press, being the model that could and did leave most archaeological evidence.

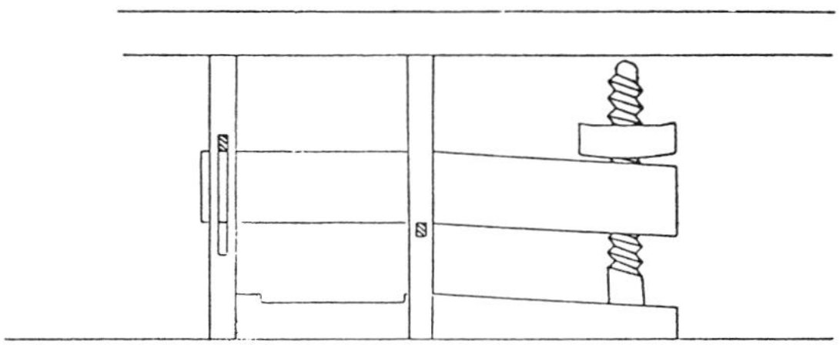

Figure 22. Model 6M: the Fenis press (Drachman 1932: fig. 13).

6C. Archaeological Evidence for Lever and Screw Presses, Screw Weights (T62): Israel and Environs

As shown above (pp. 92-93), it is very probable that the Italian presses of the circular bed type that lacked obvious evidence for the manner in which they functioned (Ts4051-1--?) were lever and screw presses raising boxes of stones (Model 6J). Similarly it was tentatively suggested that the Volubilis weights (T55111, 55412), the Cotta weights (T56911), the Francolise-Posto weight (T5541) and associated weights were screw weights (above, p. 105). In most cases, however, the evidence for lever and screw presses are unequivocal, being the presence of undisputed screw weights (Ts62; complete press Ts40------7, 40---- ---8).

In Israel and environs several different types of screw weight have been found. These have certain characteristics in common. They are nearly always cylindrical in shape and their height and diameter both average c. 0.90 m. They thus weigh c. 1.4 tonnes.

6C1. Screw Weights with Central Sockets

6C1.1. The 'Samaria Screw Weight': Screw Weight with Central Socket and Two External Dovetail Mortices (T6211). This is one of several types of screw weight (Ts621, 622, 623, 624) all having a round socket in the middle of their upper surface and additional mortices which vary according to the type of weight. All these weight types functioned according to the principles of Model H, the Arginunta press. The screw-end rotated in the socket and tenons or wedges that fitted into the mortices secured the screw-end to the weight in a manner that allowed it to revolve but not to work free. The differences between the weights is in the form of the mortices and related tenons or wedges.

In the Samaria weight the boards with semicircular indentations that held the screw-end were attached to the weight by dovetailed tenons that fitted into the external dovetail mortices.

The Samaria screw weight is the most common type both in Israel and in all the ancient world. In Israel, of the 89 recorded, 78 are cylindrical (Ts62111), three square with rounded corners (T62116), two rectangular (T62122) and one each octagonal (T62114), conical (T62123) and in trapezoid section (T62137). It is perhaps significant that two pre-industrial weights were square with rounded corners and had open mortices (T62136). Of those with exceptional mortices, 10 are with reversed-T mortices (Ts6212), a form not found in other countries. Apparently only four were with open mortices (Ts6213), one trapezoidal in section from Ḥasun, the two pre-industrial weights already mentioned (T62136) and a third pre-industrial weight cylindrical in shape (T62131). A type that is found in other regions and therefore significant is that with four mortices (Ts6214). Six examples were found.

Geographically, only 11 of the 89 were located

Photo 56. The Samaria screw weight (T62121): Tiberias, Hot Springs.

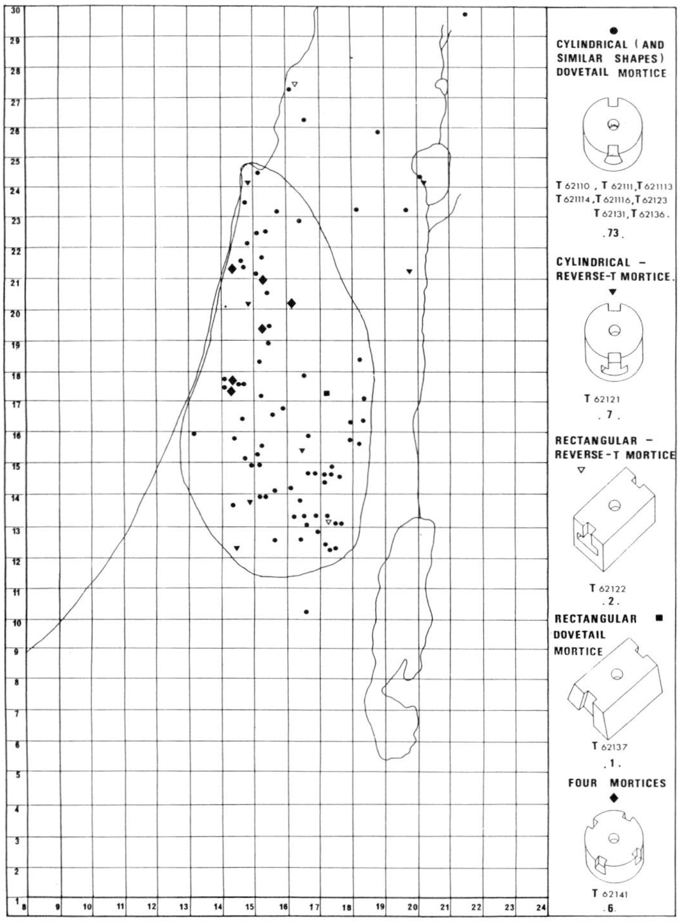

Map 19. Samaria screw weight—socket and external mortices (T621).

north of the Carmel range and only two south of 12 EW co-ordinate, making this primarily a Samarian type with a concentration also in and around Jerusalem.

6C1.2. The 'Dinᶜila Screw Weight': Screw Weight with Central Socket and Open Dovetail Mortice (T622). In the Dinᶜila weight the socket is in the centre of an open dovetail mortice that cuts across the whole upper surface of the weight. Two dovetail tenons, the ends of which were shaped to fit the spherical screw-end, were clearly introduced from opposite ends, holding the screw-end in place. This is exactly the arrangement that Hero of Alexandria described in his press B (Hero, *Mechanica* 3.15; see above, pp. 87-88).

Photo 57. The Dinᶜila screw weight (T62211): Shubeika Inst. 03.

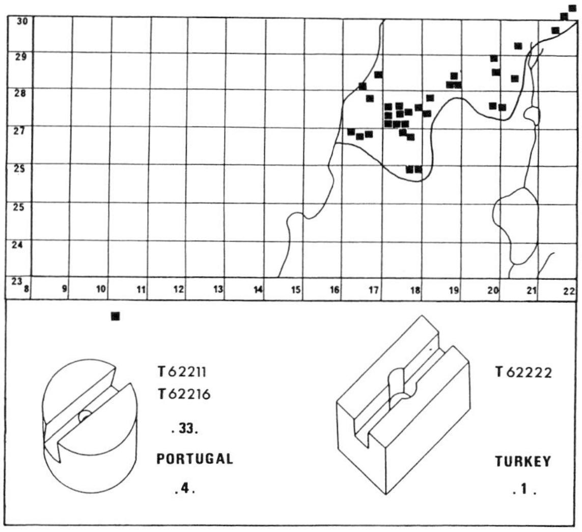

Map 20. Dinᶜila screw weight—socket and open dovetail channel (T622).

The Greek original of Hero's *Mechanica* has been lost, and Drachman (1963: 121-12) translates from the Arabic:

> So we place the end of the block into the hole that is in the stone and we make for it iron clamps which keep the block from coming out of the hole in the stone.

The question is how to understand the Arabic word *ḍb* translated as 'clamps'. Drachman shows that the same term was used for the tenons that secured the bottom of the nuts in Hero's double-screw press, his third press. Drachman suggests that these were 'boards sliding in a dovetail groove'. He also suggests that in the original Arabic text the word might have been in the form meaning 'two clamps'. If Drachman's interpretation of the text is correct, and there were two clamps that 'slid in dovetail grooves', then this is an exact description of the Dinᶜila weight. There is little doubt, therefore, that Hero knew this press weight, and he thus provides a *terminus ante quem* for its introduction. There is some controversy as to when Hero lived. It was probably first century CE but possibly earlier (see above, p. 87).

The Dinᶜila weight is extremely uniform typologically. Of the 32 examples recorded, only one is not cylindrical. Geographically, the distribution area is equally well defined and is very similar to that of the slotted piers (T42). The weight is found only in western and Upper Galilee and the northern Golan, not penetrating into Lower Galilee or to eastern Upper Galilee. Nine examples were found in square 1727, and three in square 1827, but none in square 1826 and 1926. The southern border of the distribution area is thus very clearly defined; the northern is, however, on the other side of the international border and is as yet unknown. The southern border fits to a remarkable degree the border between Phoenicia and Upper Galilee as defined by Josephus and that between Palaestina Secunda and Phoenicia in later periods.

6C1.3. Screw Weight with Central Socket, Square Channel and External Mortices (T623). Only two weights that fall into this type were recorded in Israel and there is a considerable difference even between these. That from Kabri is cylindrical and the lateral mortices are in the shape of a reversed 'T'. Kabri is in the distribution of the Dinᶜila weight (T622) and the Kabri weight combines the characteristics of this weight with those of the variant of the Samaria weight with reversed-T mortices (T6212).

The other weight of T623, that from ᶜEin el-Judeida (T6232402), on the other hand, is exceptional in almost every way. It is from a wine press in a monastery and there are very few lever and screw presses in wine presses in Israel. It is the only weight of its type so far recorded in the country. It is of exceptional shape—octagonal, or perhaps more correct, rectangular with cut-off corners. It also has four small holes for clamps (T62--2), a device not otherwise known in Israel but existing in other countries. All this suggests it to have been introduced from afar. Comparison with the weights of this type from other countries show the form of channel and mortice to be very different from those of the Rabat weight, in which the top of the mortice and the channel are apparently the same width (the weight is slightly damaged) but identical to those of the Syrian Beḥyo weight, and of the weight from the Pontus 03 in Anatolia. In both, as in the weight from ᶜEin el-Judeidah, the mortice is narrower than the channel. This suggests the ᶜEin el-Judeida weight, the Behyo weight from Syria and the Pontus 03 weight to be related. As for the origin of the ᶜEin el-Judeida weight, although holes for pins are found both in Syria and the Pontus, the Syrian weights are cylindrical, whereas those from the Pontus are rectangular. This suggests the ᶜEin el-Judeida weight to be closer to those from the Pontus and to have been brought from that or a closely related region, probably by a monk or pilgrim.

6C1.4. The 'Kasfa Screw Weight': Screw Weight with Central Socket and Internal Dovetail Mortices (T624). The Kasfa weight (T624) is related to the Dinᶜila weight (T622). In both, the socket is in a channel dovetail in section. In the Dinᶜila weight the channel is open at both ends; in the Kasfa weight it is closed. They clearly functioned similarly. In the Kasfa weight two tenons were used that were shaped in the same way as those used in the Dinᶜila weight—dovetail in section with the end shaped to fit the spherical screw-end, but in the Kasfa weight slightly shorter than the mortices. The tenons were first inserted into the mortice via the central socket. Then the screw-end was inserted into the socket, the tenons were moved to the centre to grasp the screw-end and wedged into position. The advantage of this arrangement is that the tenons cannot work themselves free, whereas in the Dinᶜila weight (T622) this could happen.

Photo 58. The Kasfa screw weight (T62411): Jerusalem Notre Dame Monastery.

Photo 59. The Kasfa screw weight (T62411): Jerusalem Notre Dame Monastery.

Only seven weights of this type were recorded and one of those, that from Jerusalem Notre Dame, was not found *in situ* nor was its provenance ascertained. The remaining six were all from the centre of the country, one from within present Jerusalem, the others all within 80 km of the city to the west and north-west. One exceptional weight is that from Pisgat Zeᶜev (T624111), in which the socket and mortices are enclosed in a rectangular depression, a characteristic typical of the Sarepta weight found in North Syria and Lebanon.

Four other possibly related weights were recorded. One is from Rosh Pina (T62421) in the north of the country, but its exact form is not clear. One from Qedumim (T62431) has similar socket and mortices to those of the Kasfa weight, but incorporates also two external dovetail mortices, thus combining the characteristics of the Samaria (T6211) and Kasfa (T624) weights, the site being within the distribution area of both types. As the internal and external mortices are not aligned, it is probable that one method replaced the other. The other two related weights, those from Bet Loya (T62441101) and Duḥdaḥ (T62451101), are very unusual. These are both very closely related to the Šeiḥ Barakat weight from Syria (T62434101) and will be discussed together below (p. 118).

6C2. Screw Weights without Central Socket

6C2.1. The 'Bet Ha-ᶜEmeq Screw Weight': Screw Weight without Socket and with Central Closed Dovetail Mortice (T625). The Bet Ha-ᶜEmeq screw weight (T625) is one of a series that lack a central socket instead of which there is a single mortice in the centre of the upper surface of the weight (Ts625, 626, 627). Two of the three main types have been found *in situ* in oil presses (T62511: Quṣeir 03–04 and Bata; T62711: Quṣeir 01) and therefore these are without doubt press weights. It could be suggested that the mortices are to secure a hook as in the Zabadi weight (T541). However, the size, shape and date of these weights all point to their being screw weights. The screw-end was clearly fixed to the weight but without allowing the screw to rotate as it does in the Samaria, Dinᶜila and Kasfa weights. In the

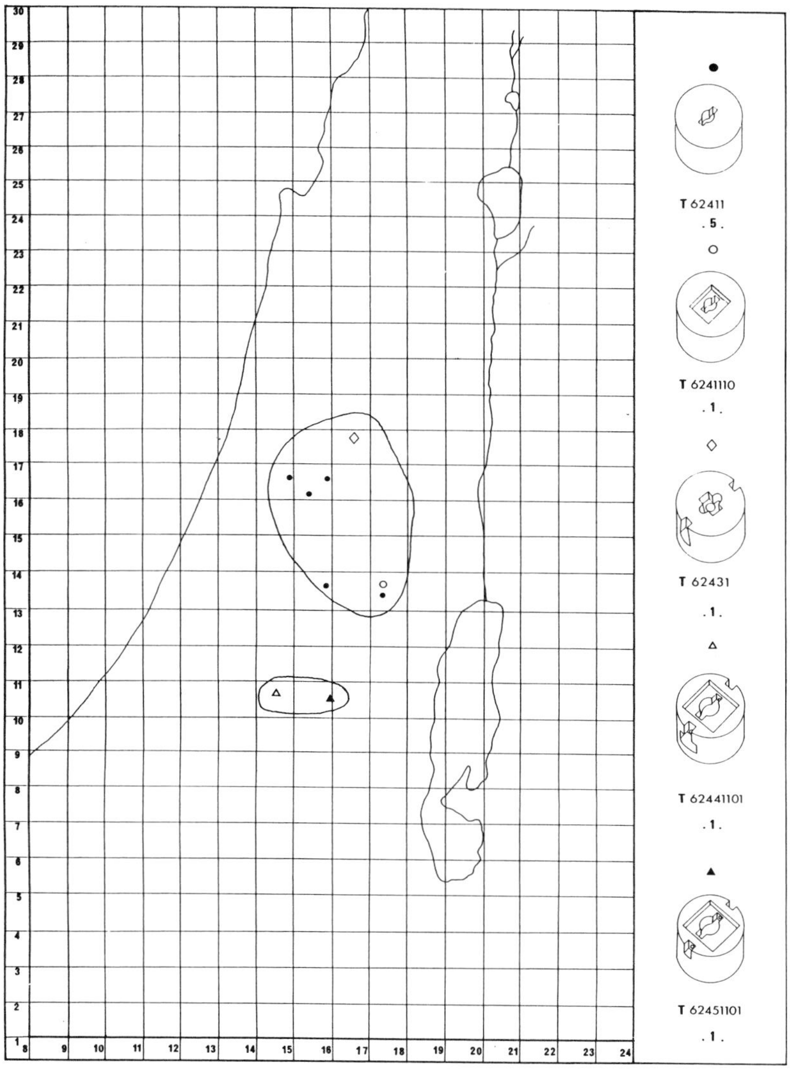

Map 21. Kasfa screw weight—socket and internal mortices (T624).

Lanquedoc press (Model 6G) the screw is similarly fixed to the weight and they rotate together. However, there is no sign on the lower side of the socketless screw weights from Israel of a pivot or a device to attach one, and clearly without a pivot the Lanquedoc press could not function.

The socketless weights almost certainly functioned in the same manner as the Fenis press (Model 6M). The screw was fixed and the nut above the beam was rotated. In the Fenis press the screw is not fixed to a weight but to the frame of the press. In a press in which the screw is fixed to a weight there is a technical difficulty that could arise. When the weight rises from the ground, the screw, nut and weight would tend all to revolve together. This could, however, be overcome with little difficulty by preventing either the weight or the screw from revolving—a hand-spake hitting a ver-

Photo 60. The Bet Ha-ʿEmeq screw weight (T62511): Quṣeir Inst. 04, press C.

tical point, for example, would have prevented rotation and allowed up and down movement.

The closed dovetail mortice of the Bet Ha-ᶜEmeq weight (T625) is rectangular at the bottom, while one half of the top is the same width as the bottom and the other half narrower. As a result, half of the mortice is dovetail in section and the other half has straight sides. The screw-end or other object to be affixed in this mortice was dovetail-shaped and was inserted into the straight-sided half, moved sideways into the dovetailed half and then wedged into position. Minute mortices of this type were used in the Levant to join ivories to their base in the Iron Age (Nimrud; Barnett 1957: 155, nos. C29, S362; 1982: 13).

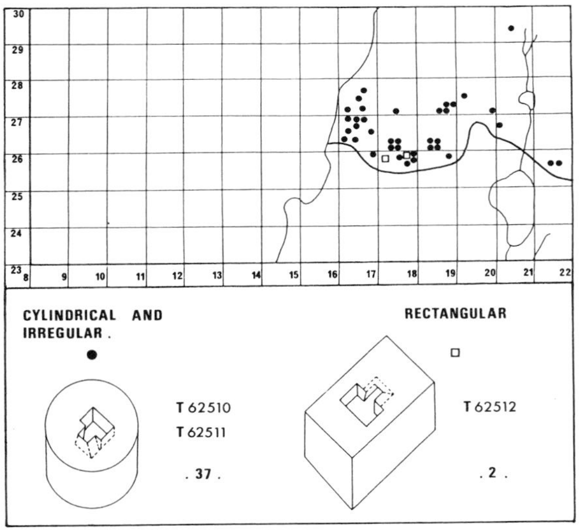

Map 22. Bet Ha-ᶜEmeq screw weight—no socket, central dovetail mortice (T625).

Hero explained how to cut this mortice as a method of lifting heavy building stones (Hero, *Mechanica* 3.6; Drachman 1963: 103-104) and it appears in several other types of installation as well as in individual installations both from Israel and other countries (all central dovetail mortices appear in T041).

Of the 39 examples of the Bet Ha-ᶜEmeq weight recorded, 32 were of the standard cylindrical shape (T62511), 7 were of irregular shape, 4 of which were re-used pillar bases and capitals, and 2 were rectangular. The distribution pattern is very similar to those of the Dinᶜila weight (T622) and of the Galilean slotted piers. As with these, the southern border of the distribution area is clearly defined, while the northern is on the other side of the international border and is as yet unknown. These weights are also found mainly in western Galilee.

Detailed comparison shows two small differences between the distribution patterns of the Dinᶜila and the Bet Ha-ᶜEmeq weights. The first is that four examples of the Bet Ha-ᶜEmeq weight appear in eastern Upper Galilee while no examples of the Dinᶜila weight were found there. The other is that in the region in the hills of western Galilee, square 1727, where eight examples of the Dinᶜila weight were recorded, only one Bet Ha-ᶜEmeq weight was found. A possible explanation is that the Dinᶜila weight reached the region at a later stage so that the lever and screw press with the Bet Ha-ᶜEmeq weight was the first to appear and reached the more open and intensely settled areas in the valleys. By the time the lever and screw press reached the more closed, mountainous areas, the new Dinᶜila weights had been introduced.

6C2.2. The 'Luvim Screw Weight': Screw Weight without Socket and with Square Mortice Widening on Two Adjacent Sides (T626). The square mortice of the Luvim weight widens in its lower section on two adjacent sides and a screw which widened similarly was inserted into the mortice pushed into the widened corner and wedged into position.

Twelve standard cylindrical weights were recorded (T62611). Of these, one was in Lower Galilee and the others on the Carmel and in the Sharon coastal plain. The distribution area does not overlap with those of the Dinᶜila and Bet Ha-ᶜEmeq weights to the north. As the data from Samaria is not complete, the distribution area could spread further to the south-east.

One variant weight from ᶜEn Hadda in Lower Galilee was square and incorporated two external dovetailed mortices (T62622).

Photo 61. The Luvim screw weight (T62611): Safti ᶜAdi.

Photo 62. The Luvim screw weight (T62611): Safti ᶜAdi.

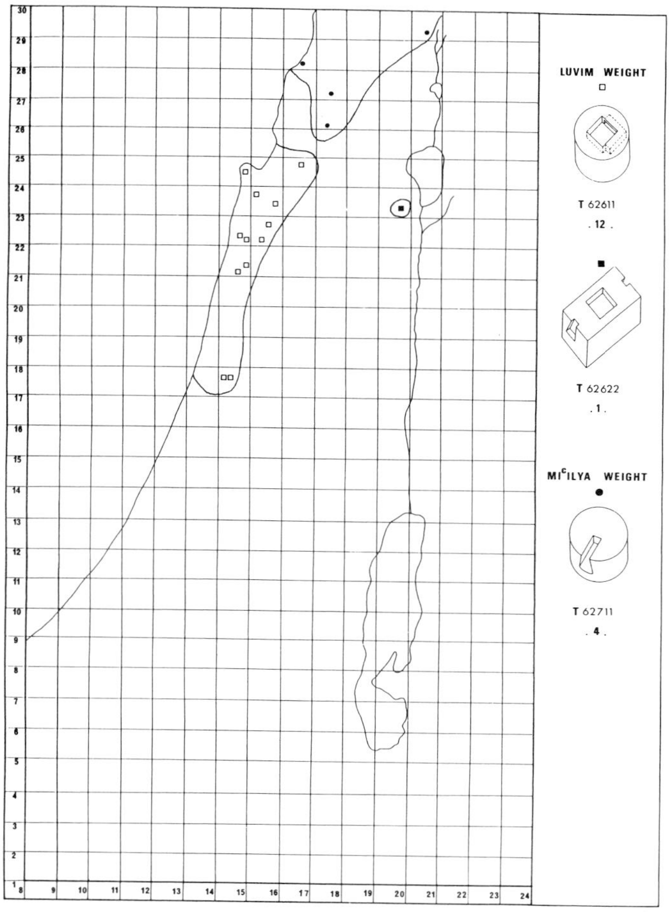

Map 23. The Luvim screw weight—no socket, square mortice (T627); The Miᶜilya weight—closed dovetail channel (T627).

6C2.3. The 'Miᶜilya Weight': Screw Weight without Socket, Closed Dovetail Channel (T6271); The 'Midrasa Weight': Screw Weight without Socket, Open Dovetail Channel (T6272). The Miᶜilya and Midrasa weights are the simplest of the screw weights, consisting only of dovetail channels cut half across the Miᶜilya weight and across the whole of the Midrasa weight. The dovetail screw-end was inserted in the channel and wedged into position.

Miᶜilya weights were recorded at four sites, at one of these—at Quṣeir 01—*in situ* in an oil press. The Midrasa weight was recorded at two sites, one on the Golan. The other and all four sites at which the Miᶜilya weights were found are in western and Upper Galilee.

6C3. Chronology of Screw Weights in Israel

Very few of the weights discussed are clearly dated. However, until now no screw weight has been recorded that dates to the Roman period, but a considerable number are dated to the Byzantine and later periods. A variant of the Samaria weight appears in pre-industrial presses—the Iksal weight, small, square with rounded corners and open mortices. Standard Samaria weights were also used—perhaps ancient weights in secondary use.

Photo 63. The Miᶜilya screw weight (T6271): Quṣeir Inst. 01, press A.

6C4. Quarrying and Dressing Screw Weights

As in the case of the simple rock-cut winery (T111; see p. 51 above), the archaeological evidence makes it possible to determine the manner in which the screw weights were quarried, transported and completed (See T03 for relevant data). Quarries have been recorded in which rounded depressions show that several weights or similar components were extracted at one site, for example, ʿAmqa North, Israel, and Puget sur Argens, France. In other cases there are signs of only one weight having been extracted at a site, leaving a round depression or a semicircular apse-like indentation in the rock face. A cylindrical weight still attached to the bed-rock was recorded at Danun, and at ʿAmqa Central on a similar stone all the stages involved in cutting were still evident. First, a small hole was drilled in the centre of the stone and then a perfect circle drawn and engraved on the rock face, presumably using dividers and then hammer and chisel. The next stage was to cut a channel round the circle leaving the four or five centimetres closest to the circle uncut. The next stage was to cut the stone to a perfect cylinder by cutting a vertical edge flush with the circle. The final stage was to free the stone from the bed-rock by undercutting.

In the centre of the base of many of the cylindrical screw weights there was a small hole averaging about 8 cm in circumference and depth. On the upper surface of the weight there is usually a socket or mortice, but some unfinished examples (e.g. Tuweiri 01) show there was originally a similar hole in the centre of the upper end of the weight also. Vitruvius (10.2.11) describes how Chersiphron transported shafts to the temple of Diana at Epheseus by attaching a frame and handle to them, somewhat like a garden roller, and thus rolling the shafts to the temple. The cylindrical screw weights were probably transported in a similar manner from the quarry to the press.

One unfinished weight (Karkara 10) found lying beside an ancient pathway lacked all signs of mortices or sockets. This confirms that the holes were used for a frame for transportation. The sockets and mortices were cut at the site of the press and were probably made to fit the wooden components available.

Photo 64. Quarrying and dressing cylindrical press component (T03): ʿAmqa Central.

Photo 65. Quarrying and dressing cylindrical press component (T03): ʿAmqa Central.

6D. Screw Weights in Other Countries (Ts62)

6D1. North Syria and Lebanon

The screw weights of North Syria have been described in detail (Callot 1984), but the manner of publication does not allow for true quantitative analysis. Nevertheless, certain conclusions can be drawn.

As in Israel, the screw weights are cylindrical, the one exception probably having been mistakenly identified as a screw weight as will be shown below. Both the Samaria weight (T62111) and the Beḥyo weight (T62321) are represented. The main type, however, is the one with socket and internal mortices (T624) which appears in three variants. These all have one characteristic in common that distinguishes them from the Israeli Kasfa weight: they have depressions around the socket and mortice which is also found in a weight from Sarepta, Lebanon. This weight in the preliminary publication (Pritchard 1978) was dated to the Hellenistic or Roman period, which would make it the earliest dated screw weight in the Levant. Including this example, there are ten with a rectangular impression—Sarepta weights (T624111, 6241112)—and one with a depression in the form of a double cross (T624112).

This raises the question of the relationship between the Syrian type, the Sarepta weight (T624111), and the Israeli Kasfa weight (T62411). The number of this type found in Israel is small; they come from a limited area in the centre of the country near Jerusalem, where there is no obvious prototype from which they could have developed, but which is a region that had many contacts with other countries. There is, therefore, little doubt that the Kasfa weight originated from another region.

In all the Syrian Sarepta weights the socket and mortices are in a depression; in the Israeli Kasfa weight this depression is lacking. It is, however, found in three weights of this group (Ts624) in Israel, one identical to the Syrian Sarepta weight (Pisgat Zeʿev, T624111), and two extremely unusual weights (Bet Loya, T62441101;

Duḥdaḥ, T62451101) which are remarkably similar to an equally unusual weight from Syria (Seih Barakat, T62434101). All three have a socket, aligned internal and external mortices, a rectangular depression and a bore going through all four mortices. The three differ only in the shape of the weight and the form of the external mortice. The weight from Beit Loya comes from an excavated oil press in which other components also point to Syrian connection, a socketless crushing basin (T343) and guider mortices (T441) (Frankel, Patrich and Tzafrir 1990).

The fact that the Sarepta weight and its variants are the dominant type in Syria and the great similarity between the Israeli and Syrian weight suggest close links between the regions. However, no weights without depression similar to the Kasfa weights were reported from Syria, whereas such weights have been found in other regions—in Anatolia, and in southern France. This shows the history of this weight to be complex and hints at links with other regions also.

One screw weight from Syria has four holes for clamps (Ksegbe 09, T6241112). Only one weight with this device was found in Israel at ᶜEin el-Judeida and this has been shown to be exceptional there and introduced from afar. Only more detailed publication can show if this is true also of the Syrian weight.

A type represented in Syria and known from several other countries but not found in Israel is that with socket but lacking mortices (Ts629); one Syrian example from Sarfud has a rectangular depression (T629111), the other, from Dar Qita, one that is H-shaped (T629113).

A square perforated stone (Kafr Nabo 10, T629424) was published as a screw weight. Perforated screw weights are known (T6293, 6295) but in these the perforation is round and they are all pre-industrial from the western Mediterranean. The Kafr Nabo stone has probably been misidentified. It is similar to water cistern covers, although a lot smaller than those usually found.

6D2. Greece, the Pontus, Kalymnos, Lesbos, Cyprus

The small number of screw weights that appear in the catalogue has made it necessary to treat this widespread region as one unit that can in the broadest possible sense of the term be called 'the Aegean'.

While the cylindrical Samaria weight (T62111) appears in Cyprus, and examples from Athens (the Agora made of architectural elements, T6211001) could be defined as cylindrical, the main characteristic that the weights from all the countries in this group have in common is that they are rectangular. Rectangular socketed weights with closed external mortices—the Arginunta weight (T62112)—appear on Kalymnos, Cyprus and in the Pontus, with open mortices (T62132) on Lesbos and from a sixth-century-CE oil press from the Laureatic Olympus or Attica. These weights are clearly related to the rectangular Delos (T5522) and Amorgos (T5562) beam weights.

A small but variegated and extremely important group of weights were published from the Pontus at the beginning of the century by Anderson (1903: 15). These provide a glimpse into the rich typology of the Anatolian screw weights. Pontus 01 (T62112) is the Arginunta weight, the rectangular version of the Samaria weight (T62111) and has already been referred to in relation to similar weights found throughout the region. Pontus 02 (T62332) and Pontus 03 (T62322) are rectangular versions of the Beḥyo weight and are closely related to the weight from ᶜEin el- Judeida (T6232402). Pontus 04 (T62222) is a rectangular version of the Dinᶜila weight (T6221) and the only example of this type in the eastern Mediterranean outside Galilee. Pontus 05 (T62412) is a rectangular version of the Kasfa weight (T62411), lacking the depression that appears in the Syrian Sarepta weight. Pontus 06 (T6291213) is a rectangular version of the Sarfud weight with socket and depression but without mortices (T629111). Pontus 07 (T62912103) is similar but is without the depression.

Both Pontus 06 and 07 have holes for clamps—Pontus 06 eight holes and Pontus 07 four holes. These were clearly to secure the boards that held the screw-end.

This group of weights hints at possible answers to questions regarding weight types already discussed, the most important being the possible distribution areas of the Dinᶜila weight and the Kasfa weight and the origin of the Semana weights.

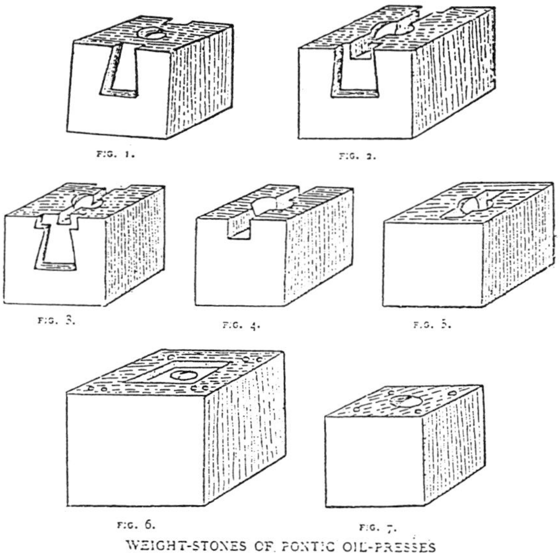

Figure 23. Screw weights from the Pontus (Anderson 1903: 15).

6D3. Italy

Little was known about screw weights in Italy until Liverani (1987) showed that a group of objects previ-

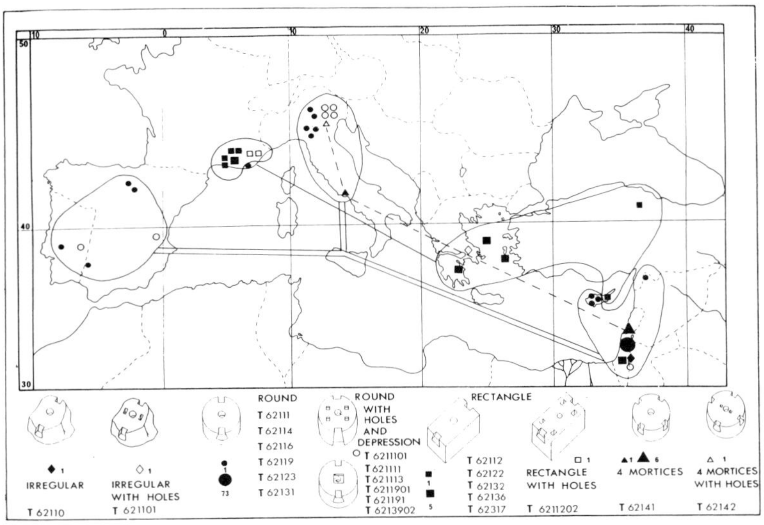

Map 24. Samaria screw weights (Ts621): Mediterranean Basin.

ously thought to be border markers of Roman centuration field systems (Alpago Novello 1957) were actually press weights. These, with some that were previously known (e.g. Francolise Posto) and others that have since been published (Capua, Carinola), provide a great variety of press weights.

The weights are of rounded shapes, cylindrical, square with rounded corners or in the shape of a rounded cylinder. They include:

a. Weights with neither mortices nor socket (T5009);
b. Weights with mortices but no socket (Ts5511, 5519), including that with four mortices from Francolise Posto, first to second century CE;
c. Weights with socket but no mortices (Ts629190, 6291901, 6291902);
d. Weights with socket and two mortices (Ts62116, 62119, 6211901, 621191);
e. Weights with sockets and four mortices (T62141, 6214901).

These weights are either of one type—that with socket and external mortices, the Samaria weight (T621), groups (d) and (e)—or can be seen as possible prototypes of this weight, stages in its development, groups (a), (b) and (c), those that lack mortices, sockets or both.

The great variety of sub-types and possible prototypes, strongly suggest that the Samaria weight developed in Italy, especially as the dominant shape of this weight in most countries of the Mediterranean is cylindrical, as are the Italian examples.

The dated weight of Francolise Posto provides a hint of one stage of development, typologically and chronologically. Whether there were more than one line of development and what were the detailed typological and chronological sequences only further research can resolve.

The fact that the weight with four external mortices (T62141) appears both in Italy and Israel suggests that the Samaria weight was a result of direct contact with Rome. It should be noted that clamp holes were common in Italian screw weights, appearing both in pairs and fours, but depressions do not appear except perhaps in T621191.

6D4. France

No screw weights have as yet been found in excavated presses in France, but many screw weights were recorded in Brun's survey of the Var region of Provence. He reached the conclusion that many, if not all, these weights were from the Middle Ages.

Several different types of screw weight were recorded in the region. Many of the screw weights with external mortices (T6211) were re-used Semana beam weights (T55121). Of those that were especially made as screw weights, however, only one was cylindrical and similar to the Israeli and Italian Samaria weight (T62111), the others continuing in the tradition of the rectangular Arginunta weight (T62112).

Another type is that lacking mortices (T6291): the Sarfud weight. The French examples, while appearing in both square and round forms, lack depressions.

Yet another type is the Ponteves weight (T6242) with socket and internal mortices. These mortices are narrower than those in the Levantine Kasfa and Sarepta weights. As in the Sarepta weight, the Ponteves weight often has a rectangular depression.

Clamp holes appear both in the Sarfud weights without mortice (T6291) and in the Ponteves weights (T6242), in one case four sets of two holes (T6291203)—a combination that appears in the catalogue only in one other case from the Pontus (T6291213).

As in the case of the presses themselves, the screw weights of Southern France are very different from those of Italy. The Sarfud weights without mortices (T6291) appear in both countries, and the one example of a cylindrical Samaria weight (T62111) probably shows direct or indirect Roman/Italian influence. The majority of the screw weights with external mortices, however, are of the rectangular Arginunta type. These are clearly a direct development from the Semana beam weight typical of the region. There is as yet no data as to the date of the transition.

Ponteves weights do not appear in Italy and those from France are almost certainly connected to the Sarepta weight from the Levant from which they probably derived, but here again the chronology is unknown.

6D5. Spain and Portugal

As opposed to the South of France, where there is much information from a small region, from all Spain and Portugal there are in the catalogue only eight ancient and seven pre-industrial screw weights. All 15 are cylindrical. Of the ancient weights, four are simple Samaria weights (T62111), one is a Samaria weight with two clamp holes (T6211101) and another a Samaria weight with a square depression (T621111). One is a Ponteves weight, cylindrical in shape (T6421) and the eighth from Turrios Berceo is of a unique type with socket, a square channel above and below, and external open mortices (T62351).

As in Italy, the dominant type is the cylindrical Samaria weight in a variety of forms. As yet, possible prototypes have not been reported, but it is likely that weights similar to the Volubilis and Francolise Posto weights will be found in the future

Three pre-industrial weights from Portugal are of the Dinᶜila type (T62211), the only examples of this type from the western Mediterranean. The questions that arise are whether they are in any way connected to the Turrios Berceo weight on the one hand, or to the Dinᶜila weights in Galilee on the other.

In Spain a pre-industrial screw weight is found, the perforated screw weight (T62931), which also appears in pairs (T62951), and that apparently does not have ancient parallels.

6D6. North Africa

It was suggested above that the Volubilis cylindrical weights (T55111, 55412) and the Cotta weight (T56911) were screw weights. The weights are very close in type and date to the Francolise Posto weight. The connection between the two will be perhaps clarified when we know more of the picture in Spain. The only hint as to the later developments in North Africa are the screw weight with socket (T62311) from Rabat and the pre-industrial perforated screw weights (T62931) that are the same in North Africa as in Spain.

Any attempt to explain the relations between the various types of screw weight, their history, development and diffusion must be related to the history of other devices and in particular the single fixed screw press (Ts81-83). The discussion of these subjects will, therefore, be deferred until the concluding chapter after the other relevant installations have been described and discussed.

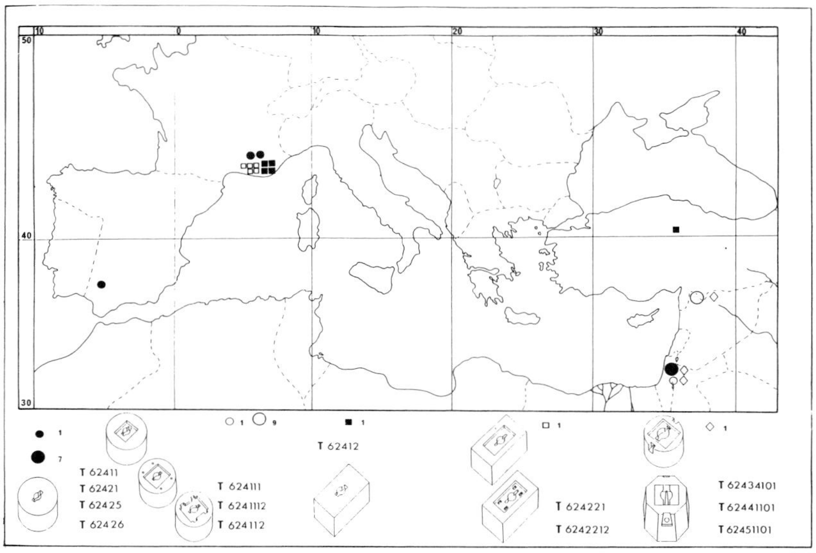

Map 25. Kasfa, Sarepta and Ponteves screw weights (Ts624).

Chapter 7

Direct-Pressure Rigid-Frame Presses: Single-Screw, Double-Screw and Wedge Presses (T7)

7A. Chronology and Typology

Technically the next stage in the evolution of the press was dispensing with the use of the beam and using the screw to exert direct pressure.

Pliny the Elder and Hero of Alexandria both describe direct screw presses: Pliny's press D and Hero's presses C and D were of this type. Pliny also provides an exact date for the introduction of the direct-screw press stating that it was 'in the last twenty-two years', which would place it in the second quarter of the first century CE.

From pre-industrial presses, the descriptions of classical written sources, and from depictions of presses in various media, and in this case also from an ancient wooden press that actually survived in dry desert conditions, we learn of many different types of direct-screw and other rigid-frame presses. The pre-industrial example are numbered: T013, single rotating-screw presses; 014, double-screw presses; 015, wedge presses; 0155, drum and weight press; 016, fixed single-screw presses.

The stone parts of these presses, which are nearly always the only archaeological remnants that survive, often allow for reconstruction of the original press in different ways, and as a result the archaeological remains of most of these types of press usually cannot be distinguished. The archaeological remains have therefore been categorized into types according to quite different criteria from those used as regards the pre-industrial presses and those used for defining the Models 7A–7G in this chapter. Archaeologically, only two main groups can be distinguished: the various types of rigid-frame presses—pre-industrial types T013–T015—and the single fixed-screw press, pre-industrial type T016. In Israel the single fixed-screw press was always a wine press, whereas the rigid-frame press was almost always an oil press. Therefore those archaeological finds that are remains of rigid-frame presses have been categorized as type T7 and the possible way they could have functioned as Models 7A–7G, while the single fixed-screw press (Pre-ind. T016) is not discussed here but is included in the chapter on components of wineries and categorized as Ts81-83.

7B. Models of Rigid-Frame Presses: Single-Screw, Double-Screw and Wedge Presses, Based on Pre-Industrial Presses, Classical Written Sources and Ancient Artistic Representations (Models 7A–7G)

Model 7A: Single Rotating-Screw Press

This is Pliny's press D: 'A shorter spar straight in the middle bearing down with full weight from above on the lid laid on the grapes'. It could be reconstructed either as Model 7A1 or 7A4 (see above, p. 79).

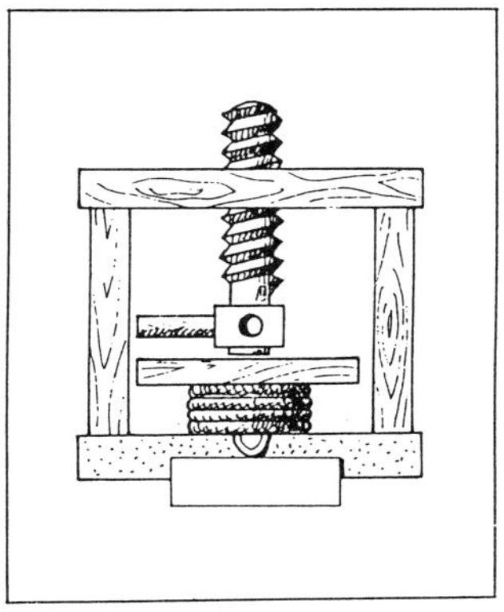

Figure 24. Model 7A1: single rotating-screw press, handle at lower end of screw.

Model 7A1: Single Rotating-Screw Press, Handles at Lower End of Screw (Pre-Ind. T0131). In this press the female thread is in the upper frame of the press. The bottom end of the screw spar below the thread, that pressing on the press board (Platen), is usually wider than the threaded part and in it are crossed holes in which the hand-spakes that turn the screw are inserted. This is the most common of pre-industrial direct-screw presses and found in most if not all the Mediterranean countries. In the catalogue there are 38 examples from 18 countries. In Israel this is the pre-industrial press most common in Galilee.

Photo 66. Screw of pre-industrial press Model 7A1 (T0131): Bet Ha-ʿEmeq.

Model 7A2: Single Rotating-Screw Press, Handles at Lower End of Screw, Vertical Auxiliary Drum (Pre-Ind. T0132). To increase the mechanical advantage, an auxiliary drum was used. There is possibly evidence for use of this device in ancient times (T790; Dabussiya 03).

Model 7A3: Single Rotating-Screw Press, Handles at Lower End of Screw, Horizontal Auxiliary Drum (Pre-Ind. T0133). This press is similar to Model 7A2, but in this case the drum is attached to the press frame and is horizontal. There is possibly also evidence for use of this device in ancient times (T7132; Ṣafṣafot).

Model 7A4: Single Rotating-Screw Press, Handles at Upper End of Screw (Pre-Ind. T0134). In this press, as in Models 7A1, 7A2 and 7A3, the female thread is in the upper frame of the press, but the handspakes are inserted in holes at the top end of the screw above the frame. This is Hero's press D (see above, p. 80). Pre-industrial examples were recorded in Germany, Rumania and Hungary.

Model 7B: Press with Two Fixed Screws (Pre-Ind. T141)

In this press two screws are fixed to a base and the two nuts that are rotated exert the pressure. This is the second most common of the pre-industrial presses. There are 15 examples in the catalogue from Israel, Rumania, Yugoslavia, Hungary, Austria, Italy, Spain, Great Britain and the islands Corfu and Karpathos. In Israel this press was found only in northern Upper Galilee. Humbel (1976: 164) terms this the 'Genoese press'.

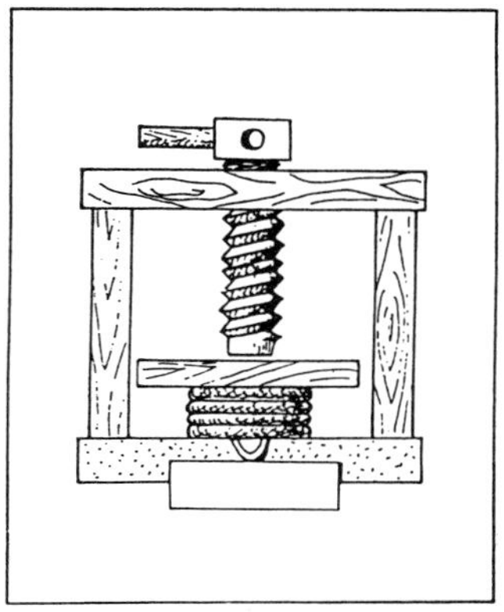

Figure 25. Model 7A4: single rotating-screw press, handle at upper end of screw.

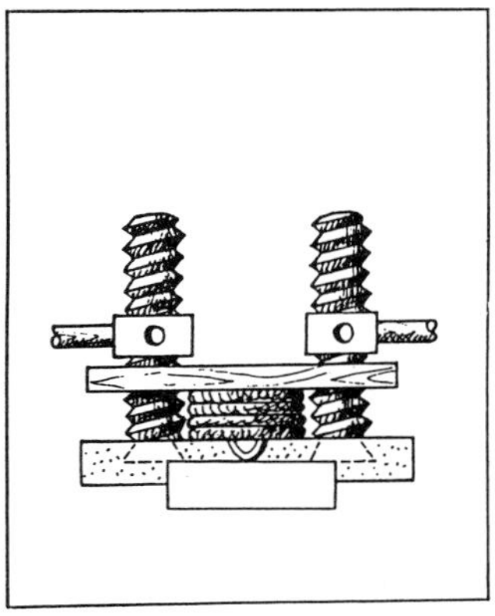

Figure 26. Model 7B: press with two fixed screws.

Model 7C: Press with Two Rotating Screws

Model 7C1: Press with Two Rotating Screws, Handles at Upper End of Screws, Female Threads in Lower Press Frame (Pre-Ind. T0142). In this press the two screws go through unthreaded holes in the pressing board and screw into female screws in the

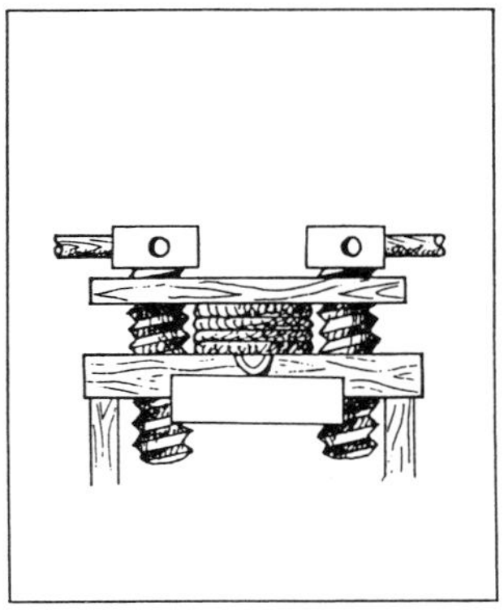

Figure 27. Model 7C1: press with two rotating screws, handles at upper end of screws, female threads in lower press frame.

lower frame of the press. A complete press of this type was found in the Fayum. Presses of this type are found in the Rhine Valley and there called 'the Worms press' (Humbel 1976: 166). There is evidence of the use of this press for cider making in Great Britain in the seventeenth century.

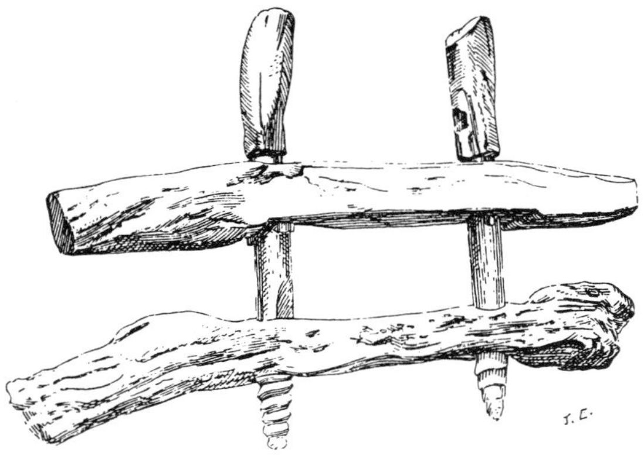

Figure 28. Model 7C1: screw press from Fayum (Billiard 1913: fig. 157).

Model 7C2: Press with Two Rotating Screws, Handles at Upper End of Screws, Female Threads in Upper Press Frame (Pre-Ind. T0143). This is a twin version of Model 7A4. Pre-industrial examples were recorded in Hungary.

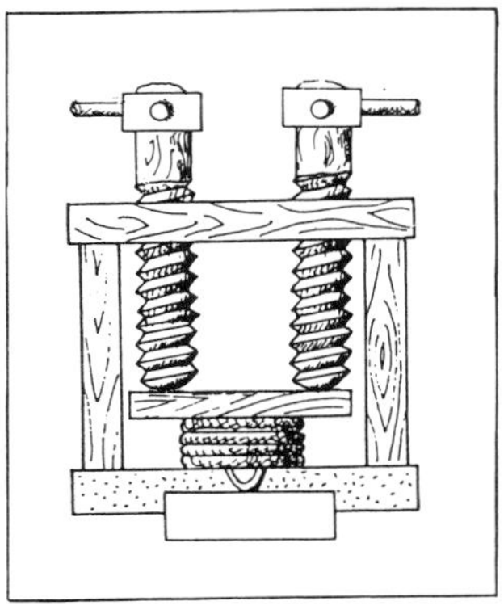

Figure 29. Model 7C2: press with two rotating screws, handles at upper end of screws, female threads in upper press frame.

Model 7C3: Press with Two Rotating Screws, Handles at Lower End of Screws, Female Threads in Upper Press Frame (Pre-Ind. T0145). This is a twin version of Model 7A1. A clothes press of this type appears in a wall painting from Pompei 05.

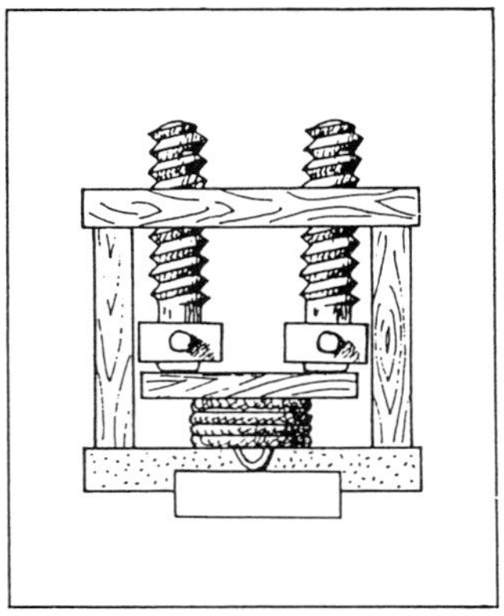

Figure 30. Model 7C3: press with two rotating screws, handles at lower end of screws, female threads in upper press frame.

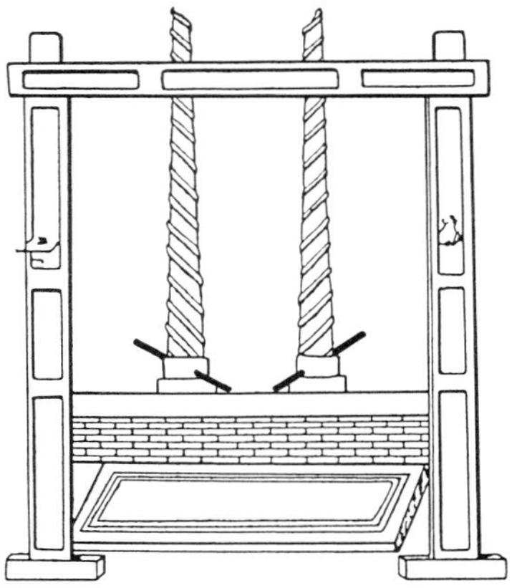

Figure 31. Model 7C3: Fuller's press—frescoe from Pompei 05 (Billiard 1913: fig. 156).

Model 7D: Press with Two Anchored Rotating Screws, Female Threads in Pressing Board

This is Hero of Alexandria's press C (see above, p. 80). In it two screws were attached to the lower frame of the press in a manner that allowed them to rotate but not to work free. Rotating the screws in female threads in the pressing board (Platen) brought it down and exerted the pressure.

There have been three different suggestions as to how to reconstruct this press, as follows:

Model 7D1: With Frame and Handles at Upper End of Screws. Nix (Hero, *Mechanica* fig. 60; Drachman 1963: fig. 50d) reconstructed this press with a frame and with handles at the upper end of the screws.

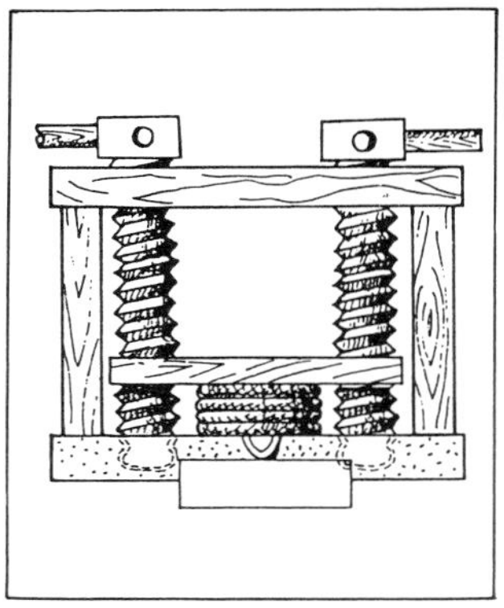

Figure 32. Model 7D1: press with two anchored rotating screws, female threads in pressing board, with frame and with handles at upper end of screws.

Model 7D2: Without Frame and with Handles at Upper End of Screws. Drachman (1932: figs. 25, 26) in his first reconstruction suggested that the press had no frame and that the handles were at the upper end of the screws.

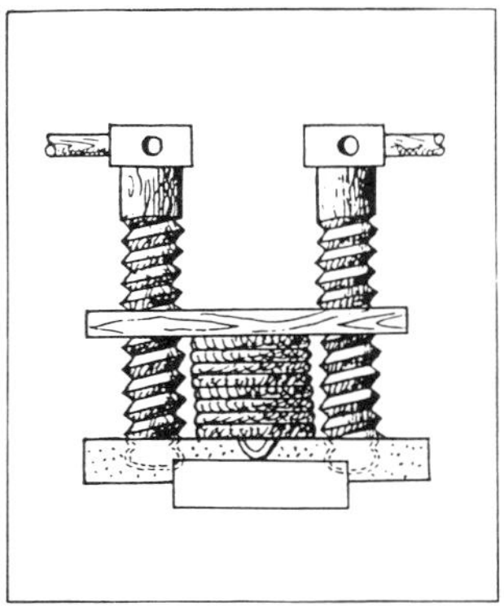

Figure 33. Model 7D2: press with two anchored rotating screws, female threads in pressing board, without frame and with handles at upper end of screws.

Model 7D3: Without Frame and with Handles at Lower End of Screws (Pre-Ind. T0144). In this press there is also no frame, but the handles are at the lower

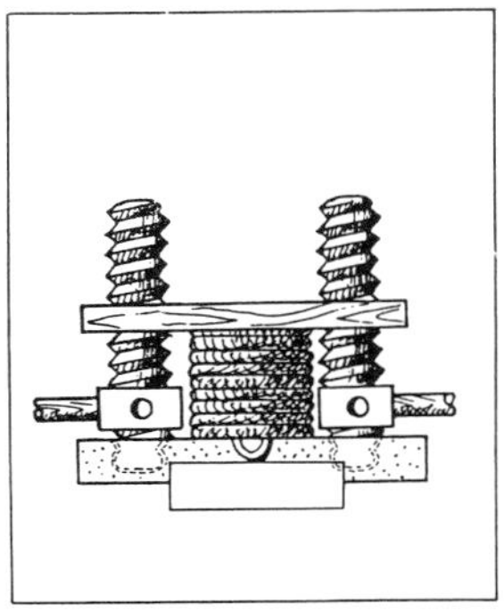

Figure 34. Model 7D3: press with two anchored rotating screws, female threads in pressing board, without frame and with handles at lower end of screws.

end of the screws. This is the way Carra de Vaux suggested Hero's press should be reconstructed, as did Drachman in his second reconstruction (Drachman 1963: figs. 50a, 50c). There are pre-industrial presses of this type both in France and Italy.

The fact that Hero's press C exists in a pre-industrial form is remarkable. The pre-industrial press is similar also to Hero's press in that both were portable. Drachman (1932: 76) has pointed out that this was a characteristic that distinguished Hero's press from that of Pliny, and the pre-industrial press from Herault was specifically called the 'pressoir ambulant'. It seems unlikely that the pre-industrial press is directly connected to Hero's press, which was probably a suggestion for an improvement of an existing press that was never actually made. The similarities between Hero's press and the pre-industrial examples are nevertheless remarkable.

Model 7E: Wedge Presses

Model 7E1: Horizontal Wedge Presses (Pre-Ind. Ts0151). From wall paintings at Pompei 06 and Herculaneum we learn of the use of horizontal wedge presses in the ancient world. Lines of pointed wedges were placed between horizontal boards within a frame so that the heads of the wedges faced in opposite directions in alternating rows. By hammering the wedges between the boards, these brought pressure to bear. Identical pre-industrial presses have been recorded from Algeria, France and Spain, and similar presses were used in Japan (Nagatsune 1974: figs. 2, 22).

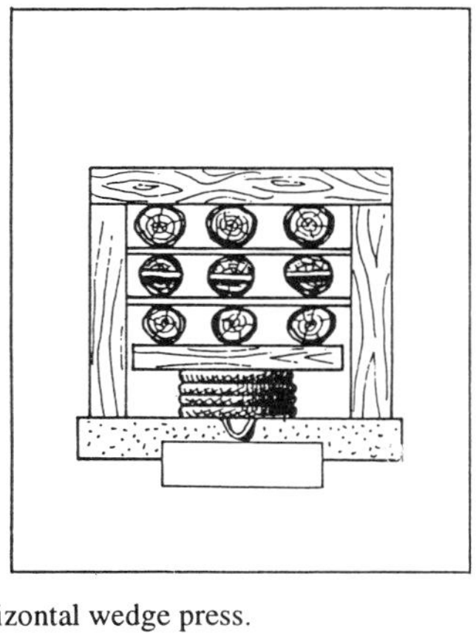

Figure 35. Horizontal wedge press.

Model 7E2: Horizontal Wedge Presses with Hinged Hammers (Pre-Ind. T0152). This improvement was recorded in pre-industrial presses in Rumania.

Model 7E3: Vertical Wedge Presses (Pre-Ind. T0153). When vertical wedges are used, the pressure applied was also vertical allowing for the use of vertical weights and hammers. Such an oil press operated by horse-power appears on a relief dated 1628 (Edam, Holland) and oil presses working on this principle but operated by wind-power were functioning until recently in northern France and Holland. Their use is recorded also in England. In several cases these vertical wedge presses operated in unison with the vertical mechanized mortar and pestle (T0184). There is apparently no evidence for the use of this device in ancient times.

Model 7F: Screw and Weight Press (Pre-Ind. T0137)

In this press a handle above the press frame turned the screw that was fixed to a weight—raising the weight. Turning the screw in the opposite direction releases the weight to exert pressure. The pre-industrial example recorded is a cheese press.

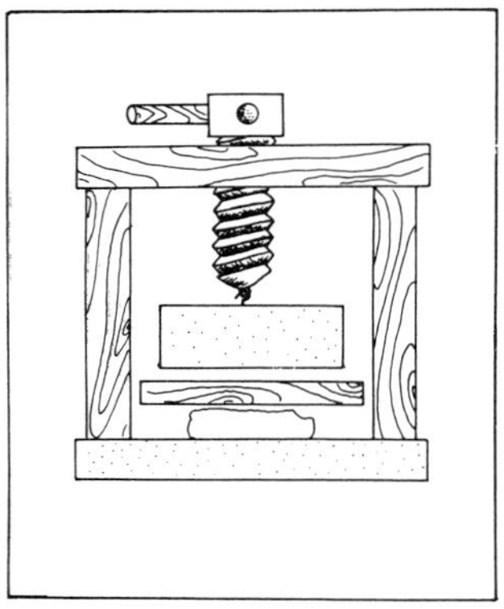

Figure 36. Model 7F: screw and weight press.

Model 7G: Drum and Weight Press (Pre-Ind. T0155)

In this press a drum raises a box of stones which is then released to exert direct pressure. In this case, too, the pre-industrial example recorded is a cheese press.

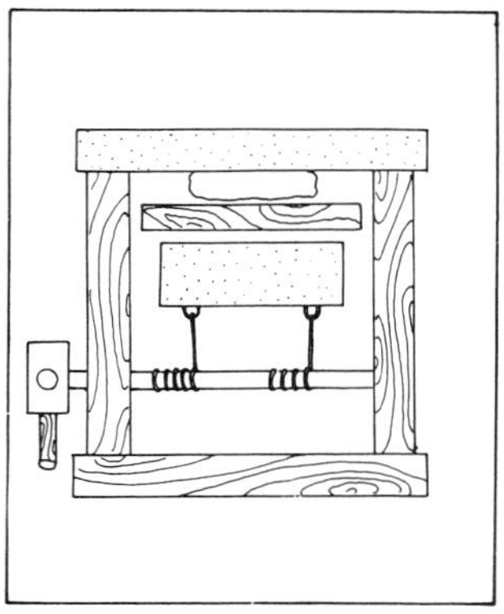

Figure 37. Model 7G: drum and weight press.

Models 7A–7G: Summary

If rigid-frame presses of any of these models (A–G) were constructed of stone bases and wooden superstructure, the stone base that would remain in an archaeological site would be similar and would consist of a press-bed with a mortice on either side which would have secured either the two posts of a press frame or two screws. Therefore, in those cases where the remains are a press base of this type (Ts73, 74), the original press could theoretically have been of any one of Models A–G including the wedge presses (Model 7E).

However, the two main types of pre-industrial rigid-frame presses of which most examples were recorded were that with single rotating screw and handles at lower end (Models 7A1, 7A2, 7A3—53 examples), and that with two fixed screws (Model 7B—15 examples). These were probably the two main types in ancient times also. For this reason Ts73 and 74 have been called 'screw press bases', although theoretically they could have been the bases of wedge presses.

Three pre-industrial press types must be mentioned here that are neither rigid-frame nor single fixed-screw presses. In Rumania a press is found in which a single screw draws two hinged boards together (Pre-ind. T0135) and in France a press is found consisting of one rotating horizontal screw (Pre-ind. T0136; in French 'pressoir à coffre'). Both these types are apparently comparatively recent inventions. A third type consists apparently of two screws anchored in screw weights (T62) and a pressing board with two female threads (T0146). This is from a sixteenth-century technical treatise and is apparently a suggestion for a new type of press. It is similar to Hero's press C and Model 7D, except that theoretically one of the two weights could have been raised off the ground.

7C. Archaeological Remains of Rigid-Frame Presses in Israel and Environs

7C1. The Grooved-Pier Press (T711)

The grooved-pier press consists of two piers with vertical grooves on the posterior and anterior faces, and on the two lateral faces mortices consisting of a short groove starting from the top of the pier and leading down to a square mortice that is connected by a bore to the identical groove and square mortice on the opposite face of the pier.

Dalman (1928–42: IV, 226-27) who was the first to show that these piers served to support a single rotating-screw press (Model 7A1), suggested that the posterior and anterior grooves served to guide the pressing board. Peleg (1981) showed, however, that this was technically unfeasible and Frankel (1981), after excavating the press at Dukas, demonstrated that the function of these grooves was to hold rods that were secured in small holes that he uncovered in the rock

Photo 67. Grooved-pier press (T711): Bet Natif (Photo: Carter Litchfield).

Photo 68. Grooved-pier press (T711): Bet Natif (Photo: Carter Litchfield).

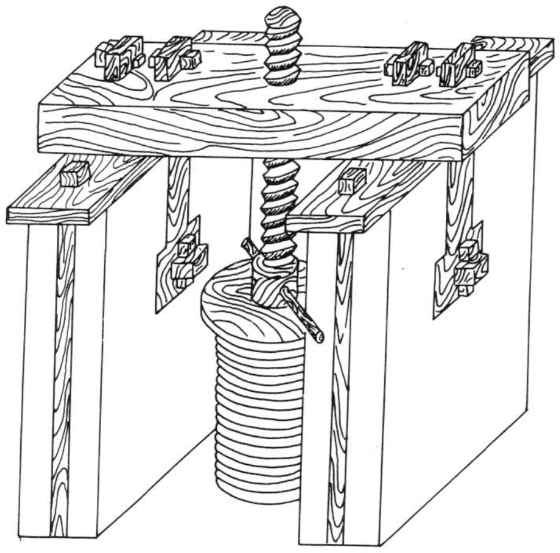

Figure 38. Grooved-pier press (T711): reconstruction.

below the grooves and in similar holes in the thread board above the piers, thus stabilizing the piers and especially acting to withstand lateral pressure during pressing. In three presses in which the pressing area was uncovered—ᶜEin Fattir, Tannim and Dukas—the collecting vat was central (T7961).

The 41 examples of the standard grooved-pier press (T711) are all from southern Israel, that is from Judaea. One only is located north of the 15 EW coordinate and only two west of the 14 NS coordinate.

The evidence suggests that this press developed directly from the Maresha press (T401110002), the lever and weights press typical of this region. As in the grooved-pier press, there were in these presses two piers, plain piers in this case, one on either side of a central collecting vat. Thus an existing lever and weights press could have been converted into a screw press by adding only a screw and suitable mortices on the piers. Piers with posterior and anterior grooves but without lateral mortices were also recorded (T4618). Of the 15 recorded, 13 were in the same distribution area as the piers with mortices. Although as yet none of these have been excavated, they were probably used in lever and weights presses and represent a transition stage between the Maresha press and the grooved-pier Press.

In two presses, Rasm el-Beida and Ḥauran, grooved piers (T[6]711) apparently served as part of lever presses rather than direct-pressure screw presses. Both presses are in caves, and in the wall opposite the piers are niches of the type that served to anchor press-beams. The screw placed between the piers apparently served to raise and lower the free end of the beam (Tepper 1988: 110-11; Kloner 1989: 72). This type possibly represents an early stage in the development of the grooved-pier press, but more probably it is a later adaptation.

Similar presses have apparently not been published from other countries, although there is a hint that similar presses perhaps exist in Lebanon. Kitchener (Conder and Kitchener 1881–83: I, 56), in reporting the existence of three presses in the region of Tyre, states:

> This is a specimen of a peculiar sort of olive-press which occurs frequently in Phoenicia and rarely in the more southern districts; there are examples at Kafr Rut, Sheet XVII and at Beit Ṣur, Sheet XXI.

Both at Burj el-Sur-Bet Ṣur and Kafr Rut grooved-pier presses were found. However, it is not certain that Kitchener defined the piers in exactly the same manner as in this book and it is also not clear what region Kitchener meant when referring to Phoenicia.

Of the few grooved piers to be dated, those from Qaṣr and Quneitra are from the Byzantine period, while that from Dukas is from the early Arab period. It has been suggested that this type of press was already in use in the Roman period (Gichon 1979–80). This, however, awaits clear stratigraphical confirmation.

Ten variant grooved piers were recorded which divide into six sub-types. All ten are located outside the distribution area of the standard type. Six, however, are located in the south on the edge of the distribution area of the main type to the south-east and north, while

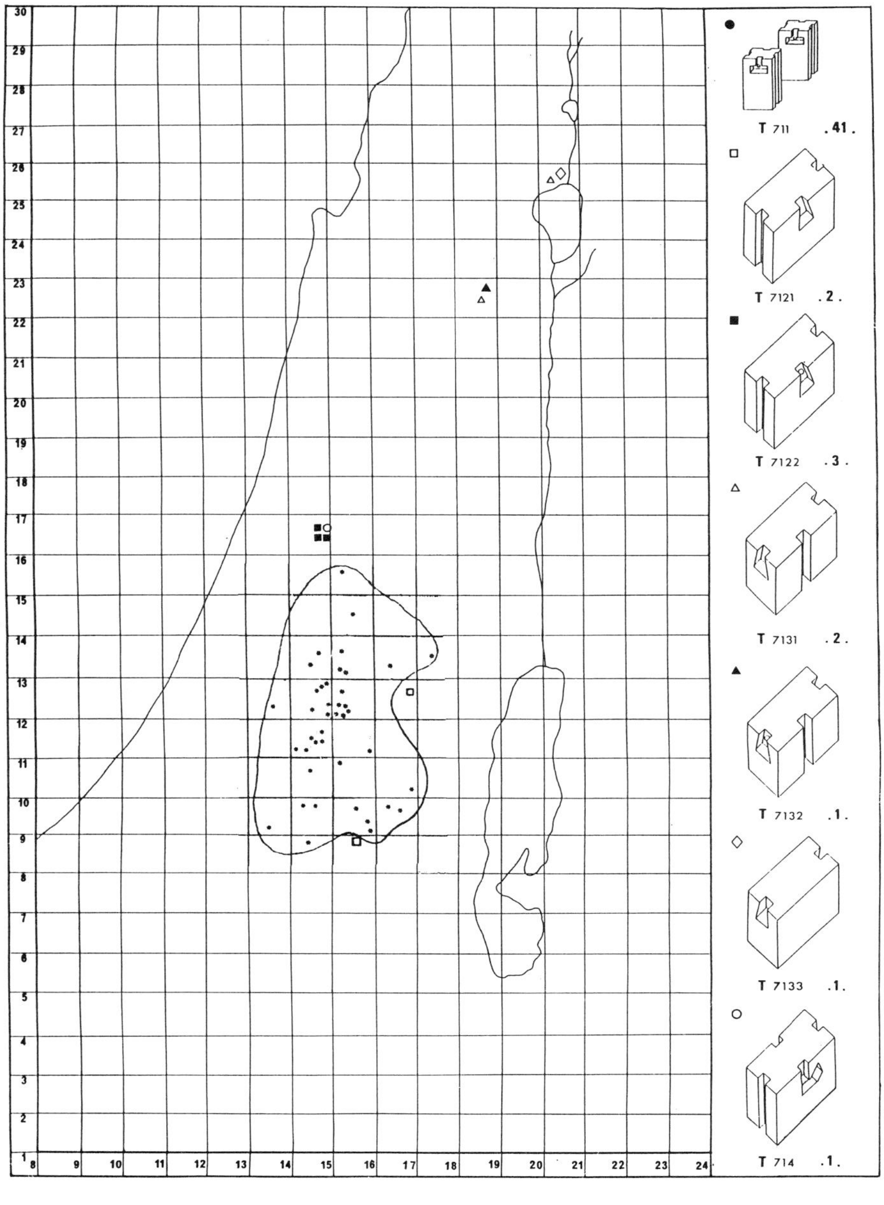

Map 26. Grooved-pier press (T71).

four are situated in eastern Lower Galilee.

These two groups are also distinct typologically. In the southern group, as in the standard type (T711), the grooves are on the narrower posterior and anterior faces of the pier and the mortices on the wider lateral faces. The variation from the standard type is in the form of the mortices. In two examples at sites to the east and south the mortice is dovetail and without a bore (T7121). In three examples from three sites very close one to the other to the north of the distribution area of the standard type, the mortices are dovetailed with a bore (T7122). The final variant pier from the south is from Kasfa, one of the same three sites, and has a mortice in the shape of an angular groove (T714).

The four variant piers from Galilee are very different. First, the mortices are on the narrower anterior and posterior faces of the piers. Secondly, in three cases there is only one groove, in two examples on a lateral face without a bore (T7131) and in one with a bore (T7132). In the fourth there is neither groove nor bore, only dovetail mortices (T7133).

The oil press from Ṣafṣafot T7132, one of the Galilean installations, has been excavated (Frankel 1988–89). There, three identical presses were uncovered. In each there was a groove on the inner lateral faces of the pier, and dovetail mortices with a bore on the anterior and posterior faces of the piers. In all the presses, however, the mortice on the left-hand pier was much longer than that on the right hand and therefore the bore was also lower. The explanation for this

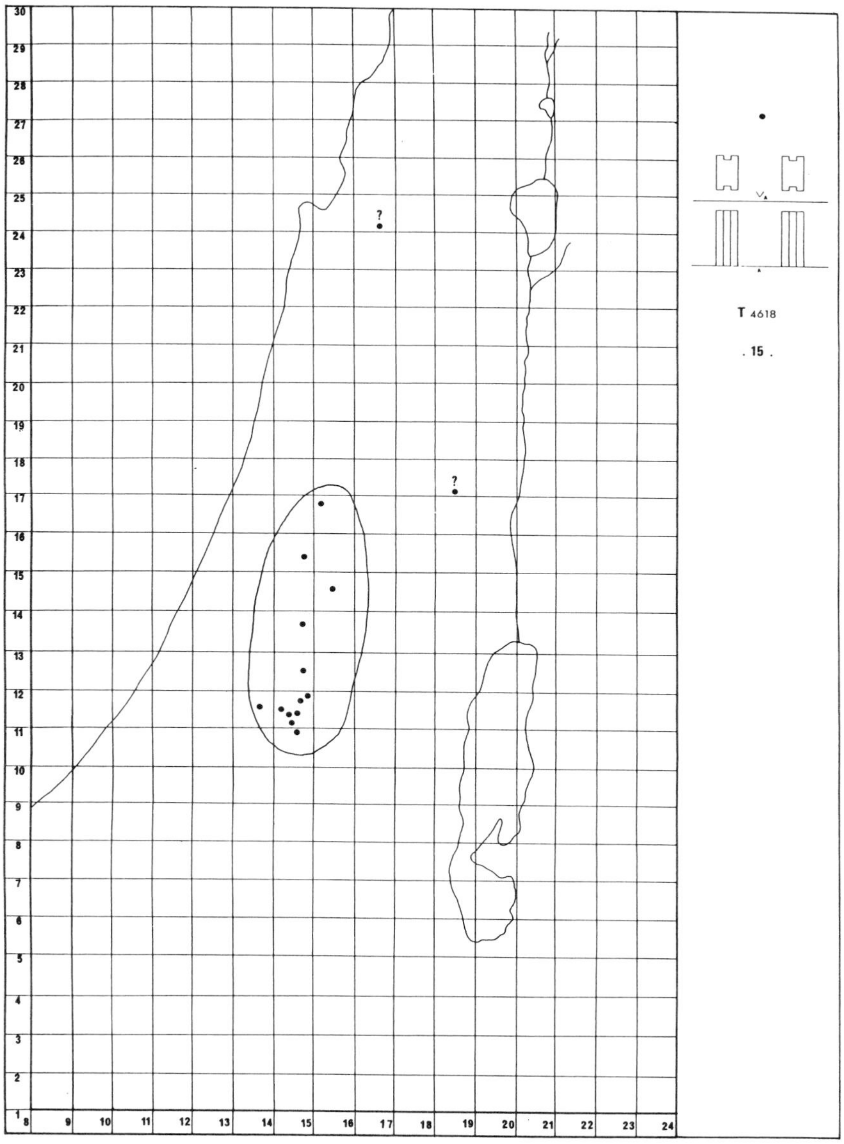

Map 27. Grooved piers without mortices (T4618).

unusual arrangement is apparently that the bores, as well as fixing the tenon that held the thread board in the mortice, also served to secure a horizontal auxiliary drum to turn the screw (Model 7A3: Pre-ind. T0133). The higher drum would have operated when the pile of pressing baskets was high and the lower when it was low. This additional function would also explain why the mortices were on the anterior and posterior faces of the piers and not on the lateral face as was usual.

It is more difficult, however, to explain the fact that mortices in the other Galilean piers are also in this variant position, on the narrow anterior and posterior faces of the piers, when in these there was no bore. A tentative explanation could be that these developed from the Ṣafṣafot type and, although the use of the auxiliary drum had been abandoned, nevertheless the variant position of the mortices was retained.

Photo 69. A grooved pier, a southern variant (T7121): Rafat.

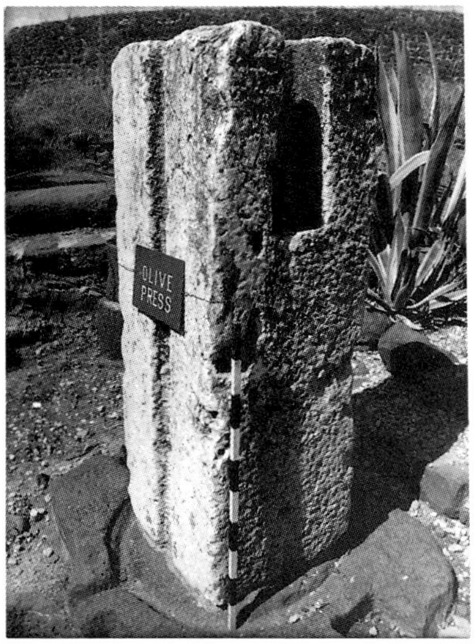

Photo 70. A grooved pier, a northern variant (T7131): Kefar Naḥum.

Photo 71. Ṣafṣafot: three screw presses and crushing basin—complete installation from north.

Photo 72. Ṣafṣafot: three screw presses and crushing basin—eastern press from west (T7132).

The distribution pattern of the grooved pier is an unusually clearly defined example of proliferation of variants in the peripheral areas around a culturally uniform nucleus.

7C2. The Cross Press (T72)

The cross press was also without doubt a single rotating-screw press with handles at the lower end (Model 7A1). It is usually found cut in the bed-rock in the walls of caves. The threaded plank in which the screw rotated was placed in the horizontal arms of the cross, the screw moving up and down in the vertical parts of the cross. In some cases the remains of a female thread can be discerned cut in the stone in the upper vertical section of the cross. Below the cross there are two connected rock-cut receptacles, the upper one in which to place the pressing frails and the lower a collecting vat.

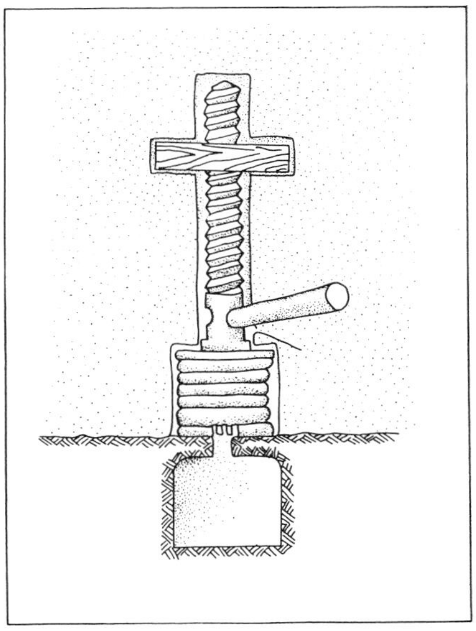

Figure 39. Cross press (T72)—reconstruction.

Twelve examples of this press were recorded in the centre of the country and one installation from Kefar Ḥananya in Galilee, which is possibly of this type but is too badly damaged for this to be certain. The distribution pattern of those recorded shows two concentrations. The main area in which nine examples were recorded includes Jerusalem and spreads to the west, north-west and north of the city. It is to the north-west of the distribution area of the grooved-pier press reaching it without overlapping. The other concentration is in the Shephala, the western foothills of Judaea, where three examples were recorded, thus in the midst of the distribution area of the grooved-pier press.

7C3. Screw Press Bases

7C3.1. The ʿEin Nashut and Ṣippori Presses (Ts73–74). Types Ts73–74 consist of a press-bed, central collecting basin or central opening, on either side of which are mortices, either open (the ʿEin Nashut press, T73), or closed (the Ṣippori press, T74). Theoretically these could be the remains of any of Models 7A–7E. It is probable, however, that those with open mortices (T73) were single rotating-screw presses (Model 7A1), while those with closed mortices (Ts74) could perhaps have been presses with two fixed screws (Model 7B).

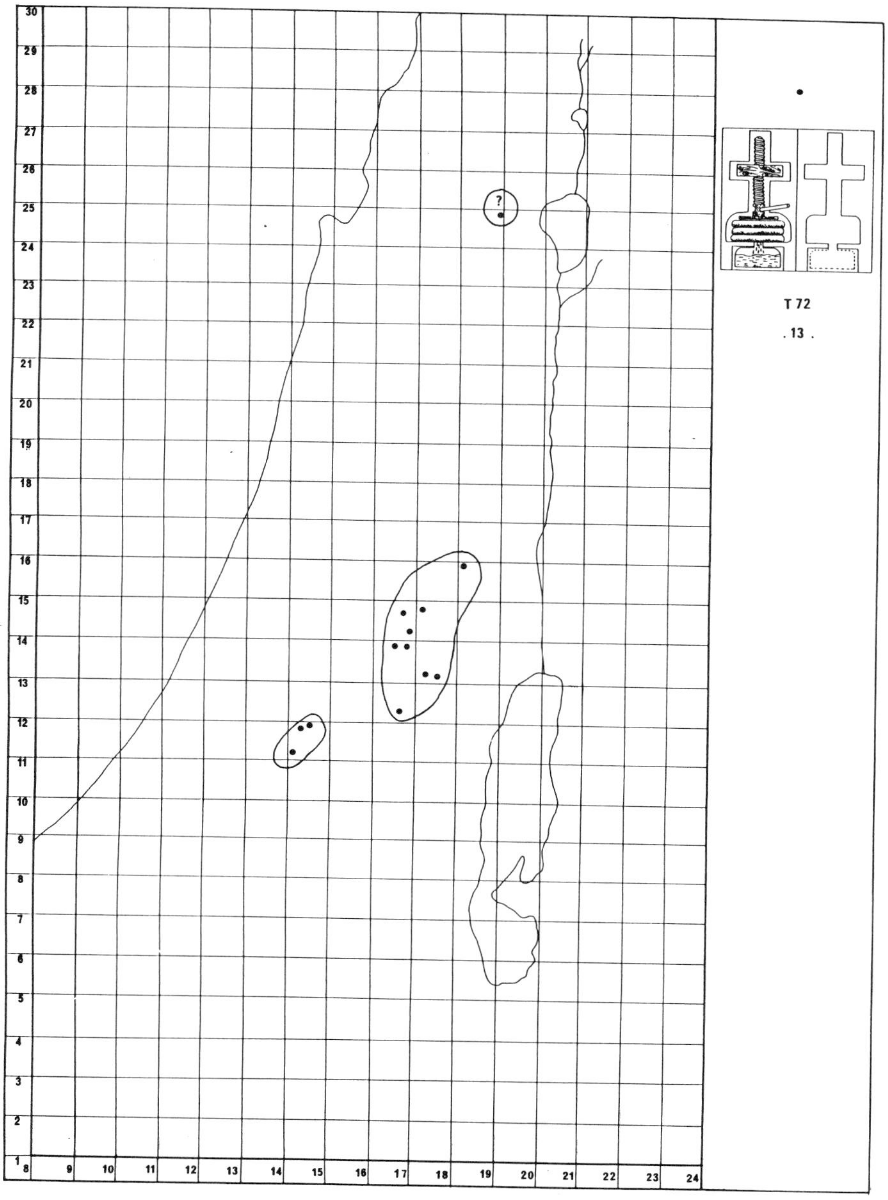

Map 28. Cross-screw press (T72).

Photo 73. The ʿEin Nashut screw press base (T732): Fakhura.

Certain generalizations can be made about the distribution pattern of this group of screw press bases as a whole (Ts73, 74 with all sub-types). First, the distribution area of this group is north of those of the grooved-pier press (T711) and of the cross press (T72) with almost no overlap. Of the 98 screw press bases recorded, only two were located south of the 15 EW coordinate. Secondly, these press bases are rare in Samaria, with ten examples, and in western Galilee, with only three examples (excluding T744; see discussion below), the two regions in which screw weights Ts62 were found in large numbers. Thus there is clear regional differentiation in the types of screw presses used: in some regions different types of lever and screw presses, in others different types of rigid frame screw presses.

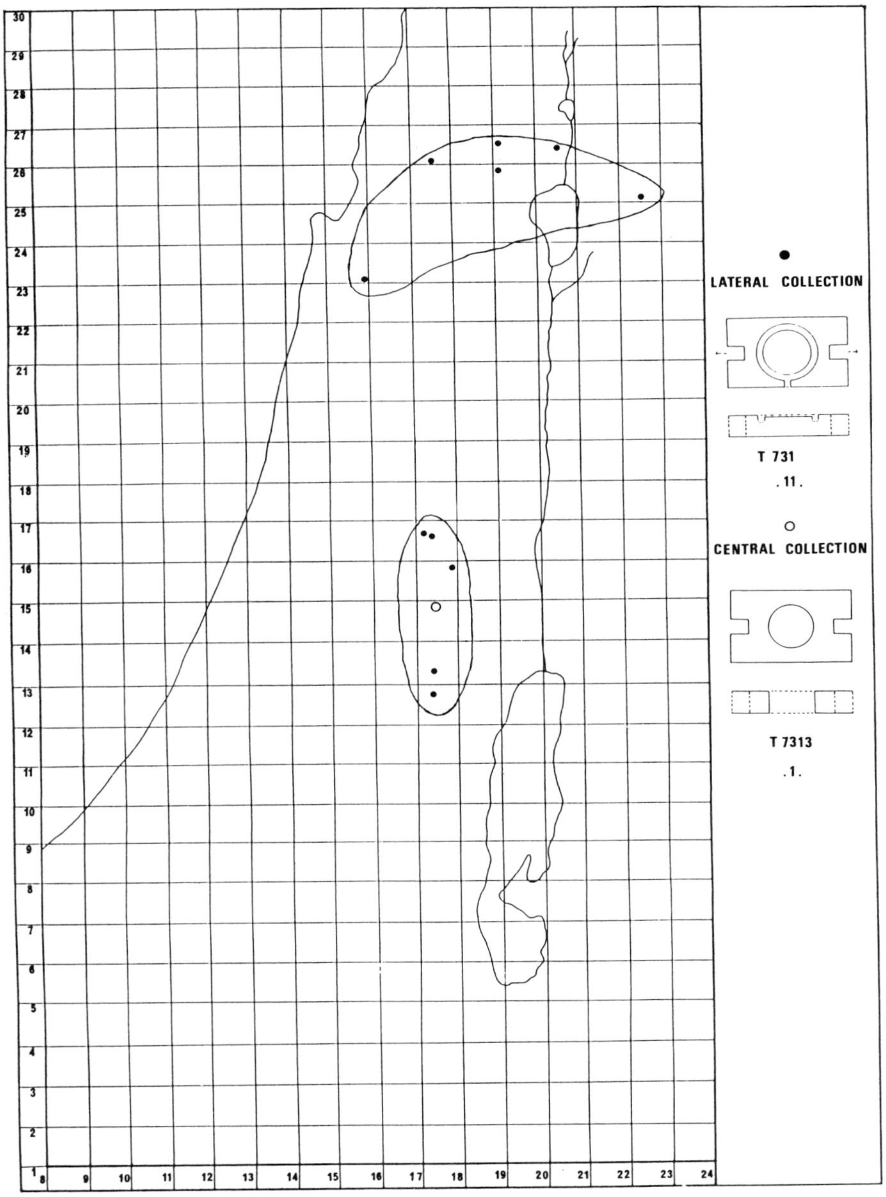

Map 29. Rectangular-screw press base, open mortices (T731).

Clear patterns can be discerned also in the distribution patterns of the main sub-types of these screw press bases. Of the 43 bases with open mortices and rounded corners, the ᶜEin Nashut press (T732), all but three were located in the regions of basalt rock in the northeast of the country, ten in eastern Lower Galilee, five on the Korazim plateau and 25 on the Golan Heights. Only five of the screw press bases found in these regions were of other types. In this case the rock formation clearly influenced the type of installation used. Basalt often comes in slabs suitable for fashioning this type of base.

Similarly, the screw bases with open mortices and square corners (Ts731) are usually of limestone, which explains the fact that this is the type found in the centre of the country.

Of the 30 bases with closed mortices, the Ṣippori press (Ts741), six were found in Samaria, eight in western Lower Galilee and seven in eastern Upper Galilee, in each case the majority of the screw bases found in the region concerned.

The majority of the dated examples of these screw press bases are from the Byzantine period. During excavations at Korazim, however, two screw press bases of the ᶜEin Nashut type (T732) and one with closed mortices (T741) were uncovered. One of the former was dated to the second century CE, at this stage in research the only screw press from Israel of any type to be clearly dated to the Roman period. The screw press base from Korazim with closed mortices (T741) was dated to the early Arab period, and therefore Yeivin (1982, 1984) has suggested that this is a late type.

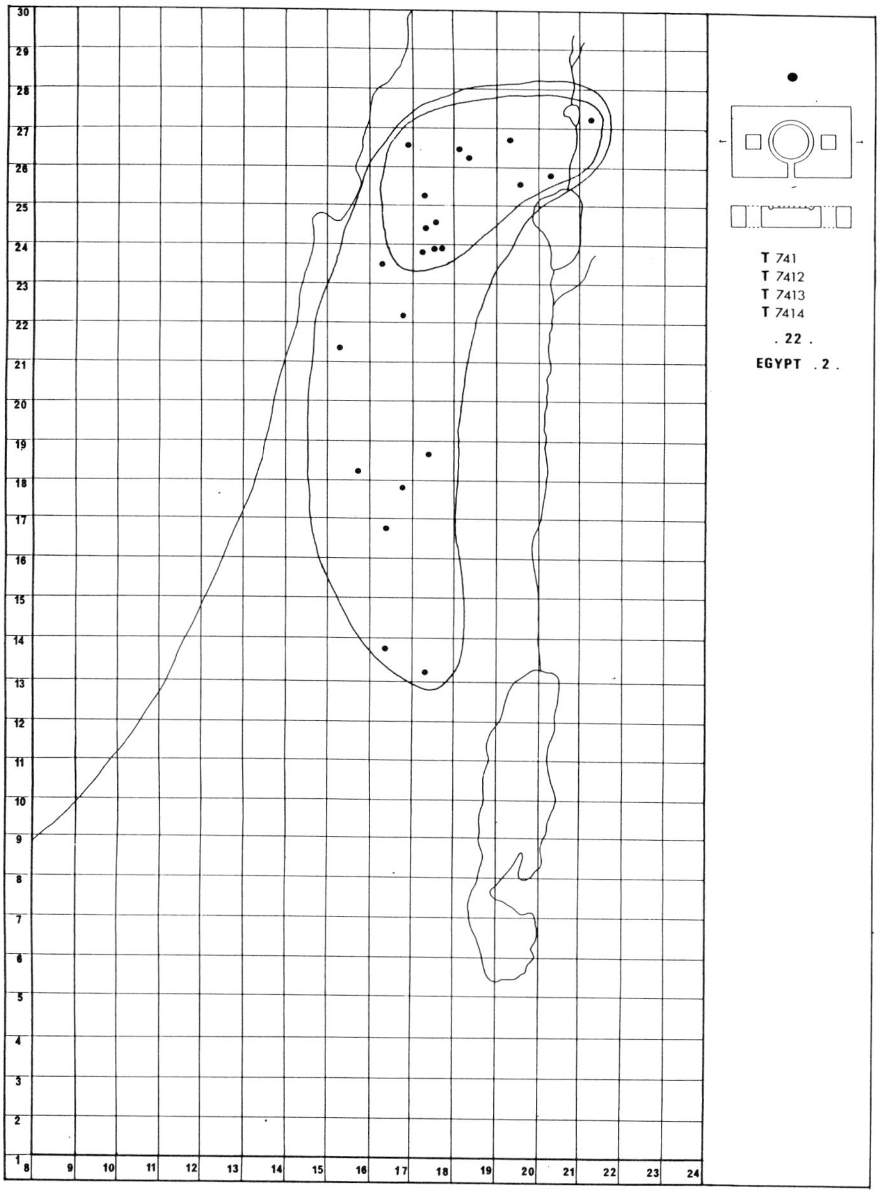

Map 30. Ṣippori screw press base, closed mortices (T741).

Photo 74. The Ṣippori screw press base (T741): Majarbin (Photo: Adam Zartal).

7C3.2. The Tabgha and Weradim Presses: Screw Press Bases with Central Collection (Ts7313, 73221, 73222, 7323, 751). In most of the screw press bases the collection is lateral, either in rectangular collecting vats (T7971) or in round collecting vats (T7972). In a small number, however, collection is central.

The main sub-type with central collection is that with open mortices, rounded corners and a central collecting basin surrounded by a circular and radial grooves—the Tabgha screw base (T73222). Nine examples of this base were recorded, all concentrated around the Sea of Galilee, the centre of the distribution area of the ʿEin Nashut screw base (T732) of which it is a sub-type. The radial grooves are without doubt related to other types of installations with this device (see T042 for all types with radial grooves). In one of the lever and weights presses from Gamla, situated not far from the Sea of Galilee, the collecting vat was of this type. This installation dated from the first century CE and therefore is certainly earlier than the Tabgha screw press base. However, several of the lever and screw presses from western Galilee, in which the collecting vat was also surrounded by circular and radial grooves were probably in use at the same time as the Tabgha bases (e.g. Bata, Quṣeir 01, 02, 03, 04).

A screw press base from Weradim (T73223) has radial grooves and a central opening to allow the expressed fluid to flow to a central vat placed below it.[1] At this same small site an unusual beam weight was also found (T534).

Photo 75. The Tabgha screw press base (T73222): Tabgha.

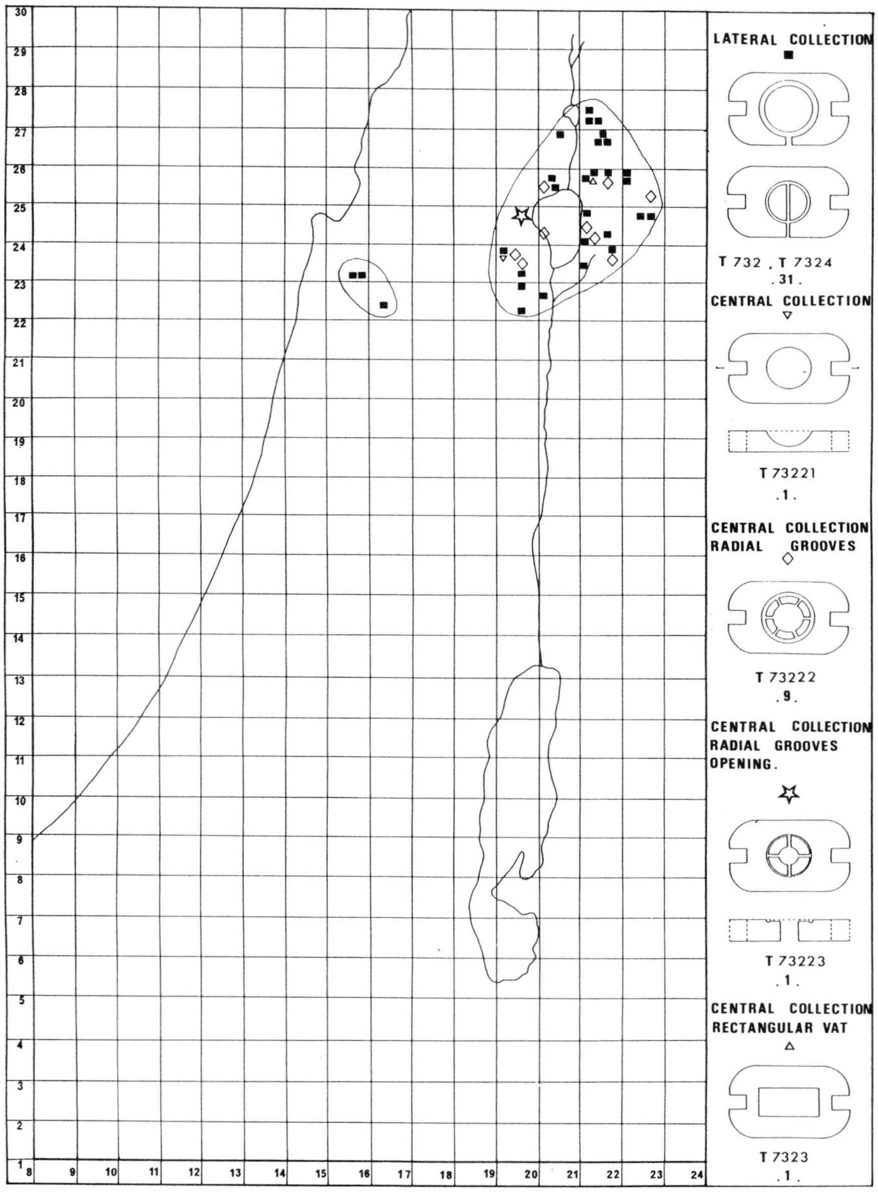

Map 31. ʿEin Nashut, Tabgha and Veradim screw press bases: rounded corners, open mortices (T732).

The concentration of the Tabqha screw press base (T73222) in a small area in the centre of the distribution area of the ʿEin Nashut screw base (T732) is a striking example of a specialized type developing in the nucleus of the distribution area of a main type, the Weradim screw base (T73223) being perhaps the acme of this pattern of development. In some ways this pattern is the reverse of that of the grooved-pier press (T71) in which the development pattern is one of peripheral differentiation.

Four screw press bases with central collecting basins but without radial grooves of three other types were recorded in the north (Ts73221, 7323, 751) as was one other base with a central opening from the south of the country (T7313).

7C3.3. The Rama and Mishkena Presses: Screw Press Bases with Mortices and Insertion Channels (Ts742, 743). Five examples were recorded of the Rama Screw press base in which the mortices are of the closed dovetail type (T743). They are from sites located in western and Upper Galilee, the region where this mortice is the main type used, both in the socketless screw weight of this region (T625), already touched upon above, and in other devices yet to be discussed (see T041 for all installations and types equipped with the closed dovetail mortice).

A related type is the Mishkena base (T742) found in three sites in Lower Galilee. The mortices used in these are adaptations of the square mortice used in the Luvim weight by the addition of an insertion channel some-

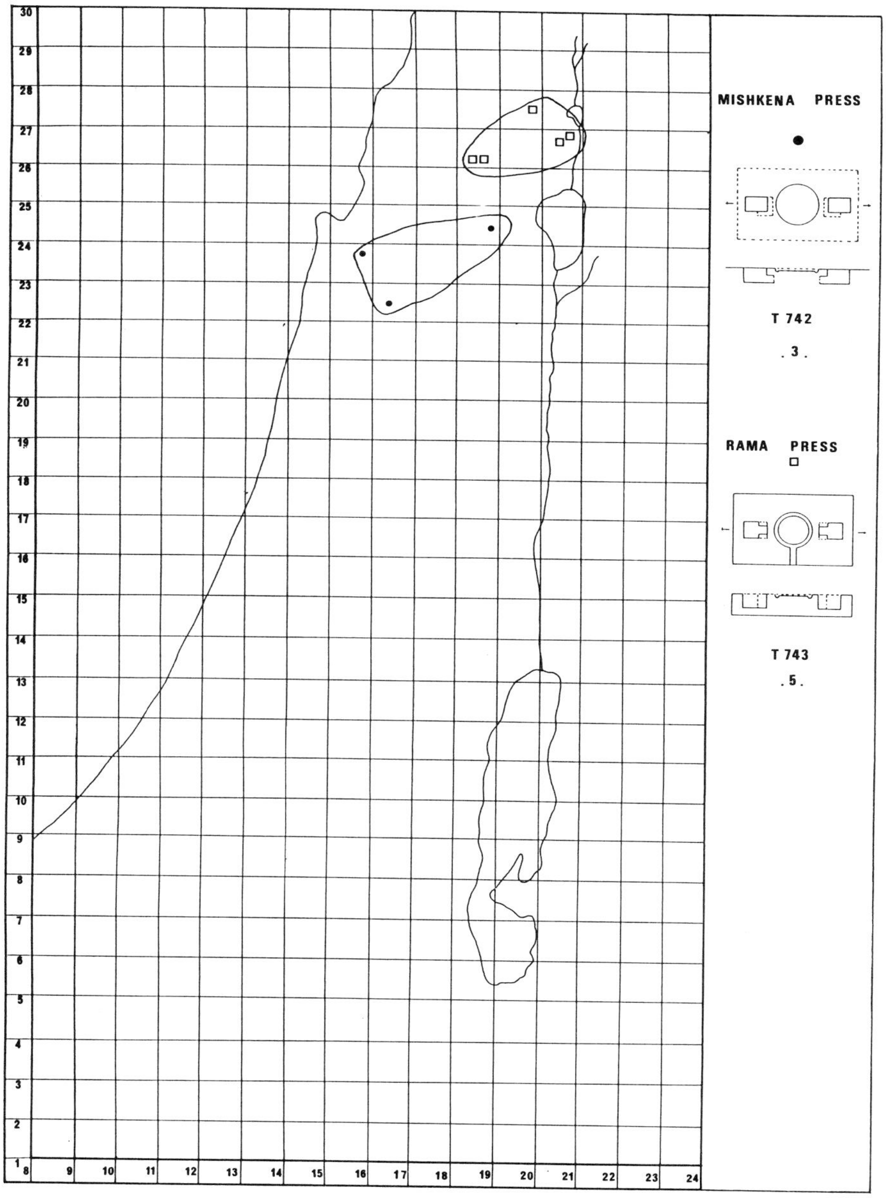

Map 32. Mishkena screw press (T742); Rama screw press (T743).

what similar to that of the closed dovetail mortice and perhaps influenced by it.

7C3.4. The Manot Press: Screw Press Bases with Mortices with Horizontal Bores for Pins (T744). The special characteristic of the Manot screw press is the form of the mortice. A narrow horizontal bore went through the mortice from one side to the other. A metal pin was presumably inserted in the bore and through a hole in the bottom of the screw frame post or screw-end, thus securing them in place. Manot screw bases were usually cut in the bed-rock on the edge of a step, and in these cases the bores for the piers in both mortices were cut from the same side. In three bases, however, where the installation was not near the rock edge and special insertion channels were cut, these approached the two mortices from opposite directions.

Eleven examples of the Manot screw press base were found in a very limited area of 250 km^2 in western Galilee. This press was used in several different types of installation. Five are connected to wineries, one is an oil press and one is attached to a Crusader sugar factory, Manot Lower (see discussion in list of sites).

The uniformity of the Manot presses and their being found only in a small area suggest them to be all from one comparatively short period. None of these presses was dated stratigraphically, but glazed pottery typical of the Middle Ages was found near most of them. This, the association with the Crusader sugar factory, and the fact that the distribution area of this press is the immediate hinterland of Acre, the capital and main port of the Crusader kingdom, all suggest that this type was introduced by the Crusaders.

The probability of this press having been introduced from abroad is strengthened by the fact that it is different in several ways from presses used in ancient Israel. First, as will be shown below, in Israel there is usually sharp differentiation between presses used for oil and those used for wine. Secondly, in western Galilee other presses used in both these processes were completely

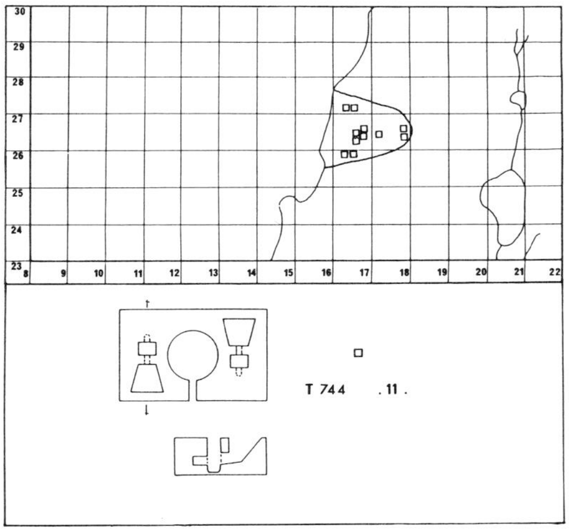

Map 33. Manot press (T744).

Photo 76. Manot screw press base (T744): Manot Lower (sugar press).

Photo 77. Manot screw press base (T744): Mana East (attached to winery).

Photo 78. Manot screw press base (T744): Judeida (attached to winery).

different from this press. The oil presses were lever and weight and lever and screw presses, and the wine presses, as will be shown, were single fixed-screw presses of type T83.

Two screw press bases from other countries similar to the Manot press (T744) appear in the catalogue, one from Cyprus and one from Sicily. It is very probable that these are connected to those from western Galilee. It is difficult, however, to determine whether these were the origin of the Israeli press or derived from it. As regards the latter possibility, it is known that after the defeat of the Crusader kingdom sugar production continued in Cyprus. In the refineries excavated, however, no presses were found (Maier and Wartburg 1983; Wartburg and Maier 1989).

7C3.5. Miscellaneous Screw Press Bases (Ts745, 751, 752). Three other types of screw press bases were recorded. Two examples were found of type T745 in which the press-bed was square and the mortices round. In type T751, already referred to in connection to central collection, and in type T752, the mortices instead of projecting from the round press-bed cut into it.

7D. Screw Press Bases from Other Countries

Only nine screw press bases appear in the catalogue from five other countries. Those of the Manot type (T744) from Sicily and Cyprus have already been mentioned. Two from Egypt are of type T741, one from Spain of type T752 and one from Italy of type T753. Three others, although each different in character, have been grouped together as T746. That from Bovillae is interesting, combining characteristics of the Tivoli pier base (T43111) with external dovetail mortices similar to those of the Samaria screw weight (T62111), both types found in Italy.

The paucity of examples does not allow for any attempt at typological or spatial analyses. It is clear from Pliny (18.74.317, press D; see above p. 83), however, that the single rotating-screw press, Model 7A, was known in Italy in the first century CE, as is confirmed by the evidence from Pompei 01 (T746). Presumably, the majority of the presses were completely of wood, as are the majority of the pre-industrial presses, and therefore left no remains.

Note

1. A similar screw base, the provenance of which is unknown, is today at Kafr Kama (Frankel Avitsur and Ayalon 1994: fig. 78).

Chapter 8

Improved Wineries: Ancillary Installations (T8)

8A. Differences and Similarities between Wineries and Oil Presses

As has been shown, it is often difficult to distinguish between the simple installations (Ts1) that served for wine production and those that served to produce oil. This is reflected in the biblical lexicon which lacks a specific term for an oil press (see Appendix 1).

Although primitive installations (T1) continued to be used for both purposes until recent times, in Israel, from the Iron Age II onwards, wineries and oil presses could nearly always be clearly distinguished. In the winery there was a treading floor with a comparatively large collecting vat and no olive-crushing device, while in the oil press the reverse was true: there was an olive-crushing device—from the Hellenistic period on, a round rotary olive crusher, a comparatively small collecting vat and no treading floor.

This clear differentiation in function is reflected also in the later Mishnaic and Talmudic literature in which there is a clear terminological differentiation, the term *gt*, referring to a winery, and *byt bd*, literally 'the house of the beam', to an oil press. The fact that direct-pressure screw presses lack a beam led to some Talmudic discussion as to the definition of the term, but even screw presses continued to be called *byt bd* and wineries with beam presses *gt* (see Appendix 1).

Only rarely is there difficulty in determining whether a complete installation from Israel is for producing oil or wine.

One group of installations where this problem arises is type T93—installations in which small collecting vats and pressing installations are found in the centre of treading floors, and no olive crushers were found.

As we have noted in many cases in other countries the problems of definition are greater, for instance in the installations from the region of Caesarea in Algeria (T407128004), which usually lack press-bed collecting vat and olive crusher.

In most cases, however, only parts of the installation are found not *in situ* and the question that arises is whether the actual presses used in wineries can be distinguished from olive presses, when other evidence such as olive crushers or treading floors is lacking.

In classical literature the distinction is far from clear. Both the lever and the screw presses described by Pliny and Hero are specifically stated to be wine presses and, although Cato describes both oil and wine presses (chs. 18 and 19), there is little difference between them. The archaeological evidence presents a similar picture.

Presses of different types appear in Roman Italy—but all are used both for wine and oil. In the terminology used by Latin agronomists, the distinction is also not clearly defined. Cato calls a wine press once *vas torculus instructus* (11.1) and in another place *vas vinarius* (19). He calls the oil press *vas olearius instructus*. The press room of an oil press Cato calls *torcularium* (18), but grapes are also trodden in a building of this name (Cato 112.3). The wine press is also called *torcular* (Pliny 18.62.230) or *prelum*, press-beam (Varro 1.54.2-3) and an oil press *torculum olearius* (Varro 1.55.7).

In Israel, however, the clear terminological differentiation of the Talmudic literature reflects the actual situation at the time. The presses used in wineries were almost always different in character from oil presses. Of the 207 presses in the catalogue from Israel and environs that were used in wineries, 199 are of types specific to wineries (single fixed-screw presses Ts81, 82, 83, and beam presses Ts85), while only eight are of types used also in oil presses (screw press bases: two of T743 and four of T744; screw weights: one of T62111, one of T6232402).

8B. Wineries: Typology

As the making of wine developed, the simple winery/treading installation (T1) changed and diversified. It grew in size and various installations and devices were added to the two basic elements, the treading floor and collecting vat. The term 'improved winery' has been used for all installations that include more than these two basic elements. Of the simple winery (T1), all examples have been included only from the sample 10 km square 1626, and from countries outside the Israel grid. However, an attempt has been made to include all examples of the 'improved winery' in the catalogue.

The ancillary installations are of various types and have been numbered T8. A: Compartments around the treading floor, Ts80. B: Rollers used instead of, or perhaps as an aid to, treading to extract the bulk of the must from the grapes, Ts86. C: Presses that served to extract the remainder of the must after the treading and/or rolling; lever presses Ts85; screw presses Ts81–83.

Often an additional intermediary vat or cavity was added between the treading floor and collecting vat to sieve the must or act as a settling tank. The arrangement, primarily of the three main elements, treading floor, collecting vat and intermediary vat, but also of the other ancillary installation, allowed for variations in the plan of the winery as a whole. These architectural plans have been numbered Ts9, and will be discussed in Chapter 9.

8C. Compartments Adjacent to Treading Floor (T80)

8C1. Small Compartments Adjacent to Treading Floor (T801)

The small compartments attached to the treading floor vary in shape. There are examples that are square, rectangular, semicircular and apsidal-rectangular with the back end semicircular. Many are paved in mosaic and some have a roof in the shape of half a cupola. They vary in size from an area of 0.25 m^2–3 m^2. They are usually at a higher level than the adjacent treading floor and some are connected to the treading floor (T8012) while others are not (T8011). There are usually two or three compartments in each winery, but there can be as many as nine. Sometimes both types appear. For instance, at ʿAlya there are several small unconnected semicircular compartments less than a metre in circumference on two sides of the treading floor and on the third two slightly larger compartments connected to the treading floor, one rectangular and one 'apsidal'.

Small compartments were recorded at 23 sites, nearly all concentrated in a small area in the centre of the country. One example, however, was recorded in the Golan Heights.

From Pliny and Columella we know that sea water, honey, pepper and a great variety of spices were added to the wine (see above, p. 43). In the Hebron mountains, today, it is usual to add marl to the grapes before treading (Ben Yaʿakov 1978: 27-29). It is probable that these materials were mixed, prepared and stored in these compartments. The fact that in so many cases there are two or three compartments suggests that two or three standard substances were added to the must.

It is also possible that the larger of these compartments were to prepare 'first must' (*prototropum, mustum lixivium*; see above, p. 42) by allowing the grapes to stand, and collecting the must that dripped out through the bore into vessels below.

8C2. Large Compartments Adjacent to Treading Floor (T802)

These compartments are much larger, varying from 6 m^2–10 m^2 in size. They are also arranged around the treading floor and are rectangular or trapezoidal in shape. They are at the same level as the treading floor and connected to it by a wide opening that, as opposed to the bores of the small compartments, cannot be stopped up. These are found in the wineries of the Negev Highlands (T92) and in each winery there are between seven and ten compartments. At Reḥovot, Duran, a site over 100 km to the north of the other sites at which these compartments were found, a winery was excavated with 11 compartments similar in shape to those from the Negev sites but slightly smaller and connected to the treading floor by openings that could be stopped up. Thus the compartments at this site combine the characteristics of both types of compartments.

These large compartments clearly served a different purpose from the small type and were almost certainly for storing grapes before treading. There are two possible reasons for such a practice. One is that this enabled regulation of the work in what were very large wineries. A more probable explanation is that these compartments served to leave the grapes for some days in the sun before treading. Columella (12.27) prescribes leaving the grapes in the sun for three days to make sweet wine and states that Mago, the Punic agronomist, gave similar instructions for making raisin wine (Columella 12.39.2). Wine made from grapes left in the sun by the name of *hylstwn* and raisin wine are mentioned in Talmudic literature (see Appendix 2) and so were known in the region. In all the wineries with large compartments (T802, Ts92), there are also two collecting vats, allowing one batch of grapes to be trodden and pressed and to ferment while the next is standing in the sun. Before the second batch is trodden, the fermented must from an even earlier batch would be drawn from the collecting vat and put in jars.

8C3. Auxiliary Treading Floor and Collecting Vat Connected to Main Treading Floor (T803)

In these wineries auxiliary treading floors are located near the main floor. The collecting vats of the auxiliary floors are similar in character to the small compartments (T8012) and are connected by small bores, which can be stoppered, to the main treading floor. Five wineries of this type were recorded, one on Mount Carmel, one in Jordan and three in the centre of the country.

It would seem that the purpose of the auxiliary

treading floors was also to leave grapes or raisins in the sun to make sweet or raisin wine as in the wineries with large compartments. The small intermediate collecting vats add one refinement not found in the Negev wineries, the possibility to collect separately the 'first must' (see above, p. 42) which would flow from the grapes during these three days. Columella specifically recommends such a practice (Columella 12.27).

In the Talmudic literature a distinction is made between '*mšṭyḥ šl ʾdmh*, an 'earth storage floor', and *mšṭyḥ šl ʿlym*, a 'leaf storage floor', the first on which grapes were stored without the specific intention of producing must, the second with that intention. Perhaps the first are the large Negev compartments (T802) and the second auxiliary treading floors (T803). Presumably in the latter, leaves and branches were placed below the grapes to aid the flow of the must (see discussion in Appendix 1).

Figure 40. Depiction of single fixed-screw press: mosaic, Tyre, Qabr Hiram.

8D. Wine Presses

8D1. Mortices of the Single Fixed-Screw Press (Ts81–83)

The single fixed-screw press was the main wine press in Israel and surrounding countries. Its purpose was to squeeze out the must left in the grape skins and stalks after treading. Representations of this device appear on mosaic pavements from two churches at Qiriat el-Mekhayat in Jordan, and from one at Tyre-Qabr Hiram in Lebanon, and possibly on an early English manuscript (Site: 32-902). From these representations alone the manner in which this press functioned could not be understood. There are, however, pre-industrial examples of this press (T016). Although these are all of metal and are apparently a modern invention, having no direct connection to the ancient press, they are of some assistance in understanding how the ancient press functioned.

All that remains of these presses in an archaeological context is a mortice, usually located in the centre of the treading floor of a winery. A wooden screw was fixed in the mortice. A hole in the pressing board made it possible to slip it over the screw and turning a large nut on the screw above the pressing board brought pressure to bear on the grape skins and stalks which had been piled up round the screw and under the board and nut.

The mortice is sometimes cut in the bed-rock (Ts811, 815, 8312), but often cut into a free-standing stone. In the latter case it could theoretically be confused with the socketless screw weights—especially the Luvim (T6261) and the Bet Ha-ʿEmeq weights (T625). There is, however, a fundamental difference between the function of the screw weight and that of the screw

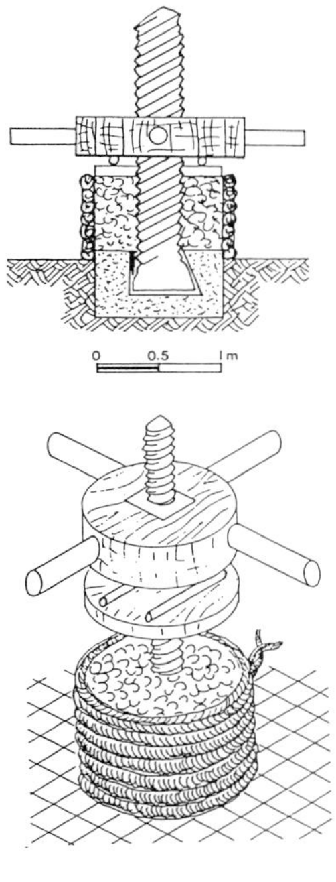

Figure 41. Fixed-screw wine press (Ts81–83)—reconstruction (Frankel, Avitsur and Ayalon 1994: fig. 82).

mortice. The screw weight serves to weigh the beam down and therefore its size is significant. In the single fixed-screw weight, the pressure of the screw acts only on the attachment of the screw to the mortice, so that the actual stone in which the mortice is cut can be small and usually is so. There are, however, some mortice stones in the Carmel region which are as large as weights or larger, but although there is some doubt as to whether these are screw weights or single fixed-screw mortices, as they are much wider than they are thick, they have been classed as the latter.

Of this press, 175 examples were recorded, 111 with square mortices, the Ayalon press (T81), one with socket and two internal mortices, the Ḥamad press (T82), and 62 with closed dovetail mortices, the Ḥanita Press (T83), and one variant (T832).

8D1.1. The Ayalon Press: Single Fixed-Screw Press with Square Mortice (T81)

8D1.1.1. Geographical Distribution of the Ayalon Press. The Ayalon press is found throughout the country, including both Lower Galilee and the Golan, but not in either western or eastern Upper Galilee.

8D1.1.2. The Form of the Press-Bed of the Ayalon Press. In all but three examples (Ts8102, 8191, 8192) the mortice was cut in a clearly defined press-bed which was raised slightly above the treading floor. In 70 examples this was square (Ts811–814) and in 42 round (Ts815–818). This variation can be explained by the differences in the devices used to contain the grape skins and stalks while pressing. On the square press-beds these were most certainly enclosed in a frame constructed of wooden boards, and in the round press

Photo 79. Ayalon screw mortice, free-standing round press-bed (T8163): Karak Inst. 05.

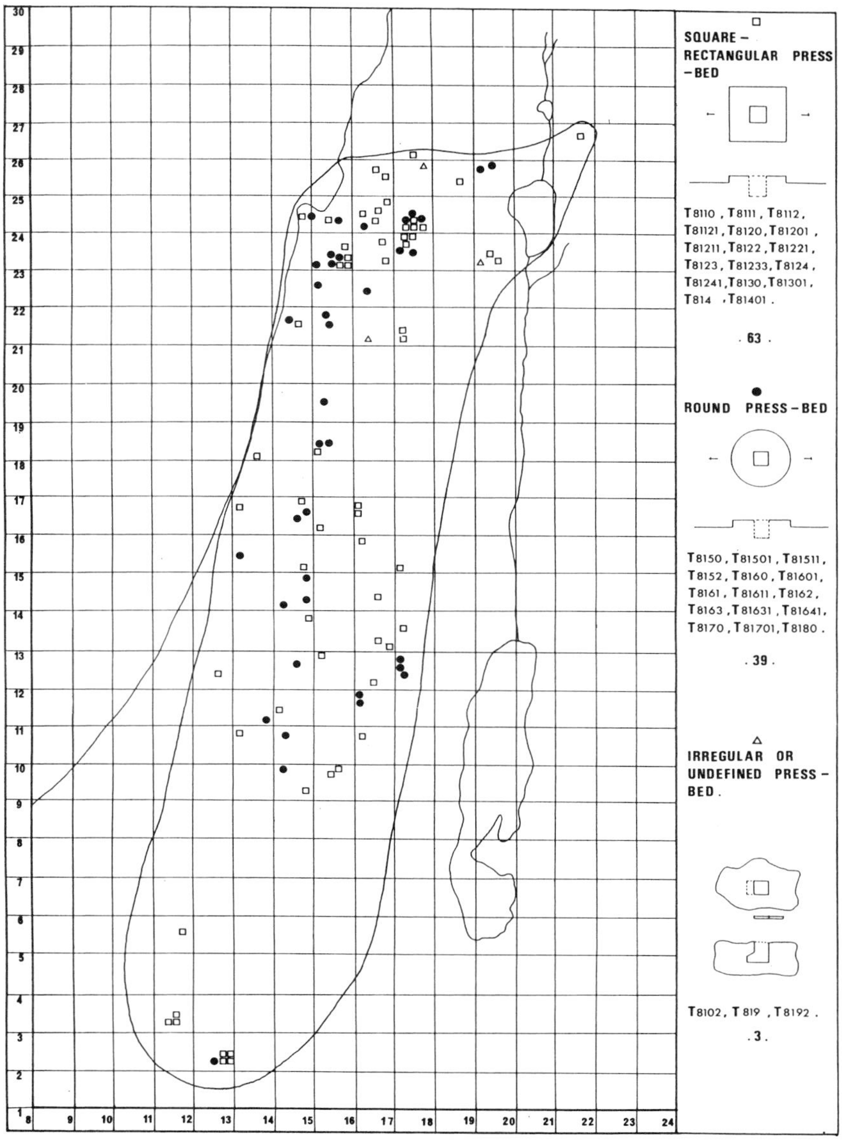

Map 34. Ayalon press: square mortice for single fixed-screw press, shape of press-bed (T81).

they were probably enclosed by a thick rope wound spirally into a cylinder (see below, pp. 147-48 for discussion on devices for containing material to be crushed). No marked differences can be discerned in the spatial distribution of these two types.

Photo 80. Ayalon screw mortice, rock-cut square press-bed and connecting channel (T81121): Usha.

8D1.1.3. The Form of the Mortice in the Ayalon Press. Four different forms of square mortices were recorded:

a. Square mortice that widens at the bottom on one side (Ts81-1).
b. Square mortice that widens at the bottom on two adjacent sides (Ts81-2). This mortice is identical to that in the Luvim screw weights (T62611).
c. Open square mortice—the square mortice cuts vertically right through the stone (T81-3).
d. Open square mortice that widens at the bottom on one side (T81-4). This is a combination of (a) and (c).

The data available were sufficient to define the type of mortice in only 45 of the 111 square mortices. Nevertheless, a spatial pattern can be discerned.

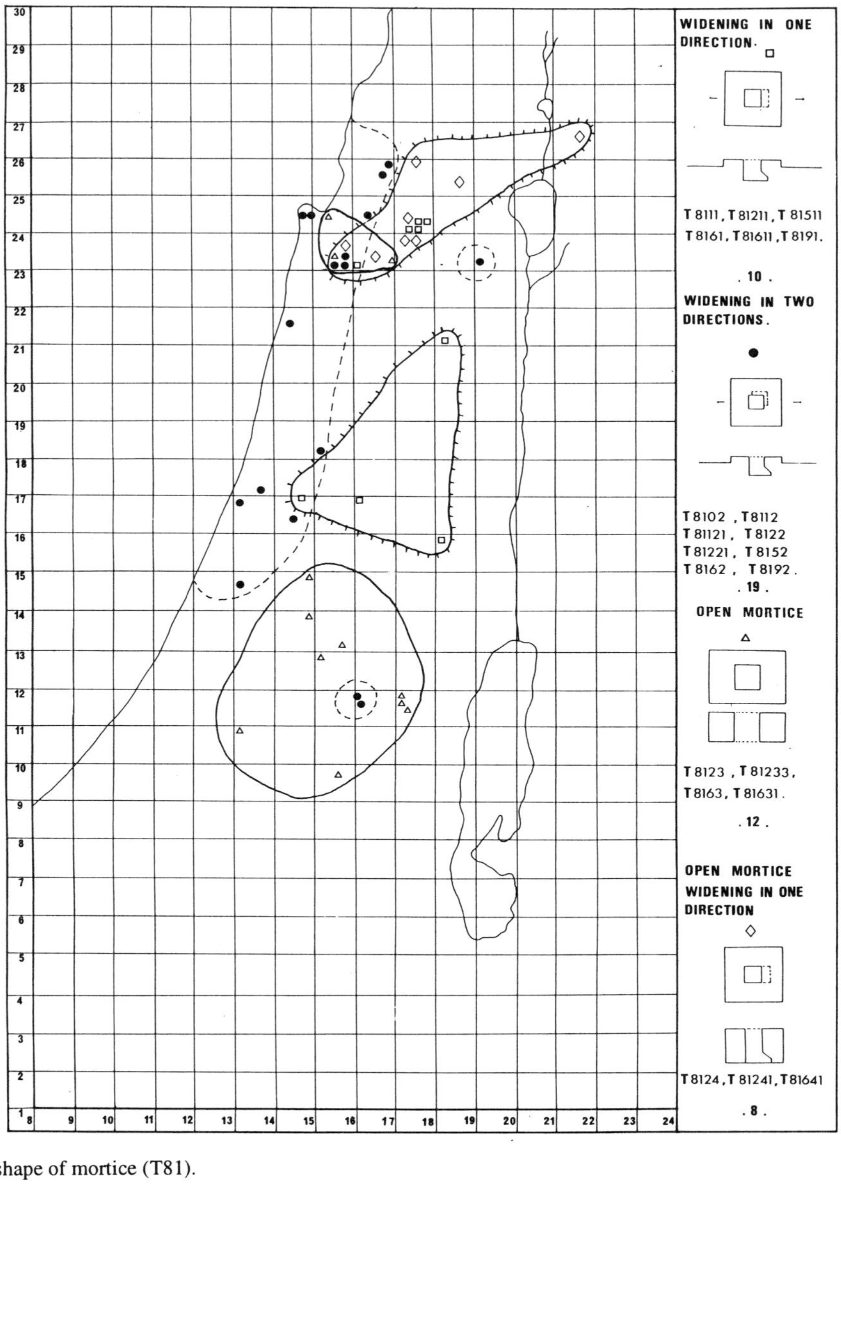

Map 35. Ayalon press: shape of mortice (T81).

South of the 14 EW coordinate, in Judaea, all 10 mortices were open ([c] Ts81-3). In the Sharon, the central coastal plain and in the foothills to the east of this region, all seven mortices widened on two adjacent sides ([b] Ts81-2). Only three examples are recorded from Samaria, but these also are of one type only, widening on one side ([a] T81-1). On Mt Carmel and in Lower Galilee the majority of mortices are of this type and of the related open mortice widening in one direction ([d] T81-4). However, in this region both the other types are also found, that of Judaea ([c] T81-3) and that of the Sharon ([b] 81-2).

This distribution pattern suggests that the three southern regions in which the mortice type was homogeneous influenced the northern where they were mixed.

8D1.1.4. Connecting Channel (Ts81---1, 813--1). In the most sophisticated wineries the mortices were connected directly to the collecting vat by a covered channel, usually of stone, in one case a pipe of lead (Tel Aviv, Eretz Israel Museum). Varro (1.54.3; see above, p. 42) distinguishes between the must extracted in the first pressing of the skins and stalks after the treading which he states should go into the collecting vat together with the must extracted by treading, and the must extracted in the second pressing (the *circumsicium* or *mustum tortivum*) which he stated should be kept separate. In those wineries in which the connecting channel ran directly into the collecting vat, the vat would have been full of must from the treading when the screw press was used and therefore it would have been impossible to separate the pressed must from that produced by treading. However, in those wineries in which the channel was connected to

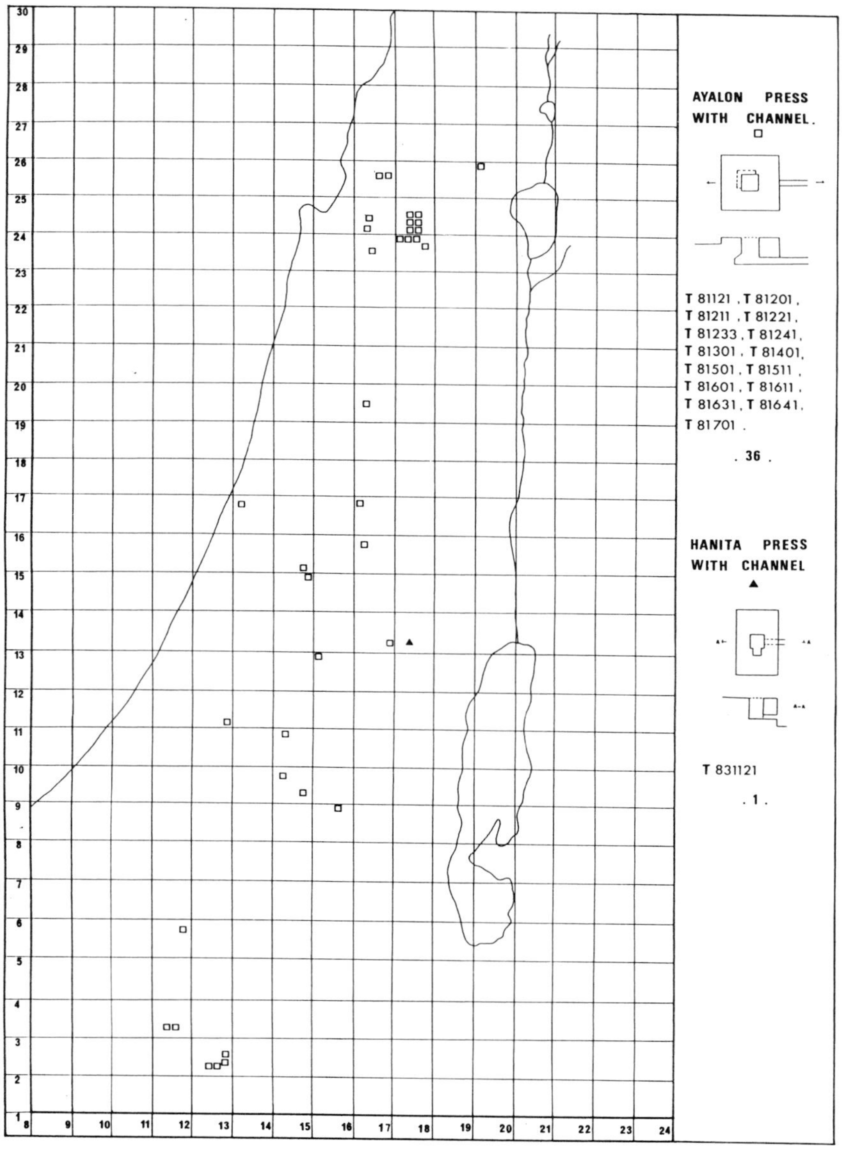

Map 36. Screw press mortices of single fixed-screw press with channels (Ts81-1, 81-3, 831121).

an intermediate vat, the pressed must could have been and probably was collected separately (e.g. the Negev wineries [Ts92], Benaya and Duma).

The wineries with the connecting channel were the technically most advanced, and their presence are evidence of a prosperous wine-producing region. The largest concentration of such presses recorded was in Lower Galilee (16 examples) and all the Negev wineries (T92) were equipped with connecting channels. Eight examples were also recorded from Judaea and six from Samaria.

8D1.2. The Ḥamad Press: Single Fixed-Screw Press, Mortice in Form of Socket and Internal Mortices (T82)

One example only of this unusual mortice was found at Ḥamad, one of the sites where the Kasfa weight (T62411) was recorded. Unless the purpose of this installation has been misunderstood, this was an attempt to use a device for a purpose for which it was neither intended nor suitable. In the Kasfa weight the screw revolves, in the fixed-screw press it does not. A similar example of a 'misplaced' device is the press-bed with radial grooves in a press with lateral collection from Dinᶜila (T4624) see p. 85.

8D1.3. The Ḥanita Press: Single Fixed-Screw Press, Closed Dovetail Mortice (T831)

8D1.3.1. Geographical Distribution of Ḥanita Press. The Ḥanita press is the only type of single fixed-screw wine press mortice found in Upper Galilee, both western and eastern. The closed dovetail mortice used in the Ḥanita press is the same mortice as that used in socketless Bet Ha-ᶜEmeq screw weights (T625) and in the Rama screw press bases (T743), all from the same region (see T041 for all installations using this mor-

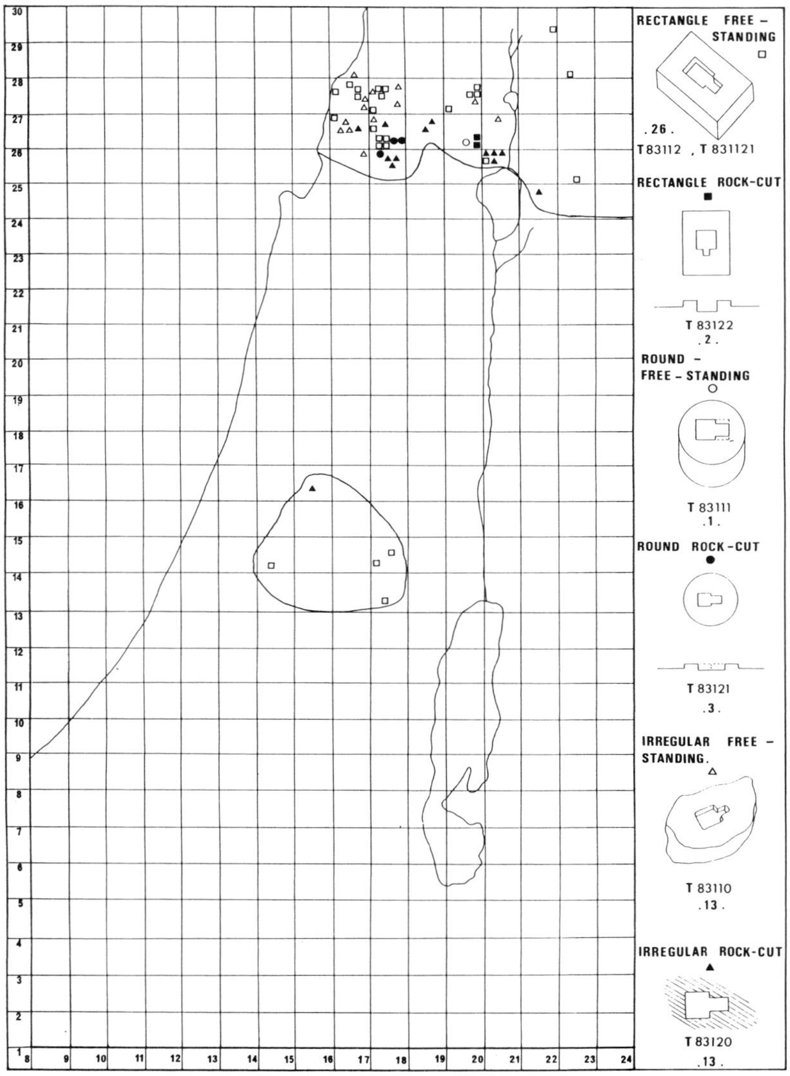

Map 37. Ḥanita screw press: central dovetail mortice for single fixed-screw press: construction and shape of press-bed (T831).

tice). However, as opposed to the southern square Ayalon mortice which did not penetrate into Upper Galilee at all, small numbers of the Ḥanita mortice are found in the south. Fifty-one examples were recorded in western and Upper Galilee, while one was recorded in Lower Galilee, four in the Golan, four in the centre of the country and possibly one in Jordan. One was also recorded from Si in Syria. Similar to other installations typical of western Galilee, this is also to be regarded as a Phoenician type. The presence of examples in the south of the country are clearly a result of Phoenician or Galilean influence from the north.

8D1.3.2. Form of Press-Bed of Ḥanita Press. As opposed to the Ayalon press, of which only three examples were found with undefined press-bed, of the 62 Ḥanita presses recorded, in 26 the press-bed was undefined (Ts83110, 83112); 28 were rectangular (Ts831-2) and only 4 were round. Three of the round press-beds were rock-cut examples from three sites very close one to the other (T83121). Only one example was equipped with a connecting channel. It is significant that this was not from one of the sites in Galilee but from a site within present-day Jerusalem (Tur, Beth Phage).

Photo 81. Ḥanita screw mortice: free-standing rectangular press-bed (T83112): Habay.

Photo 82. Ḥanita screw mortice: free-standing irregular press-bed (T83110): Muslih.

The picture reflected from these data show Upper Galilee to have been technically more primitive and less open to influence, at least from the south.

8D1.4. Mortices of Single Fixed-Screw Presses: Conclusions

In conclusion it is clear that the distribution patterns of the Ḥanita and Ayalon presses are less defined spatially than those of many other types examined up to now. The sub-types of the Ayalon press, while evincing regional diversity in the south, all appear together on Mt Carmel and in Lower Galilee. Similarly, the Ḥanita press is the only device using the closed dovetail mortice, the distribution area of which is not limited to western Galilee. This relative spatial fluidity is to be explained by the ease in which the single fixed-screw press could be made transported and installed. The mortice could be cut in a medium-sized stone building block, could be transported together with the wooden parts of the press on a donkey, and could be installed by insertion into the treading floor of an already existing winery.

This press was apparently not generally known outside the southern Levant, which explains its not appearing in classical literature, while it was almost certainly referred to in the Jerusalem Talmud (see Appendix 1). However, in an early English portrayal of Noah's sons treading grapes from the eleventh century CE (Great Britain Manuscript 32-902), behind those treading, there is a pole on which there is a spiral line, placed apparently in the centre of the treading vat. This is very probably a depiction of a single fixed-screw wine press. It is not certain that the picture represents wine production actually from England or from Normandy. This picture could be evidence for penetration of this press to the west in early times. It is more probable, however, that the illustration was copied from one from the east, similar to those in in the mosaic pavements.

The reason that this press has not survived in pre-industrial installations is because the Levant became Muslim and wine production ceased. The only pre-industrial evidence for wine production in Israel is the treading of grapes on a small scale in primitive installations to produce grape honey. On the other hand, the pre-industrial presses of this type used in Europe to press grapes (T016) appear to be a modern invention unconnected to ancient tradition.

8D2. Lever and Weights Press in Wineries (Ts85)

8D2.1. Simple Lever and Weights Presses in Wineries. Twelve wineries were recorded in which a simple lever and weights press was used to express the must from the grape skins and stalks after treading.

There is evidence for the use of this press in wineries already in the Iron Age (e.g. Ḥudash 05–07), but others are clearly from as late as the Byzantine period (e.g. Judur, Bet Hashitta).

The evidence for the use of the press is usually a round hole cut in bed-rock in the wall of the winery not more than 0.25 m in diameter.

Photo 83. Simple lever and weights press in winery (T85): Kenisa East.

8D2.2. The Marj el-Qital Beam Niche (T851). One type of beam niche apparently used only in wineries is the Marj el-Qital beam niche. On either side of a small pressing platform there are bores in which to secure the cross-piece to which the beam was attached. The bore on one side was closed, that on the other was open, in the form of a bridge with a long open channel leading up to it through which the cross-piece was inserted. Many of these were found associated with simple wineries in a small area in eastern Upper Galilee, six of which are in the catalogue. Installations working on a similar principle were found at Oshrat North and at an excavated site of the Byzantine period, Mansur el-Aqqab.

8D2.3. Large Rock-Cut Tethering Ring as Device to Operate Lever Press (T855). In two wineries rock-cut devices similar to 'tethering rings' but larger were located below the position where the beam-end would have been.

There are two pre-industrial presses that could explain these devices. In one, the end of the beam is attached to a similar device by a doubled rope (T0117). Twisting the rope lowers the beam. In the second, a rope attached to the beam is threaded through the 'tethering ring' and pulled, the ring acting as a pulley (T0118). The former is probably the method used in these presses.

8D3. Other Presses in Wineries

Of the 207 wine presses recorded, 199 were of types specific to wineries, 175 singled fixed-screw presses and 24 beam presses. Eight only were of types usually used also in oil presses. Of these, two were lever and screw presses, one from a monastery at ʿEin el-Judeida and of an exceptional type (T6232402) which almost certainly shows marks of foreign influence. The other, that from Siyar el-Ghanam, is also a lever and screw press from a monastery but using the more usual Samaria weight (T62111). Another group are the four Manot screw presses (T744), shown above to have been introduced by the Crusaders.

In a small area in eastern Upper Galilee, two Rama screw press bases (T743) were found in association with wineries, in the same district as that in which the Marj el-Qital beam niches (T851) were recorded. The use of oil press types for wine production was clearly exceptional in ancient Israel and in most cases would appear to show foreign influence. The latter group, however, are apparently a result of a local technical development in a geographical area that evinced an isolated micro-culture.

8E. Rollers in Wineries (Ts861, 862)

The use of a stone roller to crush grapes instead of treading is attested for pre-industrially in the Hebron district.

In archaeological contexts, two different types of rollers were recorded in two clearly defined regions adjacent to one another. Simple rollers (T861) were recorded from six sites, five of which were in eastern Lower Galilee. Slotted rollers (T862) were recorded at 16 sites from a region immediately to the west of the first in western Lower Galilee and on Mt Carmel. Callot (1984: 22-23, Pl. 8) has suggested that the slots served to insert levers which were used to roll the rollers. At Shiqmona, a site within the distribution area of the slotted rollers, a burial cave entrance was closed by a wheel-shaped stone rolled in the same manner, giving added support to Callot's suggestion (Elgavish 1978: 1108, photo). The distribution pattern of these two types of rollers evinces again a remarkable degree of regional diversity not only in the fact that two different types of rollers were used in two regions so close one to the other without overlapping, but also in the fact that rollers were apparently not used for this purpose in wineries in other parts of the country. This has led scholars to suggest that the installations in which the slotted rollers were used served another purpose and were not wineries. However, as already shown, clearly defined regional diversity in agricultural installations is not unusual and there is no reason to see in these rollers anything but a localized method of crushing grapes.

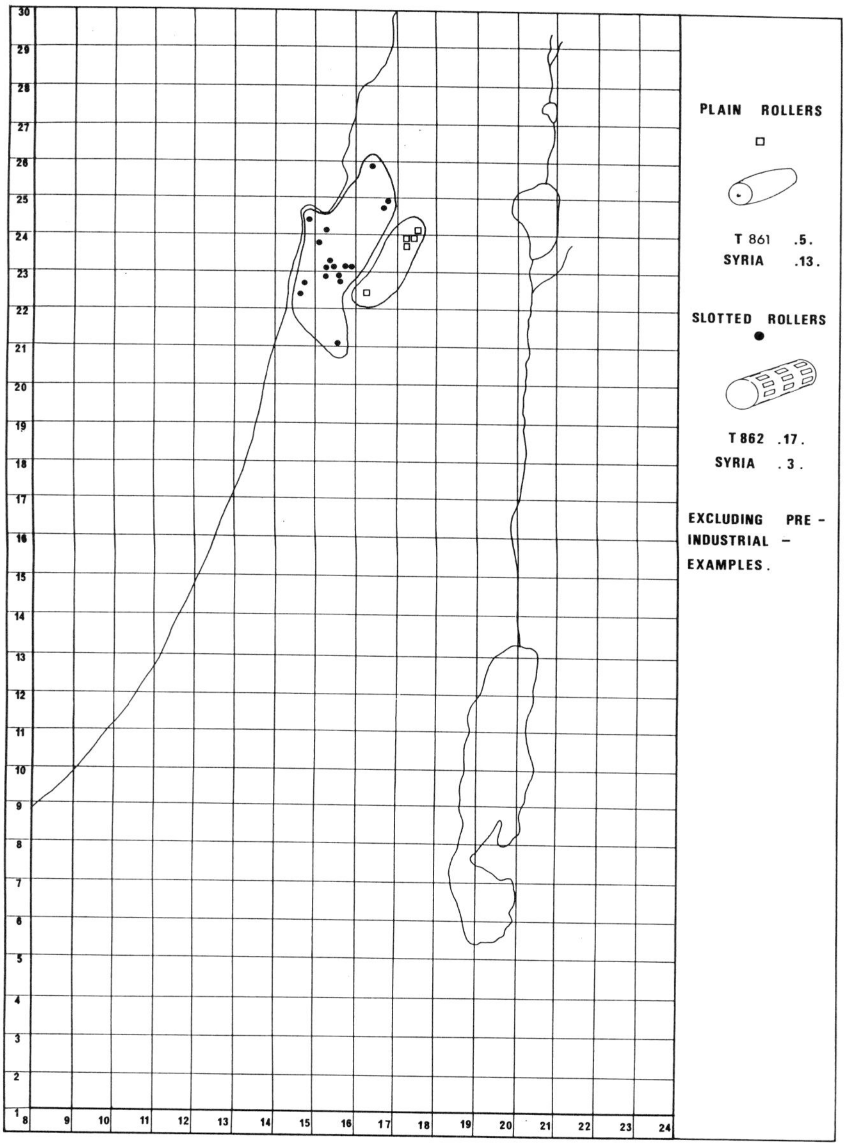

Map 38. Crushing rollers in wineries (T86).

Both types of rollers were recorded in Syria and there it was suggested that they served to crush olives (Callot 1984: 20-23, Pll. 8-11), but as shown above (pp. 88-89) the installations to which they were connected were typical wineries.

8F. Devices for Containing the Material to be Pressed

8F1. Frails: *ʿql* (Hebrew); *fiscus, fiscina, fiscellus* (Latin)

During pressing, in order to prevent the olive mash or grape skins and stalks from dispersing, these are placed in a suitable container. The contrivance usually used throughout the Mediterranean is a flat round frail with a round opening on top made of thick thread loosely woven. The material to be pressed is packed in by hand and as many as ten full flat frails are placed one upon the other. Another method used is to put the pulp or grape skins and stalks on a square cloth and to fold over the corners.

In Latin, the pressing frail is called *fiscus* (*fiscellus, fiscina*; White 1975: 88-91). Columella (12.52.22) stresses the importance of using new frails and gives instruction on how to clean them after use. First they are to be washed several times in very hot water and then kept under running water by placing stones upon them.

In Mishnaic Hebrew the pressing frails are usually called *ʿql*, although other terms are also used. In the 'Baraitha of the Lulabim' that appears in several versions in both Talmuds and the Tosefta, very similar instructions are given, distinction is only made between frails of different material. One view is that, for those of *nṣrym*, date palm fibres (wickerwork?), and *byṣbwṣ*, hemp, it was sufficient to wipe them while those of *syf*, plaited wood shavings? (reed grass?), and *gami*, reed grass (papyrus?), should be left standing—according to one opinion, until the following season. However, others said that these could also be cleaned immediately by using boiling water or watery lees or by leaving them under flowing water (see below, pp. 188-89).

Photo 84. Pre-industrial olive pulp frail: western Galilee 1980.

8F2. *Regulae, galeagra*: Wooden Frames and Ropes

In Latin literature a device called a *regula* (a board; plural *regulae*) is referred to as a substitute for the pressing frail. Cato and Varro do not mention it and Pliny (15.2.5) writes that the olives are 'pressed in frails or enclosed in *regulae*, a method recently invented'. Columella mentions this device as a substitute for pressing frails (12.52.10; 12.54.2).

Hero of Alexandria describes two versions of a device he calls a *galeagra* (Hero, *Mechanica* 3.16; Drachman 1932: 60-62; 1963: 122-26). These are four-sided boxes, open above and below, which can be easily assembled and equally easily taken apart. Hero's first *galeagra* consisted of wooden planks that interlocked at the corners and which would be built to the preferred height by adding or removing planks. The second was built up of four wooden sides kept together by three interlocking cross-pieces on each side. Hero hints that this is a new invention and Drachman suggests that *regula* and *galeagra* are different terms for the same device.

A third method used was to wind a thick rope spirally into a cylinder round the material to be crushed. Brøndsted (1928: 107) describes pre-industrial use of this method in Dalmatia, a terracotta relief in the British Museum shows grapes (?) below the press-beam wound in such a rope (Italy? Relief 27-901). and Hero speaks of 'the rope that was wound round the grapes to be pressed' (*Mechanica* 3.13).

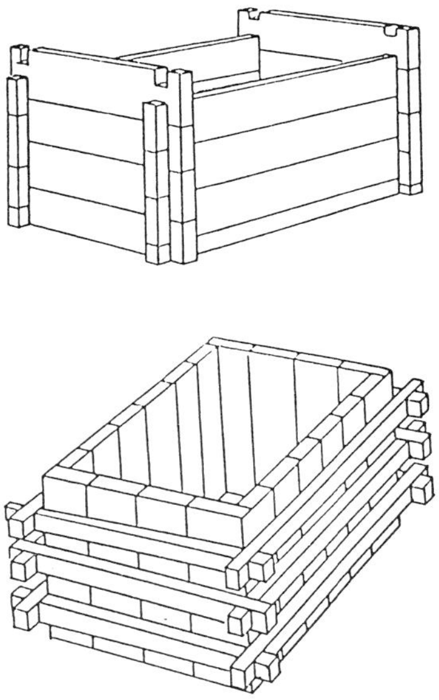

Figure 42. The *galeagra*, press frame, two types as described by Hero of Alexandria and reconstructed by Drachman (1932: figs. 18, 19).

8F3. Devices for Containing Grape Skins and Stalks in the Single Fixed-Screw Wine Press (Ts81–83)

In all the presses discussed up to now, except for the single fixed-screw press, any of the devices described above could be used and to this day the pressing frail is the most popular. In the single fixed-screw press, however, the fixed central screw prevents the use of frails or folded cloths and the only methods suitable are the wooden open box—the *regula/galeagra* or the wound rope.

The version of the 'Baraitha of the Lulabim' that appears in the Jerusalem Talmud is different in several ways from the other four versions. In the other versions both oil presses and wineries are discussed, whereas in the Jerusalem Talmud only a winery is mentioned. Of particular significance is the fact that in the latter two materials are added to the list of materials from which the frails are made, wood and ropes. It is clear that the original text has been changed and adapted to the single fixed-screw press. The only materials that could contain the grape skins and stalks in this press, wood and rope, were therefore added. As has been pointed out, the presses of this type with round press-bed were for use with the wound rope and those with square or rectangular press-bed for use with the open box—*regulae/galeagra*. It is therefore very likely that the latter device originated in this press. Hero, however, mentions the device without mentioning this type of press.

Chapter 9

Plans of Improved Wineries (T9)

The simple rock-cut wineries consisting only of treading floor and collecting vat continued in use beside more sophisticated installations until modern times. Larger versions of the simple winery, often paved in mosaics, which provide some indication as to date are also found. Neither of these types will be included in the present discussion, which is devoted only to improved wineries that incorporate elements additional to the treading floor and collecting vat. Thus, even in those cases in which a particular type of improved winery proves to be typical of a certain region, it will almost certainly not be the only type found in the region but will be accompanied by simpler wineries that were functioning at the same time.

9A. Improved Wineries in Israel and Environs

9A1. Wineries in the 'Four-Rectangle Plan' (T91)

'The four-rectangle plan' is the one found in the greatest numbers, and is the simplest, most compact and architecturally the most pleasing combination of the three primary elements of the winery. The intermediary and collecting vats are placed next to each other along one side of the treading floor and these three rectangular elements are then enclosed within a fourth rectangle. The treading floor and two vats are often all three perfect squares.

Photo 85. Winery, four-rectangle plan (T9111), small compartments (T8010), Ayalon screw mortice (T81221): Tel Aviv, Eretz Israel Museum Inst. 02 (Photo: Eitan Ayalon).

Photo 86. Winery, four-rectangle plan (T9112), compartments (T8012), screw mortice (T81301): Burak Inst. 01.

The plan is already found in Iron Age wineries (e.g. Michal 03 and Qala [T9131]) and examples have been found from the Hellenistic (e.g. Tel Aviv Ḥevra Ḥadasha and Michal 01, 02 [T9111]) and Roman period (e.g. Tel Aviv, Eretz Israel Museum 01 [T9121]). Most examples, however, date from the Byzantine period.

Comparison of sizes shows that, while the Iron Age winery from Michal 03 is comparatively small, the treading floors measuring 6.6 m^2–7.9 m^2 and the collecting vats 1.2 m^3–3.94 m^3, those from the Hellenistic period reach the large dimensions of those of later periods. In the Hellenistic winery from Michal 01 the treading floor measures 32 m^2 and the collecting vat reaches 7 m^3. In the Roman winery from Tel Aviv Eretz Israel Museum 03 the dimensions are similar—treading floor 27.5 m^2, collecting vat 6 m^3. In two Byzantine presses from Galilee Hasolelim South-East and Miṣpe Resh Laqish South the treading floors are smaller—12.4 m^2 and 13 m^2—but the collecting vats are of similarly large dimensions—6.4 m^3 and 6.8 m^3, a result presumably of different production methods.

Two-thirds of the wineries classed in this type, 35 of 54, are of the standard plan. In these, all three elements are rectangular or square and the intermediary vat clearly serves as a settling tank. The sub-types vary first in the shape of the vats, sometimes one being round (Ts912), sometimes both (T913). In others the relative size or position of the vats is varied (Ts914, 915 916). There are also variations that reflect differences in the way the winery was used. There are wineries in which there are two collecting vats and no intermediary vat (T91-2), thus allowing for treading a second batch of grapes into the second vat while the must from the first batch is fermenting in the other vat. In others, all three elements are connected allowing for various combinations (T91-3). One installation with an unusual plan is that from Jericho (T9181), in which two treading floors and intermediate vats are connected to one collecting vat. Netzer (1989), who excavated this installation, suggests it served to process dates and not grapes, which would perhaps explain the unusual shape.

Geographically, the wineries of the four-rectangle plan are those typical of the central region of the country. Only seven of the 54 recorded are in Galilee and the main concentration is in the 40 km square to the north and west of Jerusalem in which half are located. A spatial pattern can also be discerned in the distribution of the variant types. The percentage of variant types grows northwards. South of the 16 EW coordinate only six of the 26 are of the variant types; north of that line 13 out of 28.

9A2. Wineries in the 'Composite Plan' (T92)

Mazor (1981) has published the impressive group of Byzantine wineries from the former Nabataean cities Shivta, Haluṣa and ʿAvedat. Although these installations are not uniform, there are certain elements that characterize them. In some of the wineries of the group all these elements are found and in all the wineries several

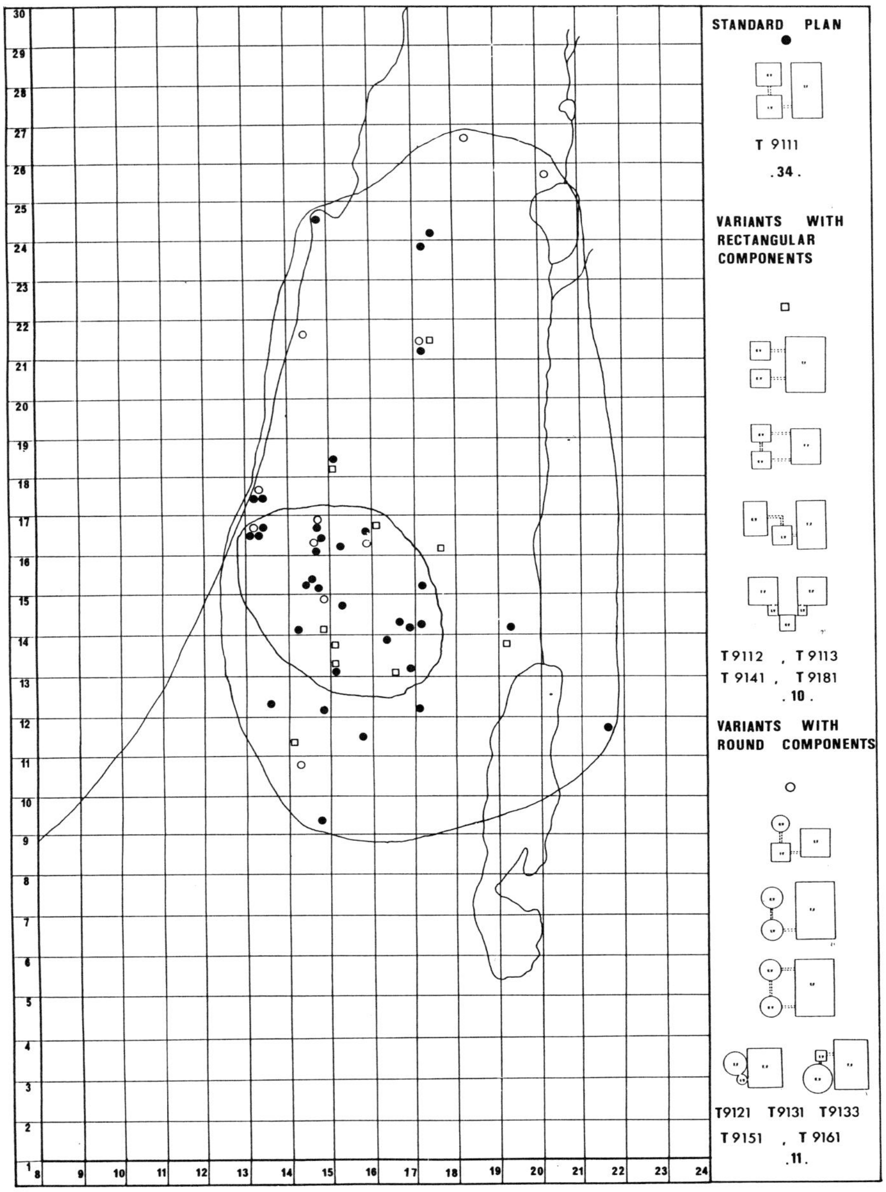

Map 39. Wineries in 'four-rectangle plan' (Ts911).

of them appear. The winery from ʿAvedat South-West incorporates all the characteristic elements. The treading floor is square, 33 m^2 in area. Around it were arranged 10 large apartments (T802). From the treading floor the must flowed into a small shallow intermediary vat from which it could continue to either of two round collecting vats, the volumes of which were 6.5 m^3 and 4 m^3. In the treading floor there were mortices for two Ayalon single fixed-screw wine presses. Both had channels, that of one joining that of the other to flow to the intermediary vat.

The layout of large compartments, two screw presses and two collecting vats allowed for the winery to be used continuously, three batches of grapes being processed at the same time. While the must of the first batch was fermenting in one collecting vat, the next

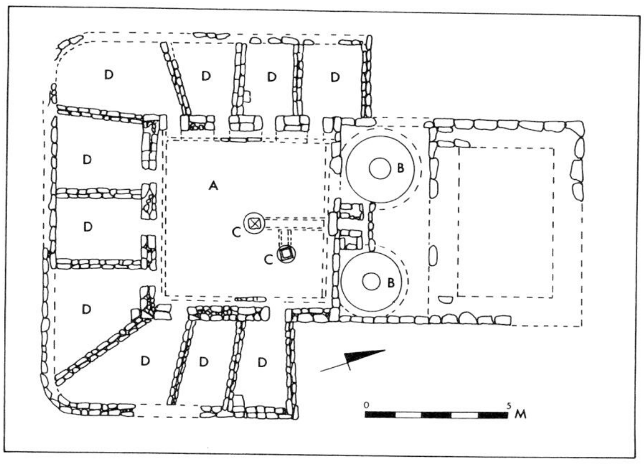

Figure 43. Winery—composite plan (T92): ʿAvedat South-West. A. Treading floor; B. Collecting vats; C. Screw press mortices with connecting channel (T82601); D. Large compartments (T802).

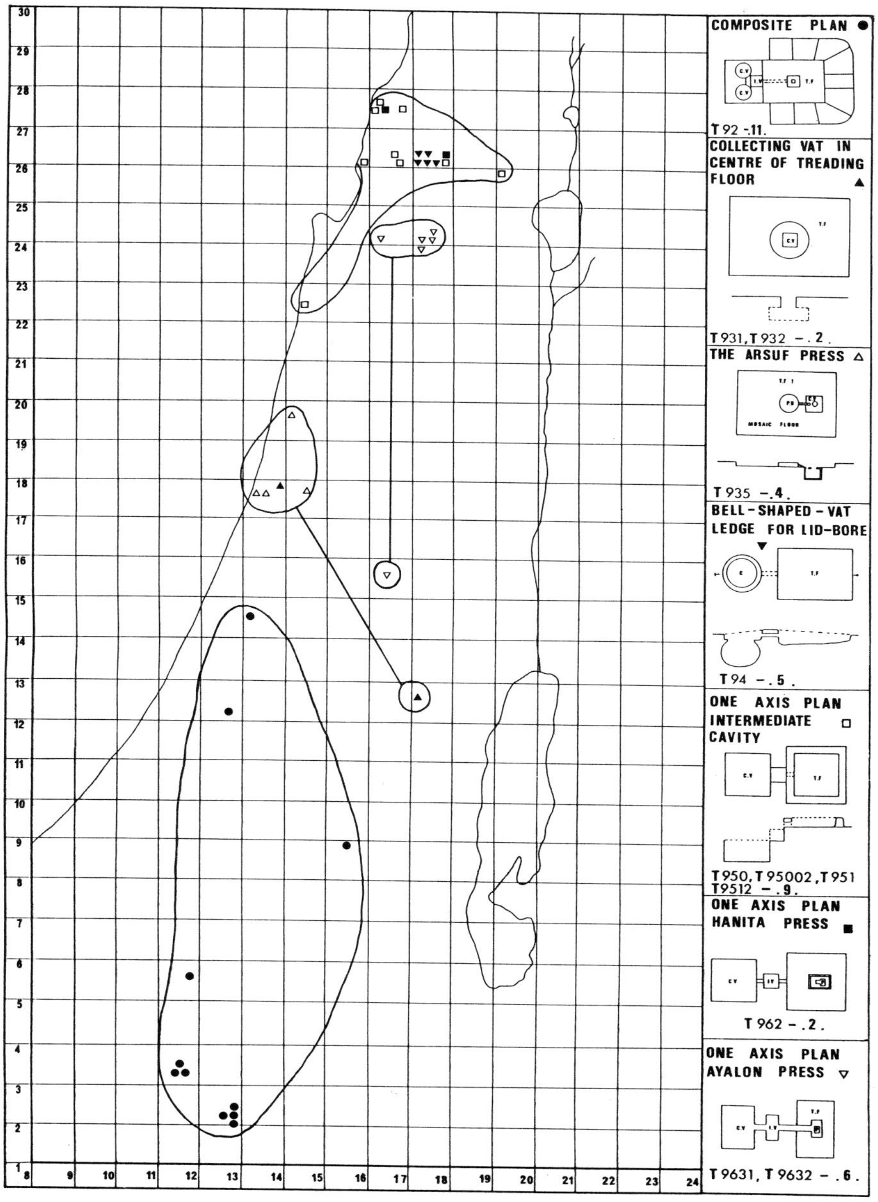

Map 40. Various types of winery (Ts9).

was being trodden, the must flowing into the second collecting vat and the remaining grape skins and stalks being then pressed and the pressed must collected separately in vessels in the intermediary vat. In the meantime, a third batch was waiting to mature in the compartments.

Most of the wineries of this group are in the Negev highlands, but three wineries situated over 100 km to the north, two in the southern coastal plan Jaladiya and Reḥovot Duran and one in the southern Judaean mountains, Samuᶜ, have certain characteristics in common with the Negev wineries. All three have two collecting vats with a small shallow intermediary vat between them. In other ways they vary considerably from one another and from the Negev wineries, but they are clearly connected to the latter, either having been influenced by them or perhaps the reverse. Additional data will perhaps clarify the question.[1]

At ᶜAvedat West 01, there is an unusual installation (T922), consisting of seven large compartments arranged in a circle around a small circular space and lacking all the other elements of a winery. A possible explanation is that it is a raisin-drying plant and that the circular space in the centre was to collect the first must that was produced.

Mayerson (1985) has placed the Negev highlands wineries into their historical perspective, showing them to have been the production centres of the famous 'Gaza wine' that was mentioned in many texts from the Byzantine period. A particular type of amphora found at sites throughout the Mediterranean has been shown to be the vessel in which this wine was exported and has therefore been called the 'Gaza Amphora' (Zemer 1977: Type 133; Peacocke and Williams 1986: Class 49; Riley 1979: Amphora Type 3; Sciallano and Sibella 1991: Amphora Late Roman 4).

9A3. Wineries in the 'One-Axis Plan' (Ts95-96)

The simplest way to arrange the three basic winery elements, the treading floor, intermediary vat and collecting vat is in a straight line—'the one-axis plan'.

In Galilee the majority of the improved wineries are of this type. Those found in this region have certain characteristics in common, and it is perhaps possible to trace stages in their development. Wineries in the one-axis plan, however, are also found in other regions and even other countries, but as this is the most obvious plan, although it is perhaps justified to name it the 'Galilean winery', this term should not be understood as suggesting that any connection of diffusion or influence should be sought between the presses of this type found in Galilee and those from other regions.

In the simplest Galilean type (Ts95) the intermediate element is not a vat but a small cavity cut in the rim of the collecting vat and connected to the treading floor by a bore. In several of these wineries there is a ledge around the collecting vat for a lid, and in these cases the intermediate cavity is cut into the ledge. The small cavity probably served to sieve the must using branches of poterum, great burnet, as a sieve. Some reach similar dimensions to those already described. In Yaᶜara West, the treading floor measures 14.8 m^2 and the collecting vat 10.7 m^3. In one example at Parod North a depression and connecting channel for a screw mortice were recorded (T9512).

Photo 87. Winery—one-axis plan with intermediate cavity (T951): Liman West.

Photo 88. Winery—one-axis plan with intermediate cavity (T951): Liman West.

Photo 89. Winery—one-axis plan with intermediate cavity (T951): Yaᶜara West.

The more common type had rectangular (T961) or round (T9612) intermediary vats. In Galilee these are very small and, as opposed to the large intermediary vats of the south which were settling tanks, the small Galilean vats, similarly to the cavities, probably served as sieving vats.

In Upper Galilee these wineries appear with Ḥanita screw mortices (winery T962) and in Lower Galilee they appear with the southern square Ayalon screw mortices (winery T963).

In eastern Lower Galilee a group of sophisticated wineries, most of which, although not all, were in the one-axis plan, had Ayalon mortices and connecting channels (Ts9631, 9632). The agricultural prosperity of this region is perhaps connected to its being in the immediate vicinity of the regional capital Ṣippori.

9A4. Galilean Wineries with Two Collecting Vats (T97)

The Negev highlands wineries (T92) have two collecting vats as have some of those in the four, rectangle plan (Ts91-2).

In the mountains of central Upper Galilee two small rock-cut wineries were recorded, neither of which had an intermediary vat, while one had two collecting vats and the other three (T971). In Lower Galilee an unusual winery with two intermediate vats and two collecting vats was recorded (T972). In these cases also, two collecting vats allow for work to continue without interruption. While the must from one batch of grapes is fermenting in one vat, the must from the next batch flows into the second vat.

9A5. Winery with Bell-Shaped Collecting Vat, Ledge for Lid and Bore (T94)

Within a very small area of less than 10 km^2 in Western Galilee, five almost identical small wineries were recorded. They had bell-shaped collecting vats, a ledge for a lid and the treading floor was connected to the collecting vat by a bore and a channel cut into the ledge to allow the must to flow under the lid. The treading floors ranged in size from 2 m^2–5.3 m^2. The collecting vats, although exact measurements are missing, were less than 1 m^3 in volume.

This is a very good example of a particularly impressive local type. The ledge for the lid and channel under the lid connect it to the similar wineries found in the same region with intermediate cavity (T952). The bell-shaped collecting vat is not found in the region. However, water cisterns in all the regions are of this shape, so it would seem more likely that its use in wineries is a local adaptation rather than a sign of connections with Syria.

Photo 90. Winery—bell-shaped collecting vat, ledge for lid, bore connecting treading floor to vat (T94): Yirka Central.

9A6. Installations with Collecting Vat in Centre of Treading Floor (T93)

Two installations were recorded in which the collecting vat was in the centre of a treading floor. In both, the collecting vat was comparatively small and in both the vat was covered by a stone lid with a small opening in the centre. In the installation at Qat (T931) the volume of the vat was 0.27 m^3 and the stone lid was an Ayalon screw mortice of a sub-type found at two other sites in the immediate vicinity (T8163). At ʿAzzun (T932) the vat was 1.6 m^3 in size and the stone cover had radial grooves. The hole in this cover was round and not square as is usual in screw mortices, and therefore the excavators were probably correct in not seeing this as a screw mortice. The cover, however, is similar to those of oil presses with central collection (see T042).

The unusual character of these two installations certainly suggests that the production methods used differed from those used in the usual Israeli wineries or oil presses. There is certainly a possibility that another fruit was processed in these installations which produced less juice than grapes or that these installations were introduced from a region with different agro-technical traditions. However, for both these possibilities concrete suggestions are lacking.

These two installations have much in common with the Arsuf press (T935). In the latter, press-beds and collecting vats very similar in character to those usually associated with oil presses were found in the centre of a mosaic pavement, presumably a treading floor. In three of the four Arsuf presses the collecting vats are of the unusual Qedumim type (T47114) and in no case was an olive crusher found. All four sites at which Arsuf presses were found are in the Sharon coastal plain and located not far from ʿAzzun.

The exact function of these installations remains extremely difficult to determine. It would seem, however, that the installation at Qat is a winery and those from the Sharon, the Arsuf presses and that from ʿAzzun perhaps unusual oil presses.

9B. Wineries in Ancient Artistic Representations and Classical Literature (Ts02)

In the discussion on simple wineries (T1), it was shown that both the archaeological and early pictorial evidence suggested that different methods of wine production were used to the north of the Mediterranean than those used in the south. Instead of collecting the must in the large vats of the Levant or Egypt, in the north it was collected in *pithoi*. The question that arises is to what extent this difference continued into later periods.

From later pictorial and written evidence a very similar picture emerges. Of the 16 later representations of treading grapes from the Hellenistic, Roman and Byzantine periods, only two showed vats, one of these from Egypt (Ts02124, 0213), while ten showed collection in *pithoi*, these varying in number from one to four.

Most of the representations are very uniform stylized sarcophagus reliefs in which the grapes are trodden in a vat that looks like a free-standing bath, often with everted sides. A very realistic representation is that from the mosaic pavement at St Romain en Gal showing two men treading and a third playing the flute. The treading floor appears to be large with a low parapet around it. The must flows through four pipes directly into four *pithoi*. There are other points of interest in the pictorial representations. In six cases the must flowed out through lion-head spouts (Ts02-4), including in a tomb painting from Egypt from the Hellenistic period (Hermopolis, Tomb of Petosris), probably the earliest in this group of pictorial representations in the catalogue. In this Egyptian painting the men treading kept their balance by holding on to a horizontal rod above their heads, one of the traditional Egyptian methods. On an Egyptian silver goblet from the Roman period (Egypt, Alexandria Museum, Goblet) a vintage scene shows grapes being trodden in a vat with a lion-head spout. The workmen kept their balance by holding on to ropes, another traditional Egyptian method. In the representations from other countries, balance was kept in other ways, in three by placing hands on each others' shoulders (Ts02-3), a method also appearing in Egyptian tomb paintings, but in seven cases by holding sticks either on the ground like walking sticks or in the air—methods not appearing in Egypt (Ts02-4). The fact that the methods of keeping balance in depictions from Egypt are identical in Pharaonic and Classical periods shows this to be not an artistic convention but a picture taken from everyday life.

There are many references to treading and pressing grapes in classical literature, and many more references to the preparation of wine, but there is apparently only one detailed description of a winery, that of Palladius, which is part of the paragraph *De Cella Vinaria*, 'On the Wine Cellar' (Palladius 1.18). The treading floor (*calcatorium*) was raised between two collecting vats (*lacus*). Three or four steps led up to the treading floor. Channels or earthenware pipes took the must from the collecting vats to *dolia* outside the walls (Plommer 1983: 9 n. 2 suggests correcting *extremos muros* which would have meant 'the edge of the walls' or 'the end of the walls' to *externos muros*, 'outside the walls'). The following section gives two alternative suggestions as to where to place the barrels for storing the wine. In the second suggestion it is stated that these should stand in such a manner that, if they burst, the must should flow into the collecting vat. Plommer (1973: 10, fig. 1) understands this to mean the same collecting vat. However, all the archaeological evidence suggests that Palladius meant that there should be a special vat for this purpose.

9C. Wineries in Other Countries (Ts98)

9C1. Wineries in Egypt

A group of 11 small rock-cut wineries were published from Upper Egypt, partly in modern Sudan (Adams 1966). They have intermediate vats and are in the 'one-axis plan' (Ts981). The treading floors measure 3 m^2–7 m^2, and the collecting vats 2 m^3–5 m^3. These

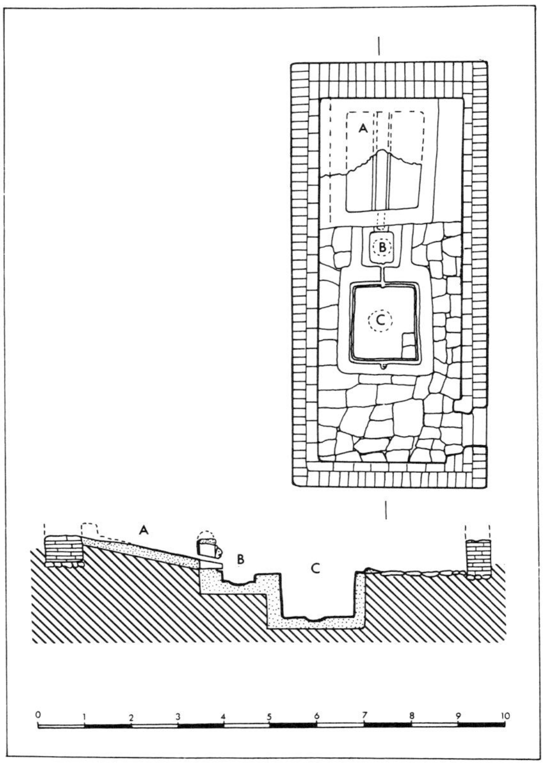

Figure 44. Winery of Roman period from Meinarti, Egypt, based on Adams (1966: fig. 2). A. Treading floor; B. Intermediate vat; C. Collecting vat.

installations date to the fourth century CE. Over half of them have lion-head spouts (T98103).

The winery from Crocodopolis Theadelphia, which was part of a temple complex, is built so that the floor of the collecting vat is at ground level and its walls above ground. Its volume is 6 m^3. Steps lead up to the treading floor, which measures 15 m^2. Here too there is a lion-head spout. The winery at the City of Menes is similar. The collecting vat is also built above ground, steps lead up to the treading floor and the must flows through a lion-head spout. That from Aswan (Simons Kloster) is similar but lacks the lion-head spout.

The Egyptian wineries are very similar to those from Israel. The main differences being that in Israel there are no examples of collecting vats standing above ground while Egyptian wineries lack presses.

9C2. Wineries in North Africa

An impressive winery was published from the villa at Tipasa. There are two treading floors the larger measures 10 m^2 and is connected to the collecting vat by a spout. The other measures 3.9 m^2 and is connected to the collecting vat by a lead pipe. As at Crocodopolis, Aswan (Simons Kloster) and the City of Menes the collecting vat is built above ground and steps lead up to the treading floor. At Tipasa the fact that the floor of the collecting vat is at ground level allowed for an outlet to be made at the bottom for cleaning.

The group of rock-cut installations published from the Grande Kabylie in Algeria by Laporte (1983) as oil presses are very similar to wineries from the Levant and were probably for wine (Ts98, 9802, 980002, 982). One at Azeffoun also has rock-cut steps leading up to the treading floor.

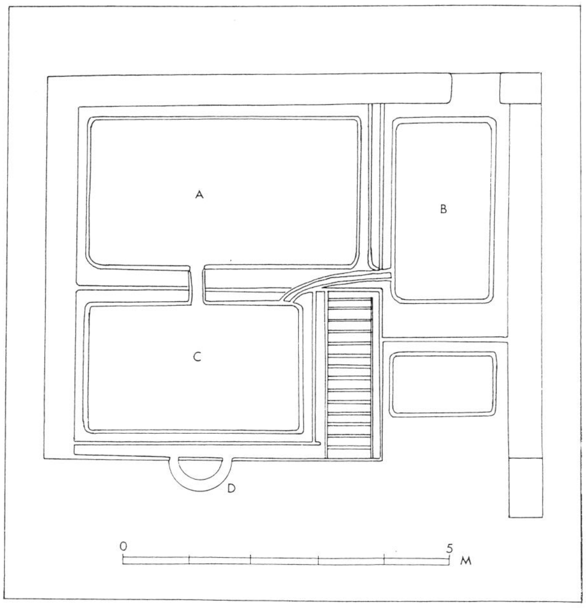

Figure 45. Winery from Tipasa, Algeria, based on Gsell (1984: fig. 55). A and B. Treading floors; C. Collecting vat; D. Draining sump.

9C3. Wineries in Syria

It has been suggested above (pp. 88-89) that installations in North Syria with large pressing platforms (Ts4661-4662) which were published as oil presses (Callot 1984) are actually wineries. The treading floors vary in size from c. 5 m^2 to c. 16 m^2 and the very uniform and characteristic bell-shaped collecting vats range in volume from c. 1.5 m^3–2.5 m^3. Many of these wineries are equipped with lever and screw presses. In some there are small cup-marks connected to the treading floor (T805), and in many more cup-marks on the rim of the collecting vats (Ts476-1, 2, 3, etc.). Both apparently served to add substances to the must. Both plain and slotted rollers (Ts861 and 862) are also found. Three examples of wineries with auxiliary treading floors were recorded (T982) and the small lateral basins next to the collecting vats doubtless served a similar function to the intermediate vats from Israel and Egypt.

The Syrian wineries are similar to those from Egypt and Israel, comparing in size to the smaller Israeli examples (e.g. T94). The most significant difference is the use of lever and screw presses in the Syrian wineries and the lack of the single fixed-screw press, thus giving some indication of the northern extent of the distribution area of the latter type of press.

9C4. Wineries in the Crimea

Gaidukevych (1958) has presented a very interesting and important group of wineries from the Crimea that date from the third to second centuries BCE to the third to fourth centuries CE. They have certain basic characteristics in common, all consisting of rectangular treading floors and collecting vats situated inside buildings. Three distinct types of winery can be discerned, however, which largely correspond to a chronological sequence (Gaidukevych 1958: 363-64).

The first group, which are all dated to the third to second centuries BCE, are all simple wineries placed in long rectangular buildings. Rectangular treading floors (15 m^2–35 m^2) occupy the whole of one end of the building, square or almost square collecting vats (7 m^3–8 m^3) are placed in the centre of the building and a large space at the other end of the building remains as a work or storage area (Tiritake 01; Karrantinnaya Slobodka 03; Partizani 01).

The second group of wineries, ranging in date from the third to second centuries BCE to the first BCE to first CE, are twin wineries in which, in each case, in one winery there is a beam press, the evidence for which being either a press-bed, a weight or both. These have been discussed in previous chapters (pp. 91, 104). The press-beds in these wineries vary in form, including round, spouted and pear-shaped, square with spout, and one consisting of blocks of stone. The size of the treading floors suggests that in these wineries the grapes were first trodden on both treading floors and then the remaining grapes and stalks were pressed under the press in one (Mirmiki 01, Stage A; Mirmiki 04; Tiritake 02; Chersonesos).

The third group, which dates from the first to second centuries CE to the third to fourth centuries CE, consists of triple wineries. In two cases (Tiritake 03; Tiritake 07) the presence of both press-bed and weight show that the central of the three wineries was equipped with a beam press and the same was probably true of the others in this group (Mirmeki 01, Stage B; Mirmeki 02; Mirmeki 03; Patrasia 01). In all the presses of group three, the press-bed was square, and in two there are small intermediate vats (Tiritake 03; Tiritake 07). The differences between the second and third groups are demonstrated by the changes that took place in winery Mirmeki 01, between the first stage dated to the third to second centuries BCE and the second dated to the first to second centuries CE. In the first stage (A) this was a twin winery. On the smaller of the two treading floors there was a rectangular stone press-bed with spout which was almost certainly part of a beam press in which the beam was aligned to the short west–east axis of the building. In the second stage (B) the winery was rebuilt, three treading floors replacing the previous two. The press-bed of the first stage was covered and replaced by a square press-bed located in the central of the three treading floors in a position suggesting that, at this stage, the beam was aligned to the long axis of the building, north–south, allowing for it to have been up to 12 m in length as compared to a length of 3 m or 5 m in the first stage. In the triple wineries in which the treading floors were of equal size, grapes were probably trodden on all three, but in the latest and most sophisticated of these wineries, Tiritake 07, the small central treading floor was clearly for pressing only.

The Crimean weights are of the Semana group (Ts5532, 55321, 55326), rectangular with external mortices. Gaidukevych (1958) follows Drachman's (1932: 97, fig. 32) reconstruction of the Semana weights suggesting the Crimean weights to be screw weights, but Christofle's (1930b) reconstruction of the Semana weight as a beam weight with the drum attached to the beam is to be preferred in the case of the Crimean weights also. The Crimean weights all have straight open mortices, and of the four in the catalogue three have an upper groove. The exception, the Chersonesos weight which lacks an upper groove (T5532) is also not only the earliest of the Crimean weights but the earliest of the whole Semana group of weights (Ts551, 552, 553). The significance of this weight and the other Crimean weights for the understanding of the history of the Semana weight has been discussed above.

The Crimean wineries are basically very similar to those from Israel, Egypt and North Africa, but inside buildings.

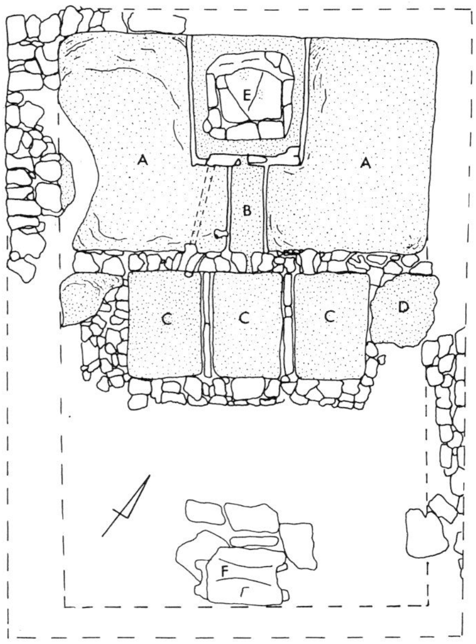

Figure 46. Triple winery with beam weight: Tiritake Inst. 07, Crimea (based on Gaidukevych 1958: fig. 68). A. Treading floors; B. Intermediate vat; C. Collecting vats; D. Fragment of raised working area; E. Press-bed; F. Press weight.

9C5. Wineries from the Greek Mainland, Mikonos, Delos, Crete and Cyprus

From the Greek mainland there is pre-industrial evidence for treading grapes in a wooden tub (Greece Photograph, T024) and also for must flowing from a treading floor directly into *pithoi* (Stavros). There is also archaeological evidence for direct must collection in *pithoi* from two sites (Athens Acropolis and Olynthos).

The picture in the Greek islands appears to be different. At Mikonos pre-industrial wineries have been published with treading floor and collecting vat and archaeological evidence for this type of winery comes from Delos, Dhioros in Cyprus and Knossos in Crete, from the Roman and Late Roman periods.

9C6. Wineries from Italy

The two main types of press from Roman Italy, the 'platform' lever and drum press of the region of Pompei and the 'circular bed' presses were both used as oil presses and in wineries.

In some and perhaps all the wineries associated

with platform presses, the must flowed through small collecting vats and then on through pipes to *pithoi* that stood in rows in an adjoining wine store (T985).

In the 'circular bed' presses, this was apparently not the case. At Sette Finestre 02 the oil press and winery were on the second storey and there was an opening in the centre of the treading floor through which the must apparently flowed directly into *pithoi* placed below.

The installation at Posta Crusta Foggia 02, although considered by the excavator to be an oil press, was probably a winery. The must from a treading floor (room 15) and from a press (room 17) flowed into a collecting vat placed between them (room 16). From there the must was then apparently transferred by hand to the *pithoi* in the adjacent room (18).

A unique element found at Posta Crusta Foggia and at three other Italian wineries is a round raised platform placed in the centre of the treading floor (T84). The grape skins and stalks were clearly placed on this platform during the final stages of the treading process. After the first treading, when the treading floor is full of must that has not yet separated from the grape skins, treading is very difficult. Placing the grape skins on the platform would avoid this difficulty. It is surprising that this improvement remained localized in central and southern Italy.

9C7. Wineries from Yugoslavia

The remarkable winery of Verige Bay is the largest in the catalogue (T985). An L-shaped treading floor 124 m^2 in extent drains through a channel and short pipe to a collecting vat 16 m^2 in area (18 m^3? in volume) and a short pipe connects the collecting vat to the wine store, which is 224 m^2 in extent. A very small winery was also recorded at Banjole.

9C8. Wineries from France and Spain

In France a number of wineries have been published but no two are exactly alike. The one at Cadillac is a simple very small winery—treading floor 2.8 m^2; collecting vat 1.25 m^3. At Allas la Mines there are two installations, near each other. The larger is a winery with a treading floor 8.9 m^2 in area and a collecting vat 3.4 m^3 in volume. The smaller installation has been reconstructed as a press. At Lambesc an installation published as an oil press is possibly a winery—treading floor 8.9 m^2; collecting vat 3 m^3. At Donzere a large winery was uncovered in which six rows of 34 (total 224) *pithoi* were reconstructed. It was suggested that there were two treading floors, one at either end of the wine store and on either side of each a collecting vat. Next to one of these treading installations and probably also next to the other there were also beam presses. The plan suggests strongly that the vats were connected to the store by pipes (T985), but no mention is made of these in the short report. This winery is without doubt that most similar to the *calcatorium* described by Palladius so far discovered, a treading floor with collecting vats on either side leading to *pithoi*.

At La Roquebrussanne Le Grand Loou I there is a winery—treading floor 17 m^2 collecting vat 3.2 m^2. Next to it is a platform with two lever presses connected to two collecting vats each 4.6 m^3 in volume. It is suggested that the two installations are part of one winery. It is more probable, however, that the presses were oil presses.

Recently a simple winery similar to those found in the Levant has been excavated at Benimaquia-Denia in Spain, in a context dated to the seventh to sixth centuries BCE and showing Phoenician influence.

9C9. Wineries from Germany

Three simple wineries from the Roman period were published from the Mosel region of Germany: from Losnich (treading floor 27 m^2; collecting vat 2.6 m^3); from Neumage (treading floor 8.8m^2; collecting vat 2.4 m^3); and from Maring Noviand (treading floor 8.4 m^2; collecting vat 1.33 m^3). One more complex improved winery was excavated at Piesport Trier, which was a twin winery. In each winery there was a treading floor, 20 m^2, an intermediate vat, 5 m^2, and a collecting vat, 2.8 m^2. A weight of the Semana group (T5522) provides evidence for a beam press.

9D. Improved Wineries: Conclusions

It has been suggested above that the pictorial and archaeological evidence from pre-Roman periods point to two basically different methods of producing wine—the first collection and fermentation in large collecting vats, a system usual in countries south of the Mediterranean; and the second collection and fermentation directly in *pithoi*, the method practised to the north of the Mediterranean.

The wineries from countries to the south of the Mediterranean from Hellenistic and later periods continued the earlier methods. The collecting vats from Israel, North Africa and Syria are large, often reaching 5 m^3 or even double that volume and there is no evidence for collection in *pithoi*. The wineries from the Crimea are of the same type, although those are in buildings and not in the open, while those from the Greek islands, both ancient and pre-industrial, present a similar picture. As regards wine production in countries to the north of the Mediterranean in later periods, the data presented in the present chapter have shown that the picture is not so clearly defined.

The pictorial representations from the Roman Empire fit the model. In almost all of these, the must is shown as flowing into *pithoi*. Similarly, in the ancient and pre-industrial wineries of the Greek mainland, the few data available also point to a tradition of collection into *pithoi*.

The winery described by Palladius would, at first glance, appear to incorporate elements from both techniques, the must flowing first through collecting vats, *lacus*, and then into *pithoi*. The collecting vats of Palladius, however, were in actual fact intermediary, serving as settling tanks so that clear must would flow on into the *pithoi*.

The wineries of Italy and Yugoslavia (e.g. Verige Bay) and that from Donzere in France accord almost exactly to that described by Palladius, by which the must flowed through a collecting vat and then through pipes into *pithoi* in a separate room. The winery closest in plan to that of Palladius, in that the treading floor is placed between two collecting/intermediary vats, is that from Donzere in southern France.

In other wineries in France and Germany, however, the must was collected in vats without flowing on into *dolia*. The collecting vats were much smaller than those in the Levant, reaching a maximum of 3 m^3 as opposed to 5 m^3–10 m^3 in the east, but nevertheless they are certainly similar to the smaller wineries of the east and it is very probable that in these also the must fermented in the vat. The question is whether this technique developed in this region or where it was introduced from the east, as is hinted at perhaps by the recent finds from Benimaquia in Spain where a simple winery was found at a Phoenician site dated to the seventh to sixth centuries BCE.

Another aspect worthy of attention is the lion-head spout, which appears on so many of the Roman pictorial depictions. Archaeological evidence for this element has apparently been found so far only in Egypt, and the earliest pictorial representation from the Hellenistic period is also Egyptian. It is most unlikely, however, that the standard motive on Roman sarcophagus reliefs originated in Egypt. Vitruvius (3.5.13) refers to lion-head spouts as a standard element in temple architecture, and the reality on which the bath-like treading vat with lion-head spout was based must surely be sought in Roman Italy.

Note

1. After the text catalogues and lists of this book were closed, two wineries almost identical to those from the Negev highlands were excavated by Y. Israel at a site 5 km north of ancient Ashqelon, some 60 km north-west of those of the Negev highlands.

Chapter 10

Pre-Industrial Installations (Ts01–02)

10A. Presses for Purposes other than Oil and Wine Production

Crushing and pressing, the two main techniques used in producing wine and oil, were used also in the processing of other foods and in various crafts, and in the catalogue examples of such presses have been included. The different functions often effected the form of the apparatus used.

10A1. Clothes Presses

Presses were used to squeeze out the water after laundering and at the same time to smooth the clothes as is done today by a clothes iron. One of the two examples of T0145, the press with two rotating screws with handles at the lower end, is a clothes press from a wall painting at Pompei 05. This press is a variation on the most common screw press type—that with rotating screw and handles at the lower end (T0131). The use of two screws enabled a larger area of cloth to be pressed without folds.

10A2. Cider Making

The equipment for cider making is very similar to that used for making olive oil and includes a crusher and a press. The rotary crusher turned by a horse was used in Normandy, France, western England and Wales until recently. The difference between this crusher and the one used for olives is that in the apple crusher the crushing wheel ran in a narrow channel around the crushing basin. This was apparently because the apple pulp is more watery than the olive mush, therefore needing the channel to ladle the pulp out. This variant has not been numbered separately in the catalogue. The presses used in cider making included many of the types used for olives and grapes, no types being specific to the process (Bowyer; French 1982; Quinion 1982; Williams-Davies 1984; Worlidge 1689).

10A3. Cheese Presses

Some cheese presses are simple lever and weights presses, for example, Granvas Lozére, France (T01113). However, several other types are unique to cheese making and are alike in that the benefits of mechanical advantage are exploited only in raising the heavy weight which was then released to exert direct pressure on the cheese. In the screw and weight press (T0137), a screw fixed to the weight is used to raise it and it was then released; in the screw and drum press (T0165), a drum raises a box of stones which is then released; and in T01112 a lever raises a hinged board on which a heavy stone is placed. The latter is in theory a lever and weights press, but the hinged board is so short that it provides next to no mechanical advantage and it is the direct weight of the stone that does the pressing.

The special character of the cheese presses is because a long continuous period of pressing is needed. Therefore all the various types of direct-screw presses are unsuitable. The various types of lever and screw press would be suitable, but are clearly much too large, and even a lever and weights press takes up much room, which is why these unique direct-pressure weight presses were developed.

10A4. Printing Presses

The early printing presses were all single rotating-screw presses with handles at the lower end (T0131), which is the most universal of the press types. It is also the ideal press for this purpose, allowing great pressure in a short time. Printing presses have not been included in the catalogue.

10B. The 'Mystic Press'

A very important source of information on mediaeval presses from the Christian world are the representations of the 'Mystic Press' in manuscript illustrations and sculpture (sites: 28-901; 29-901; 29-902; 30-901; 23-0-0649-00-001). The Mystic Press represents Jesus in a wine press and is a concept connected to the wine of the Eucharist and to numerous biblical references (Alexandre Bidon 1990; Isa. 5.1-7; 27.2-9; 63.2-3; Pss. 75; 80; Jn 15.1-2; Rev. 14.19-20; 19.15; etc.) It again is based on the misunderstanding of the biblical term *gt*, usually translated as 'wine press', 'pressoir'. This rich repertoire of pictorial representations of mediaeval presses provides an invaluable chronological link between the archaeological remains and the pre-industrial presses.

10C. Pre-Industrial Presses: A Regional Survey

Only some aspects of the subject of pre-industrial installations will be touched upon in this chapter. The stress will be laid on presses and on those regions from which enough data is available to allow for attempts at spatial or chronological analysis.

10C1. Pre-Industrial Presses from Israel and Environs

Twenty-eight pre-industrial presses from Israel and environs appear in the catalogue. Ten of these are Arginunta lever and screw presses (Model 6H), all of type T012161, lacking central piers, with beam consisting of one tree trunk, and with Samaria weights (T621). These were often re-used ancient weights, but those specifically made for these presses were apparently usually of the small Iksal type (T62136), rectangular with rounded corners and with open mortices. Of these presses, seven were from the centre of the country and three from the Carmel and Lower Galilee. The gap between these two areas is probably a result of lack of data, and presumably the same type was also in use in the intervening region.

Six single rotating-screw presses with handles at the lower end were recorded in Israel and environs (Model 7A1; T0131). Four of these were found in western and Upper Galilee, one in Lower Galilee and one in the southern coastal plain.

Three presses with two fixed screws were recorded (Model 7B; T0141), all in the extreme north-east of the country.

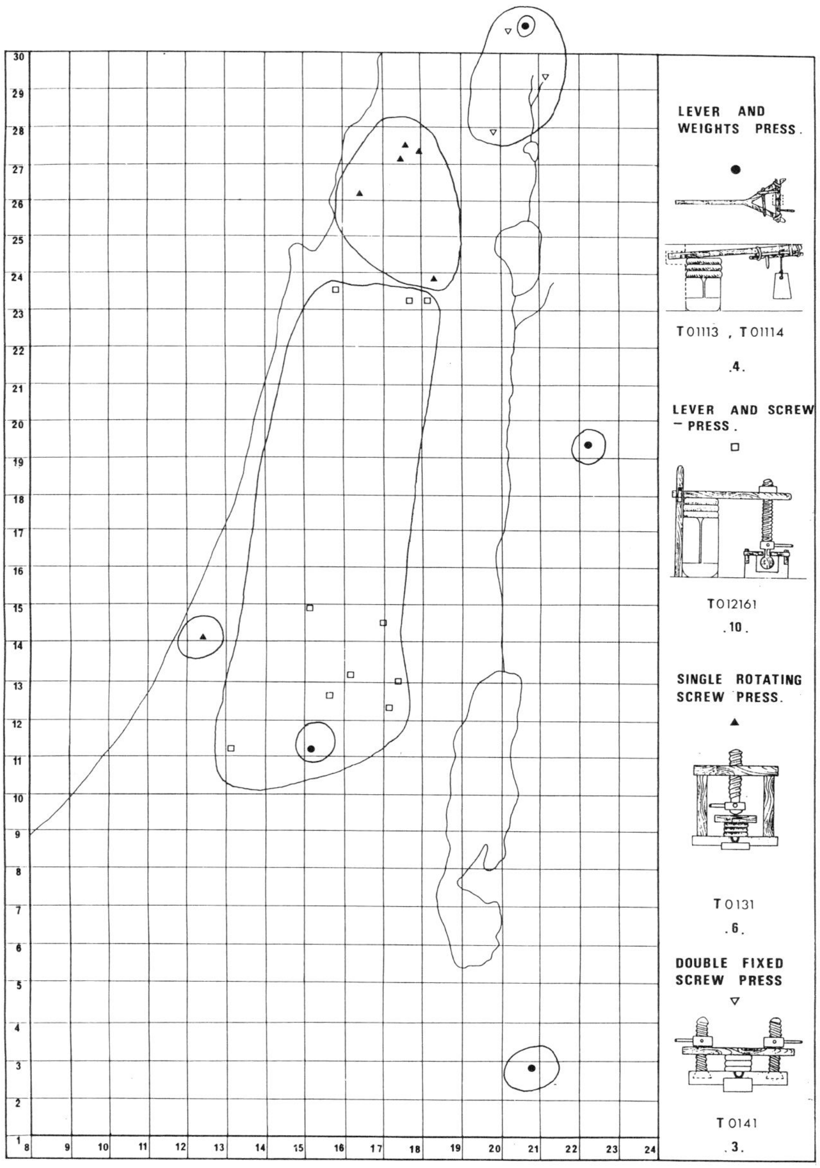

Map 41. Pre-industrial presses (Ts01).

Six lever and weights presses were recorded (Ts01110, 01113, 01114), one from Judaea, one from Samaria, two from Jordan, one from the Golan and one from southern Lebanon, to which can be added a second from Kfar Hai. In four of these, the weight was raised by a drum attached below or above the beam. It is significant that all three of the main types of beam weight used in presses in this region in late antiquity appear also in the pre-industrial presses—that with horizontal bore (Ts5122, 5123) at Bet Guvrin and ʿAjlun, that with a reversed-T bore (T5311) at Ṭafila, and that with metal hook at Kfar Hay (T541). Remarkably, there is also pre-industrial evidence for central oil collection from Ṭafila.

Thus all the ancient presses used in the country continued to be used side by side well into the twentieth century. The degree to which they had changed was remarkably small. The ancient lever and screw presses in Israel usually lacked central piers. Similarly, the pre-industrial presses also lacked such piers, as opposed to the European presses of this type in which these piers usually appeared. Pre-industrial beam weights and screw weights were both similar to the ancient types. Because of this great degree of continuity, it can be presumed that in ancient lever and weights presses the weights were also raised, as they were in pre-industrial presses, using a drum attached to the beam.

The pattern of spatial distribution of these presses manifests clear connections to the ancient patterns. The lever and screw press remained the dominant press in Samaria, having apparently spread further north and only one of the various types of ancient screw weights having survived. The direct-screw presses of the north continue those that were found in eastern Upper Galilee and perhaps also the type that was introduced into western Galilee in the Middle Ages, the Manot press (T7444). In the ancient presses the single- and double-screw presses could not be distinguished with certainty. However, the pre-industrial double-screw presses suggest that the screw press bases with closed mortices (Ts741, 743, 744) were perhaps double-screw presses.

The lever and weights press, the simplest of these presses, was not concentrated in one region.

10C2. Pre-Industrial Presses from North Africa

In North Africa two types of lever presses are found in different regions. The lever and weights press in which the drum is attached, not to the beam as in the Levant, but to the weight (T0112) is found in Algeria and Tunisia. This is clearly basically the same press as was used throughout ancient North Africa (T40------4) with the typical Semana weight (T55121). In Morocco the pre-industrial press was the Fez press (T01218050), a lever and screw press with perforated weight. As in Israel, the spatial distribution of the types of pre-industrial presses reflects the situation in late antiquity—in North Africa lever and weights presses in the east and lever and screw presses in the west.

The sophisticated pre-industrial oil separator using simultaneous overflow and underflow decantation (T0194) is clearly a continuation of the Madaure separator (T47132). Similar pre-industrial separators are used in southern Spain (Ts0191, 0192) and Ibiza (T0193).

10C3. Pre-Industrial Presses from Europe

In Europe the variety of pre-industrial presses is usually greater and their distribution patterns much more complex than is the case in Israel and North Africa. This is demonstrated by three surveys that have been published, two of wine presses, one from Hungary and one from a small part of France, and one of cider presses from western England.

A detailed survey of the pre-industrial wine presses of Hungary was published by Vincze (1959). It included 114 presses divided into four main types. (The definition of the types was, as is to be expected, not identical to that used in this book and therefore only those of which illustrations appear have been included in our catalogue.) The diffusion map (Vincze 1959: fig. 32) evinces extremely interesting spatial patterns. The two main types were: (A) beam presses—mainly but not only lever and screw presses of various sorts (42 examples); and (C) the single rotating screw with handle at the lower end (37 examples). These two types were found in all parts of the country, but in each case there were small areas where this was the only type found. The other two types were: (B) the press with two fixed screws (8 examples); and (D) the press with one rotating screw and handles at the upper end (27 examples). These two types were found each in one limited area, but in both cases they were not the only type in that area. This spatial pattern developed apparently mainly during the last one or two centuries. It is probably to be explained in that the latter two types arrived on the scene later than the other two.

Humbel (1975: 109, fig. 20) has published an interesting map of the types of presses found in the wine-producing regions of the Saone valley in France. In the southern region, the Beaujolais, the presses are mainly single rotating-screw presses (Model 7A1; T0131); in the central region, the Maconnais, the presses were mainly 'Grand Point' lever and screw presses (Model 6A; Ts012111); and in the northern region, the Chalonnais, they were of the 'Taissons' lever and screw type (Model 6D; T012120).

Preliminary notes on presses used in cider making

in Great Britain published by Quinion (1982) hint at both chronological change and regional diversity. He states that there was a development from presses with one rotating screw (T0131), using a wooden screw, to that with two fixed screws (T0141) using metal screws, while in the south-west, in Devon, the press used was the type with one fixed screw (T160). In English seventeenth- and eighteenth-century manuals more unusual presses appear: that with two rotating anchored screws (T01441; Hero of Alexandria's press C); and the Fayum press (T0142), that with two rotating screws with handles at the upper end and the female thread in the press-bed. It is very difficult to explain their appearance in England.

The numbers of examples in the catalogue for other regions do not allow for similar detailed spatial analysis. Attempts will therefore be made to determine only wider patterns. More detailed data, however, could well show these to be inaccurate.

The press that was in use in the greatest numbers in Europe and throughout the Mediterranean was the single rotating-screw press (T0131; Model 7A1) and it is found in almost every country from which pre-industrial data is in the catalogue. For this reason, without the aid of dated evidence ancient and later, it is unlikely that it will be possible to construct a diffusion model of this type or to try to put forward an argued hypothesis as to its history. The paucity of ancient remains is presumably to be explained by the fact that many of these presses were completely of wood.

Several types of pre-industrial lever and drum presses are known. In two of these the drum lifts a weight. In the Isfahan press the weight is lifted by a pulley and the drum remains connected to the weight throughout the pressing process (T0113). In the type found in the islands of Ibiza and Corsica, and appearing in an eighteenth-century English manual, the weight is attached to the beam and the drum raises both together. During pressing the drum is disconnected (T0114).

There are, however, two types of pre-industrial lever and drum press that function without a weight. In one found also in Corsica, the drum is attached to the beam and the rope to the ground (T0116). In the only type found on the European mainland, the drum is attached to the ground and the rope to the beam (T0115). Five examples are found spread over a considerable region in north and central France and this type is known there as the 'Casse Coue' press. These are very similar to the presses that Cato described, to Pliny's press A (see above, pp. 83-84) and to those found in the Roman platform presses of Campania (T4091-9--9). It is of great interest that in at least two of the French presses, Saint Lauren de la Plaine Layon and Cheillé, the beam is anchored in a single slotted standard as it was in the Campanian presses. There is therefore every reason to see in the French presses a direct continuation of the Roman presses.

Photo 91. Pre-industrial lever and drum press—'Casse Coue' (T0115): Saint Lauren de la Plaine, France (Humbel 1976: Pl. LXIV).

The 'Grand Point' press, the lever and screw press in which the screw is anchored in the press frame (Model 6A; T01211) is found over a large area in north France, but also in Hungary which suggests that it will be found in other regions as well. Manuscript illustrations provide evidence for its existence at least from the eleventh century CE. This press would, however, not leave identifying archaeological remains so that we have no tools with which to reconstruct its earlier history.

The 'Taissons' press, the lever and screw press in which the screw is anchored between two posts fixed in the ground (Model 6D; T012120) has apparently a much more limited distribution area in north-east France. There is possibly one example from Hungary, but nevertheless the history of this type is apparently shorter, although it is perhaps related to the other lever and screw press fixed to the ground—the fixed-block press' (Model 6E; T012122).

The fact that two types of lever and screw press found in France, the 'Grand Point' and the 'Taissons' presses, lack screw weights while lever and screw presses in other parts of the Mediterranean Basin always have weights is almost certainly to be explained in that the French presses developed directly from the Roman lever and drum press, Cato's press and Pliny's first press, which also functioned without a weight.

This also explains why lever and screw presses in

abroad as the presence of the unusual weight (T62441101), guider mortices (T441) and crushing basin (T343) suggest (Frankel, Patrich and Tsafrir 1990). It would seem, therefore, that Judaea did not go through the stage of the lever and screw press, but passed directly from the Maresha lever and weights press to the direct-pressure screw press. The grooved-pier press, the screw press typical of Judaea (T711), developed directly from the Maresha press. In both, there are two piers on either side of a central collecting vat. In the first case a screw board and screw were probably fitted directly on the piers of existing Maresha presses. The distribution area of the standard grooved-pier press (T711), that with perforated square mortice, is limited to Judaea only.

Thus the three presses—the simple Bet Mirsham lever and weights press, the improved Maresha lever and weights press and the grooved-pier direct-pressure screw press—are all in one line of development and found in one region, Judaea.

The other main screw press found in this region is the Ayalon single fixed-screw wine press, which in this region is equipped with the square mortice of the open sub-types (T81-3). This square open mortice is very similar to the square perforated mortice of the grooved-pier press, which justifies defining the square open mortice as the Judaean mortice. It is significant that very few examples of the many types of installations that use dovetail mortices (Ts041) are found in this region.

One other press that appears in small numbers in Judaea is the cross press (T72). It probably reached the region from slightly futher to the north, having the great advantage of having been cut in the walls of caves—of which Judaea had many.

11A2. Phoenicia: The Northern Culture

The northern Iron Age oil press, although as yet found only in smaller numbers, is also clearly defined and very different from the southern Bet Mirsham press. The oil collection is lateral, and the characteristic Rosh Zayit press-bed (T262211) is round free-standing with one or more circular grooves. The olives were usually crushed in a round mortar (Ts1152, 134) and not in a rectangular basin and the beam weight was a field stone with natural bore (T5110) and not fashioned. The small number of these presses are found in Galilee and as far south as Balata-Shechem in northern Samaria.

The characteristic northern improved lever and weights press, the Zabadi press, can be seen as a direct successor of the simple Iron Age Rosh Zayit press. The collection is lateral and the round press-bed very similar in character to the Rosh Zayit press-bed. The main development is the change in the method of anchoring the beam: the use of slotted piers (Ts. 421). In both examples excavated, there are also twin round collecting vats allowing for overflow oil decantation. The southern border of the distribution area of the slotted piers is very well defined and corresponds almost exactly to the border between Phoenicia and historical Upper Galilee as defined by Josephus (*War* 3.3.240; Avi-Yonah 1977: 133; Ilan 1984), a border that later became that between Phoenicia and Palaestina Secunda (Avi-Yonah 1977: Map 9). Thus the Zabadi press is clearly Phoenician. As pointed out above, the Phoenician Zabadi and the Judaean Maresha presses are different in almost every way—in the method of anchoring the beam, of collecting and separating the oil, and although there are piers in both presses these stand in different parts of the press and serve different functions.

The transition to the use of the screw in western and northern Upper Galilee is also very different from that in Judaea. If in Judaea the lever and screw press is very rare, and the one used is a direct-pressure press, in western Galilee the reverse is true—the lever and screw press is the main press and three different types of screw weight are found.

This difference is, paradoxically, to be explained by the similarity in the transition to the use of the screw in the two regions. In Phoenicia, as in Judaea, the press previously in use was converted to use with a screw. The Phoenician Zabadi presses with the almost monumental gate-like slotted piers could only be converted to lever and screw presses, as apparently actually happened at Karkara 01.

The Din^cila, Bet Ha-^cEmeq and Mi^cilya screw weights (Ts62211, 62511, 62711) are similar technically, all being based on a dovetail channel in their upper surface and all three are found in the same region as the slotted piers, Phoenicia. They are therefore not only Phoenician, but probably also derived one from the other or have a common origin.

As in Judaea, in western and Upper Galilee also, the mortice used in the single fixed-screw wine press is of similar type to mortices used in the region in other devices. The Ḥanita fixed-screw wine press mortice (T8311), the Bet Ha-^cEmeq screw weight and the Rama press base (T743) all use the closed dovetail mortice (T041). All these are Phoenician devices, making it very likely that the *coagmenta punicana* mentioned by Cato in his description of the oil press is the dovetail mortice (Cato 18.9; see Drachman 1932: 119, fig. 39; Sleeswyk 1980; and Basch 1981 for other suggestions). Thus the closed dovetail mortice is one of the many devices described by Hero of Alexandria (Hero, *Mechanica* 1.6; Drachman 1963: 103-104) that is of Phoenician origin. It is perhaps also of significance that many centuries earlier in the Iron Age

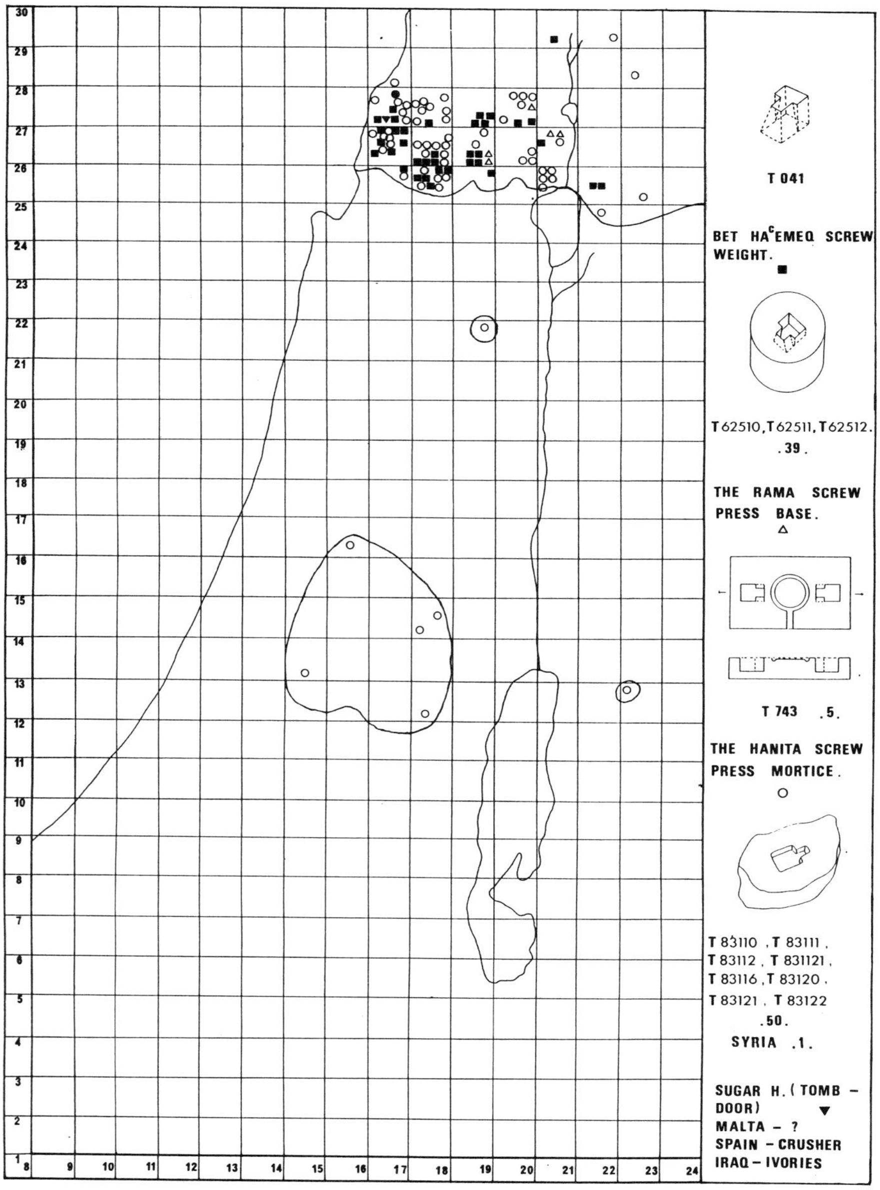

Map 42. Closed dovetail mortice (T041).

minute mortices of this type were used in the Levant to join ivories to their bases and Barnett suggested that this was a Phoenician device (Barnett 1957: 155 nos. C29, S362; 1982: 13).

The closed dovetail mortice appears in the catalogue on one device from Ecija in southern Spain and one from San Paul Milqi in Malta and possibly one from Taourienne Leveau site 197 in Algeria. Meister (1793: fig. 1) apparently knew of it. These data are insufficient to attempt to reach conclusions as to the wider diffusion of this device. It is sufficient, however, to show that it has a more complex history which awaits to be unravelled.

The developed wineries found in Galilee as a whole are nearly all in the one-axis plan (Ts95–96). However, this is so obvious a plan that more evidence is needed from more northerly regions of Phoenicia before defining this also as Phoenician.

The two technological cultures of Judaea and Phoenicia, although different in almost every way, have certain basic characteristics in common. In each region there is at each stage one type of installation for each function (the different types of screw weights in Phoenicia are here regarded as of different technological stages), and those of each stage developed from that of the previous stage. These two cultures can both be described as closed integrated cultures showing great uniformity and internal development, and both are clearly connected to entities that are ethnic, cultural and political in character.

11A3. Samaria and the Sharon Coastal Plain

The quantity of data from Samaria in the catalogue is less than from other parts of the country, while from the Sharon, the coastal plain of Samaria, the quantity of data is fuller.

The Iron Age Bet Mirsham oil press of Judaea is found also in Samaria. In fact, the centres of the distribution area of the type as a whole and of the main sub-type—that with circular groove and bore (T261111)—were in southern Samaria. Samaria was for some of this period within the Northern Kingdom of Israel and for a part an Assyrian province. However, the fact that Samaria was separated from Judaea politically did not leave its mark in the spatial patterns of the oil presses.

As opposed to the uniformity of the improved lever presses in the two nuclear cultures of Judaea and Phoenicia, in Samaria and the Sharon these appear in a great variety of sub-types. Only one press, Tira, was recorded with simple central vat and plain piers as in the Maresha press. In the majority the oil collection is lateral, usually without plain piers and in several cases with the unusual Qedumim collecting vat with perforated lid found only in this region (T47114). Of those in which the collection is central, several are with radial and circular grooves (Ts4612, 042), a characteristic hardly found in Judaea but common further north. Two other types are found almost only in this region. One is the guider mortices found at two sites in Samaria and at Bet Loya in northern Judaea, but common in Syria and North Africa (T441). The other is the Arsuf press found always surrounded by a mosaic pavement (T935).

Several types of presses are also found in this region that are found in other regions also. However, the main types are specific to the region. The typical press is the lever and screw press with rotating screw and the Samaria weight (T62111), while that with fixed screw and the Luvim weight (T6261) has been found mainly in the Sharon and Mt Carmel.

The square mortice of the fixed-screw Ayalon wine press appears in two sub-types: in the Sharon coastal plain in the form that widens in two directions (T81-2), while the three examples from the mountain areas of Samaria widen in only one direction (T81-1). Screw press bases with both open and closed mortices have also both been recorded in the region in small numbers (Ts73, 74).

In spite of this great variety of types, it is nevertheless possible to define two characteristics as being typical of the region: the Samaria screw weight (T62111) and lateral collection.

Our knowledge of the history of Samaria shows that it was to some extent a separate region ethnically and culturally the land of the Samaritans. There was, however, always a close relationship with the adjoining regions of Judaea and Galilee.

11A4. The Carmel, Lower Galilee and Eastern Upper Galilee

The examples of the Iron Age Rosh Zayit oil press in the catalogue are all from these regions. These have been shown to be the predecessors of the later Phoenician presses which were limited to western Galilee, to the north-east of the regions discussed in this section.

In the Carmel, Lower Galilee and eastern Upper Galilee a great variety of presses are also found. The majority of the lever presses have central collection and, of these, the majority have radial and circular grooves. Examples of presses with plain piers were found in all three types, in the one press with lateral collection from Yodfat, and in presses with central collection both with and without radial and circular grooves.

The main characteristic of the presses in Lower and eastern Upper Galilee that distinguished these regions from Samaria and the Carmel to the south and Phoenicia to the north is the rarity of the screw weights and therefore of the lever and screw press. The direct-pressure screw press, however, appears in a greater variety of forms than in any of the other regions. The Ṣippori press bases with closed mortices (T741) and the ʿEin Nashut press bases with open mortices (T731) are types found in other regions also. However, the Rama press with two closed dovetail mortices (T743) and the Mishkena press with rectangular mortices (T742) are found in this region only. The various types of the grooved-pier direct-pressure screw press with dovetail mortices (Ts713, 7132, 7133) are found in this region only but are clearly sub-types of the Judaean grooved-pier press (T711).

The mortices in the single fixed-screw wine press are of different types in the three areas that have been grouped together in this section. In Lower Galilee the vast majority are square Ayalon mortices, widening in one direction. On the Carmel several types of mortices are found, while in eastern Upper Galilee the mortice is of the Phoenician Ḥanita type with closed dovetail mortices.

One device found only in Lower Galilee is the crushing roller used in wineries—on the Carmel the slotted roller (T862) and in eastern Lower Galilee the plain roller (T861).

Thus the presses of the Carmel, Lower Galilee and eastern Upper Galilee evince connections with Phoenicia, Samaria and Judaea. The main connection with Phoenicia is the dovetail mortices. These are most marked in eastern Upper Galilee where the Ḥanita and Rama presses are found, but also appear in the Galilean grooved-pier presses found in Lower Galilee (Ts713).

The connections with Samaria and the Sharon are felt mainly on the Carmel where the Luvim screw weight is found, but the central vat with radial grooves

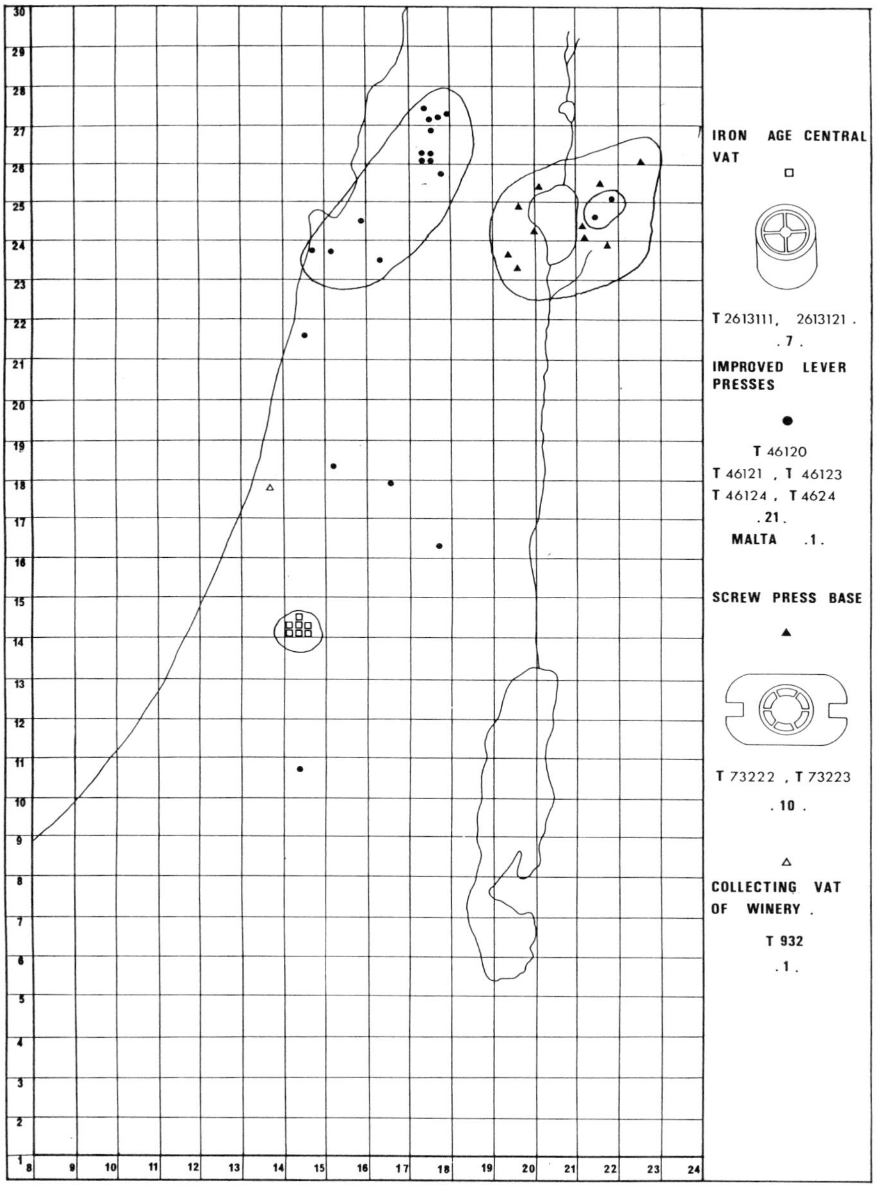

Map 43. Central vats with radial grooves (T042).

(T042) and the Ayalon mortice widening in one direction also connect Lower Galilee to Samaria and the Sharon.

The connections to Judaea are dominant, however. The percentage of central vats is greater than in Samaria: the grooved-pier press (Ts71) is a clear link, while the most important element is the lack of lever and screw presses and the dominance of the direct-pressure screw press. The latter two factors are true of Galilee only and not of the Carmel.

Thus, while Mt Carmel has much in common with Galilee, the factors linking these regions to Samaria are stronger on the Carmel and those linking them with Phoenicia and particularly to Judaea are dominant in Galilee.

This special connection between Galilee and Judaea, in spite of Samaria's separating the two regions geographically, without doubt reflects the historical situation in which Galilee and Judaea were both Jewish while Samaria was Samaritan. This situation existed throughout the Roman period, but probably had roots in earlier periods (Frankel 1992b: 893-94).

The most marked aspect of this special connection is the fact that both in Judaea and Galilee the direct-pressure screw press was used to the almost complete exclusion of the lever and screw press, whereas the latter was the main press of Samaria and of Phoenicia. What demands an explanation, however is the fact that the direct-pressure screw press is of different character in Judaea and Galilee. In Judaea there is one type only, while in Lower Galilee there are six or seven different types of this press. It can only be suggested that Lower Galilee was both farther from the uniform nucleus of the distribution area of the grooved-pier press in Judaea, and also in a region more open to outside influences, two factors that made it easier to desert orthodox methods and thus allow for experimentation and the development of a great variety of press types.

11A5. The Golan Heights

The picture in the Golan Heights is very similar to that in Lower Galilee, the main difference being that there is less variety in the types found.

Two lever and weights presses have been excavated: that at Gamla with central collection and that at Givᶜat Hayeᶜur with lateral collection, while many uniform beam weights with reversed-T bore have also been found (Ts313).

Screw weights are found in the region of Mt Hermon in the north, which was apparently part of the Phoenician sphere of influence but in the south, in the Golan proper, screw weights do not appear, only screw press bases. These are the very uniform ᶜEin Nashut bases with open mortices (T732), while the more sophisticated Tabgha and Weradim sub-types were found around the Sea of Galilee (Ts73222, 73223). The preference for the open mortices and rounded corners is surely a result of the basalt rock of this region. The preference for direct-pressure screw presses and the lack of lever and screw presses links this region to Galilee and Judaea.

11A6. The Jerusalem Area

Jerusalem is exceptional in the great variety of types of installations found in the city and the region around it.

Certain types, of which considerable numbers have been recorded, are found mainly in this region or centre upon it. Two examples are the Kasfa screw weight (T624) and the cross press (T72). Other types of which one or very few examples have been recorded are also found only in this region. The ᶜEin el-Judeida screw weight (T6232402; one example), the Qat screw mortice (T8163; three examples in the region) and the Tur mortice, the only closed dovetail wine press screw mortice with channel (T831121), all come under this category. Other types are found both in other regions and in the Jerusalem area. These include: the Samaria screw weight (T621111): press bases (Ts731 and 741); wineries with small apartments (T8011); auxiliary treading floors (T803); and Ayalon screw mortices (Ts8123, 81301).

This phenomenon can certainly be partly explained by the intensity of the archaeological research carried out by institutions located in Jerusalem, the Department of Antiquities, the Hebrew University, and the many institutes of Archaeology and Biblical Studies from other countries located in Jerusalem. Many excavations and surveys have, as a result, been carried out in the Jerusalem area and have enriched our catalogue. And yet, it would seem that this explanation is not sufficient. The explanation is to be sought in the unique character of the city. It is not only the capital of the country and on the border between different regions, but was also a site for pilgrimage and a Holy City which drew people from many lands to come to settle in it and around it. Many of the installations enumerated above were found in monasteries and almost certainly were brought to the region by foreign monks who brought new techniques from their home countries.

11A7. Israel and Environs: Summary

The history of the ancient wine and oil technologies of Israel can be summed up as consisting of two primary closed integrated technical cultures, Judaean and Phoenician.

In early antiquity—the Iron Age—the Phoenician sphere of influence included all Galilee and the Judaean included Samaria.

In later antiquity the manifestations of the Phoenician culture were evident primarily in western Galilee, while in other regions the spatial pattern modified and Samaria, Galilee and Golan developed local sub-cultures. These, while being closely connected to each other, were connected primarily to Judaea. Galilee and Golan, however, were technically closer to Judaea than was Samaria and also closer to Judaea than to Samaria.

This complicated mosaic reflects the political and cultural relations in the region, in which a Samarian culture that was basically very close to that of the Jews was wedged between two Jewish areas, Judaea to the south and Galilee and Golan to the north.

It should be stressed that these regional differences continued into periods when the country had become largely Christian and even later when it had become largely Muslim.

The special character of Jerusalem as a focus for foreign influence affected its technical culture throughout these periods, resulting in an eclectic culture in many ways the reverse of the integrated technical cultures of Judaea and Phoenicia.

11B. Oil and Wine Presses in the Mediterranean Countries

In this section the discussion will be limited to oil and wine presses and oil separators. These are the only subjects that demanded an attempt at examination in a broader perspective. As regards olive crushers and wineries, it was felt that there was no need to add to what has already been written in the relevant chapters above.

11B1. Beam-Anchoring Devices

The primary characteristic that distinguishes the different lever presses is the methods in which the beam is anchored. These methods can be divided into three main categories: anchoring in a niche; anchoring in a standard; and anchoring between slotted or perforated piers.

Anchoring the beam in a niche in the wall of a building or in a rock face was the method used in all the early presses from the Bronze Age and Iron Age in the Levant and apparently in presses of all periods in the Aegean. This remained the main method used in southern Israel throughout the later periods and was still that used there in pre-industrial lever and weight and lever and screw presses. At this stage of research, it is not clear whether all these presses are connected and have a common origin. This is possible but not certain, and if this is so the time and place of origin have certainly not been determined.

A standard is in effect a free-standing niche. This is demonstrated by the one example of an anchoring standard found in an Iron Age press in Israel at Gezer (T2112), which remained an isolated example and did not develop further. Many of the beam niches of North Syria (presses T404; niches Ts413, 414) are of this type, as are the Cypriot monoliths (T427), the Campanian anchoring standards (presses T4091; shafts Ts439) and the North African dovetail beam niche (press T4032; mortice T416). The slotted niche (T422) should perhaps also be included in this group and will be returned to below. This method of anchoring the beam survives in several pre-industrial presses in France.

Although there is a possibility that some of these types are connected, it is improbable that they all were. The Cypriot and Syrian examples are not distant from each other geographically and apparently also not chronologically. They are, however, considerably different in form. The Syrian free-standing niches are often secured in a dovetail mortice, as are the North African examples, but the North African niche was of wood and the Syrian of stone, while the Campanian standard has a very unusual underground anchoring shaft.

In conclusion, it can be stated that the step from a niche to a free-standing niche or standard is certainly one that could and probably did occur independently at various places.

The third method is anchoring the beam between two slotted or perforated piers. This was the usual

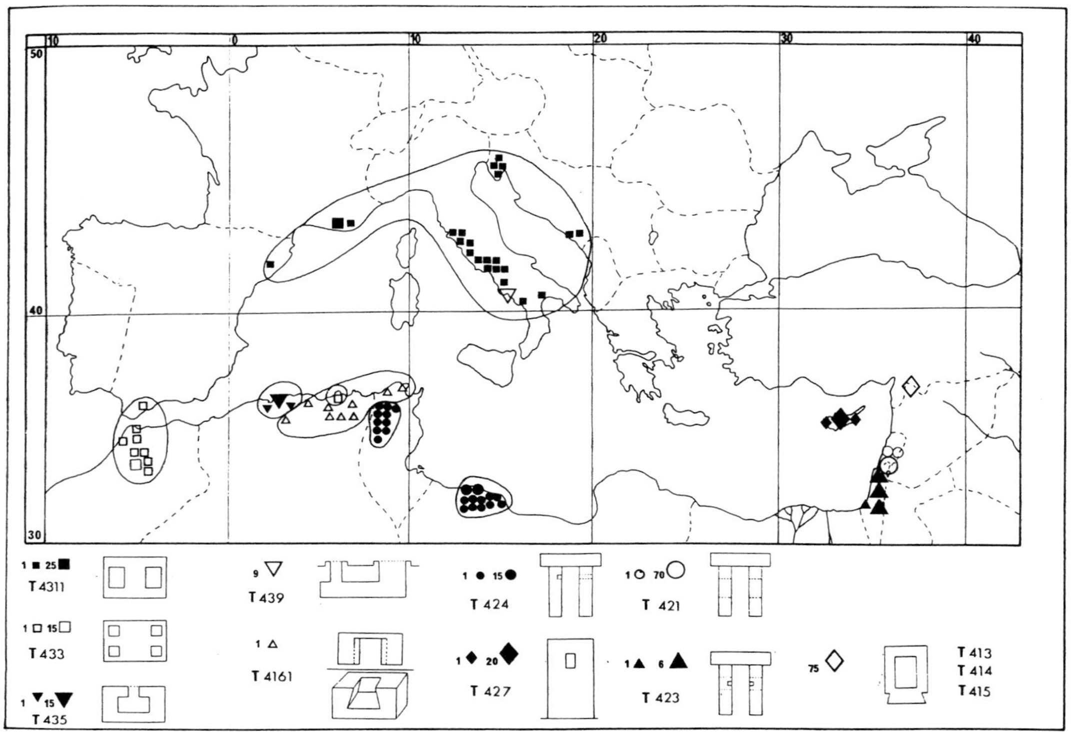

Map 44. Beam presses—anchoring devices: Mediterranean Basin.

method in late antiquity throughout North Africa, southern Europe and Phoenicia. Wooden piers of this type are very common in pre-industrial presses in southern Europe. The remains of the anchoring device from ancient presses are either stone bases for wooden piers or actual stone piers.

Throughout southern Europe a uniform Tivoli pier base is found, the rectangular pier base with two mortices (T4311; press T4051). Forty-nine examples of the standard type of this pier base have been recorded in Yugoslavia, Italy, France and Spain. It was not recorded outside southern Europe.

In North Africa, two types of pier base and two types of stone pier were recorded: in Morocco, a pier base with four mortices (T433; press T4061), found also in southern Spain; in the region of Caesarea in Algeria, a T-shaped pier base (T435; press T4071); in Tunisia, slotted stone piers (T4248, 4249; press T4022); and in Tunis and Libya perforated stone piers (T424; press 4032).

In Phoenicia slotted piers (T421; press T4022) and slotted niches (T422; press T4023) were recorded, and throughout Israel perforated piers (T423; press T4031).

In the case of the slotted piers, there is at our disposal clear dated historical evidence. The slotted piers are the *arbores* of Cato's press and the stone pier base is the *forum/pedicinus* of that press (Cato 18). Cato is clearly referring to the Tivoli pier base (T4311), showing it to have existed in Italy in the first half of the second century BCE. The geographical distribution of this pier base shows clearly that it is a Roman installation diffused by Roman influence.

The slotted and perforated pier, as opposed to the niche and standard, are almost certainly devices too complex to have been invented independently at different places; therefore the question that arises is: what is the connection between the slotted and perforated piers of North Africa, Phoenicia and southern Europe?

The slotted niche from Umm el-ʿAmad, Lebanon (T422), from the Hellenistic period is both the only installation that is a possible prototype of the slotted piers and also the earliest dated example of an anchoring installation with slots. This installation predates the Roman influence in the east and the slotted piers of western Galilee clearly represent the southern tip of a very large concentration of this type of installation. All this suggests that the slotted piers originated in the east. In North Africa they are found only in the Punic areas and they probably reached North Africa as a result of contacts between the Punic states and the Phoenician homeland which continued long after the political ties were cut. The slotted piers probably reached Rome from North Africa. The works of Mago of Carthage were greatly respected by Roman agronomists and the contacts between Rome and Carthage were not only ones of hostility.

11B2. Methods of Applying Force to the Press-Beam

11B2.1. Lever and Weights Press: Beam Weights. Simple beam weights with horizontal bore of various shapes (Ts511) have been found in Cyprus and Syria in the Late Bronze Age and in later contexts and in Israel in Iron Age and later contexts. In France they have been found in pre-Roman contexts and in the island of Chios in late Roman contexts. As with the simple niche, if there is any connection between the various appearances of this simple weight, there is at present no way of ascertaining or demonstrating it.

However, with the development of more sophisticated weights, two completely different types of weight emerged, each found in different regions. The weight with reversed-T bore (T531, 532) is found in Israel, Lebanon and Cyprus and, although it is rare in western Galilee, apparently not having penetrated to this area, can nevertheless be regarded as the Phoenician weight. The rectangular Semana weight with two external mortices (T55121) and all its sub-types is found in the Aegean, France, Spain, North Africa, Cyprus, Malta, Germany and the Crimea, and it has been suggested above that it is of Aegean or Anatolian origin. The Taqle weights found in North Syria (Ts535, 536) are a hybrid of these two types, hinting perhaps at where the distribution areas of these two types met.

Pre-industrial evidence demonstrates that these two types of weight also operated in different ways. The Phoenician weight was raised by a drum attached to the beam (Pre-ind. T01114) and the Aegean by a drum that was attached to the weight (T0112).

11B2.2. Lever and Drum Press. It is of significance that there is no archaeological evidence for the existence of lever and weights presses in Italy, which is confirmed by the written evidence, neither Cato or Pliny mentioning a press of this type.

Only one ancient lever and drum press from outside Europe appears in the catalogue in a small rock-cut installation from el-Ma Ougelmine in Algeria. Sixteen have been recorded from southern Europe, 10 of which are from Campanian 'platform' presses in Italy (T4091----9). The six others are in presses equipped with Roman Tivoli pier bases (T43111; presses Ts4051----5, 4051----6). Three of these are from Yugoslavia, one from Italy and two from France.

The lever and drum press is that described by Cato (18), and is Pliny's first press, and the fact that this type of press is found almost only in southern Europe and in two types of press that are both Roman culturally clearly also classes the lever and drum press as a Roman press.

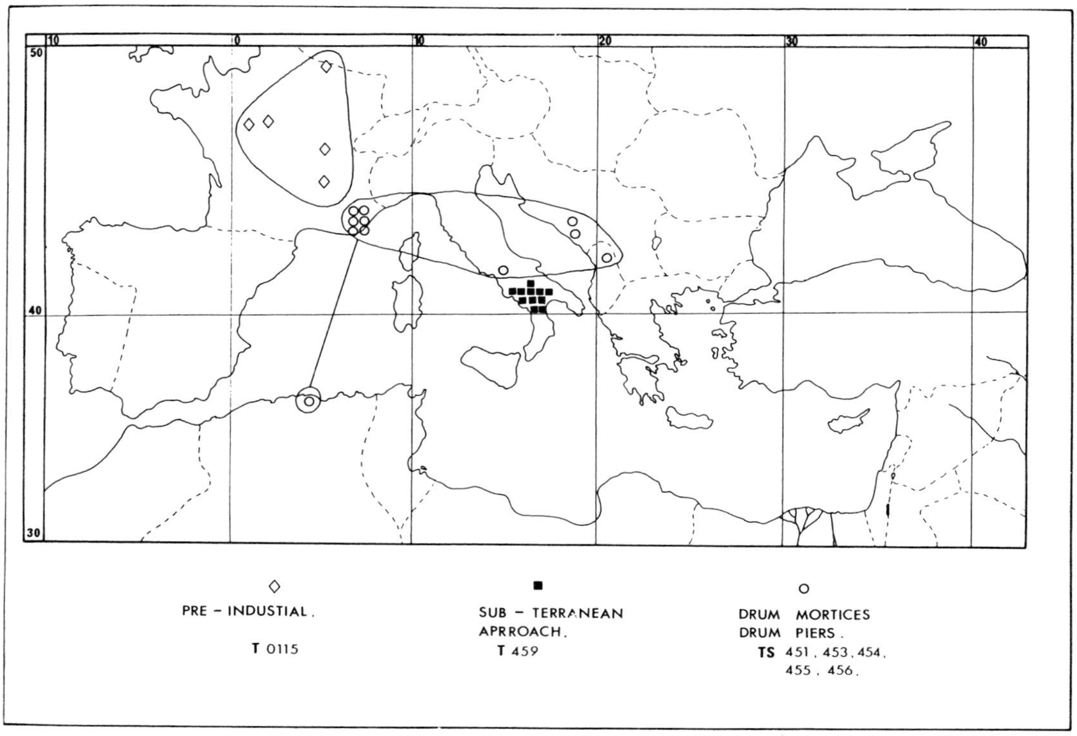

Map 45. Lever and drum presses.

The question that must be asked is: can its origins be traced? The Aegean Semana weight, to which a drum is attached in the first stage of pressing, before the weight is lifted from the ground, is in effect a drum base. There is one site, San Paul Milqi, Malta, where two large Semana weights have been joined and converted into a drum base (T5622). It is therefore very probable that the Roman lever and drum press derived from the Semana weight in one of its forms, from North Africa, from southern France, or from another region.

The French pre-industrial lever and drum presses (T0115) are of the type that was described by Cato and used in Italy in the Roman period and are almost certainly a direct continuation of this Roman press.

Hero of Alexandria's first press (*Mechanica* 3.13-14; see above) is a lever drum and weights press, in which the drum is fixed to the ground and a pulley attached to the beam-end. Drachman (1932: 67) regards it curious that Hero does not mention the lever and drum press. The explanation is that he wrote in the east and, although his presses are all clearly suggestions for improvements on existing installations, they are based on those that he knew, presses used in Egypt and the Levant. The question should therefore be not why he does not mention 'the simple lever and drum press', but from where did Hero's drum derive. Did he perhaps have knowledge of Roman presses, perhaps even that of Cato, and invented a press that combined elements from the Roman and eastern presses. There is also the possibility, however, that presses similar to Hero's first press did exist, because a pre-industrial press of this type has been recorded from Isfahan (T0113). A slightly more primitive pre-industrial lever and drum press has been recorded from the islands of Corsica and Ibiza and Great Britain. In this press the drum raises the beam with the weight attached to it, but is disconnected during pressing (T0114).

11B2.3. Lever and Screw Press. As shown above, there is little doubt that the Roman lever presses from Italy of the first centuries of the present era that lack evidence as to how they operated (Ts4051----?) are lever and screw presses of the type that Pliny (18.74.317; see above, pp. 86-87) described, in which a screw raised a box of stones (Pre-ind. Ts012162, 012163).

The pre-industrial evidence also shows that there were many types of lever and screw press that operated without screw weights. Nevertheless, the archaeological evidence for the lever and screw press consists primarily of screw weights, and therefore the main subject to be discussed here will be the various types of screw weight and the technological and historical relationships between them.

The most common screw weight throughout the Mediterranean is the cylindrical Samaria weight (T62111). It is found in central Israel, North Syria, Italy, France, Spain, Portugal and Cyprus. In Israel no weight was found that could be seen as a prototype for the Samaria weight, and it is clear that this type was introduced from abroad. It is basically very similar to the Semana beam weight (T5512). In both, there are external dovetail mortices and there can be little doubt that it was influenced by this weight. However, not only is the Samaria screw weight found in Italy in the greatest number of variants, all cylindrical in shape, and not only is this the only type found in that country, but among these variants are several that could represent stages in the development of the Samaria weight (Liverani 1987: fig. 1). Some have neither sockets nor mortices; some have mortices but lack a socket; and others have a socket but lack mortices. One of these, a weight with four mortices but no socket (T5541), was found in situ at Francolise Posto in a first/second-century context and probably represents a stage in the development of the Samaria weight. The present evidence suggests, therefore, that the Samaria screw weight is the Roman screw weight and from Italy reached other regions of the Mediterranean. When this occurred is not yet clear. In Israel no Samaria screw weights have been dated to before the Byzantine period. The weight from Camino de Pago in Spain has been dated to the Roman period.

The Kasfa weight (T624) is found in Spain, North Syria, Lebanon and Israel in a cylindrical form, in the Pontus in a rectangular form, and in southern France in both cylindrical and rectangular forms. In Israel it is found in small numbers and was clearly introduced to the country in its final form from another region. In North Syria this is the main type and it appears there in several variants. The Syrian Sarepta and the Israeli Kasfa weights are not identical, however, the latter lacking the rectangular depression that appears in all the Syrian weights. One weight from Israel, however, is of the Sarepta type and two others, those from Bet Loya and Duḥdaḥ, are extremely similar to the Syrian Seiḫ Barakat weight, showing close ties between the Syrian and Israeli weights. The rectangular example from the Pontus also has no depression and hints at a wider distribution area and a more complex history. At the present stage of research it can only be suggested that the Israeli Kasfa weight and the examples from southern France originated in the northern Levant or Anatolia.

However, as opposed to the Samaria and Kasfa weights that were clearly introduced into Israel from afar, all the evidence suggests that the three Phoenician weights, the socketed Dinᶜila weight (T62211) and the fixed-screw Miᶜilya (T62711) and Bet Ha-ᶜEmeq (T62511) weights all developed in the region. All three are related typologically: all are based on a dovetail channel and all are found in the same region. Furthermore, the Bet Ha-ᶜEmeq weight and the similar fixed-screw Luvim weight (T62611) are clearly both closely related to the single fixed-screw wine presses of the Ḥanita (T831) and Ayalon (T81) types respectively. All these six types, in addition to being technically very similar, are found in one region, the southern Levant, and only in this region, except for one example of the most developed type—the Dinᶜila weight, found in the Pontus, Turkey (T62222)—and pre-industrial examples of the latter that are found in Portugal.

The development apparently began with the fixed-screw wine press. This device involves an extremely simple use of the screw, is particularly suitable for use in simple rock-cut wineries (T111) and has the added advantages of easy transport and the possibility of being set up in existing installations. The socketless Bet Ha-ᶜEmeq and Luvim screw weights then developed from the Ḥanita and Ayalon mortices to which they are very similar. The socketed Dinᶜila weight then developed from the Bet Ha-ᶜEmeq or from the closely related Miᶜilya weight. The example of the Dinᶜila weight in the Pontus suggests that its distribution area stretched northwards. The fact that in the mountaineous area and closed areas of western Galilee the Dinᶜila is the only one of the Phoenician weights to be found suggests also that the lever and screw press penetrated to this region at a late stage.

The Dinᶜila and Kasfa/Sarepta weights are also very similar, the latter being in effect an improvement on the former. The open channels of the Dinᶜila weight are closed in the Kasfa and Sarepta weights, thus preventing the wedges that keep the screw-end in place from working free, as could possibly happen in the Dinᶜila weight. Both weights are found in the same region, the Levant, and it is therefore very probable that the Kasfa weight developed from the Dinᶜila weight.

In this case also, dated written sources are of great assistance. Hero of Alexandra's description of the screw weight to be used in his second press, the lever and screw press, corresponds exactly to the Dinᶜila weight (Hero, *Mechanica* 3.15; see above, pp. 87-88, 113). This shows that Hero knew this Phoenician weight and shows that it was in existence before the middle of the first century CE. This fits well with the dating of the weight from Sarepta to the Hellenistic/Roman period.

If our conclusions are correct, the socketed screw weights developed separately in two different areas and at slightly different times, the Roman Samaria weight in Italy and the Phoenician Miᶜilya, Bet Ha-ᶜEmeq and Dinᶜila weights and the Syrian Kasfa/Sarepta weight in the Levant. It is therefore possible that similar independent developments took place in other regions. The socketed rectangular screw weight with external mortices (T62112) could well have developed separately in the South of France and in the Aegean and Anatolia.

Another screw weight type is that with socket but without mortices in which the boards that secured the screw were held in position by clamps that fitted in the superior holes that appear on the majority of these weights (Ts6291). These are found in several countries, North Syria, the Pontus in Turkey, Italy and France. There is, however, no data on which to base a history of this type. Superficially it is similar to the pre-industrial perforated screw weight of Spain and North Africa (Ts6293, 6295). The manner in which the screw is attached to the weight differs, but it is possible that the pre-industrial weight developed from the ancient one.

The model suggested here, however, does not imply that the socketed screw weight was 'invented' at different times, nor that there was not contact between the different technical cultures. The process envisaged is one of new ideas and techniques being gradually integrated into existing technical cultures.

11B3. Oil Separation

In modern oil presses the oil is separated from the watery lees by use of centrifugal force. All ancient and pre-industrial methods were based on the fact that the lighter oil rose by force of gravity to float on the lees. Four different methods of oil separation are documented: skimming, overflow decantation, underflow decantation, and combined overflow and underflow decantation.

11B3.1. Skimming. The simplest method was to skim the floating oil off the lees using a ladle or similar utensil. Cato (66) and Columella (12.52.11, 12) both describe a similar routine of pouring the expressed liquid from one vessel to another. Columella speaks of a row of 30 vessels. Each time the liquid was allowed to stand and the oil skimmed off. Columella explains 'the more often it is aereated by being transferred from one pan to the other, and as it were, kept moving, the more transparent it becomes and freer from dregs'. The workman who separated the oil was known as a *capulator*. Cato speaks of him using a shell, *conca*, for skimming the oil, while Columella speaks of 'iron shells', *conchae ferrae*, ladles. In the oil press of the Roman period from Zabadi, a fragment of the sea shell *Tonna Galea* was found. This type is particularly suited to serve as a scoop. The whole shell could have contained 550 cc of liquid. Attached to many oil presses there were very large separating vats several cubic metres in capacity in which the expressed liquid was left in order to allow the oil to rise (T4821). At one site in Tunisia, Thuburbo Majus, and one in Italy, Francolise Posto, a small adjunct projected from the large vat in which there was a sump. The corner formed between the adjunct and the sump constituted a convenient working platform (T4822). Simple skimming-off of oil was the method used in a pre-industrial press at Fassuta.

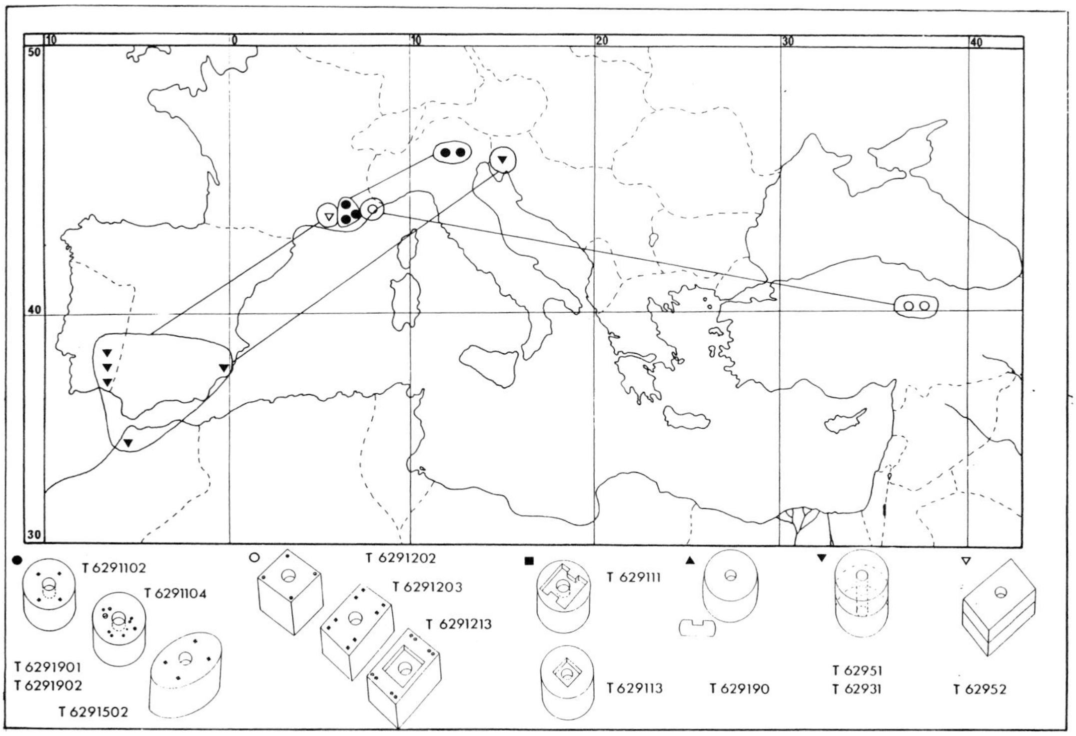

Map 46. Screw weights without mortices (Ts629).

11B3.2. Overflow Decantation. In overflow decantation an opening just below the rim of a vessel or vat allowed the floating oil to flow out, while the watery lees remained.

The pottery separators with a spout from Rasm Harbush (T182) from the Chalcolithic period worked on this principle, as did the single shallow stone Iron Age vat from Rosh Zayit and the twin round collecting vats of the Zabadi and Karkara presses from the Roman and Byzantine periods (T4722). Similar twin collecting vats, but square, were characteristic of the oil presses in Syria and found also in presses in Italy, France and Malta (T47121). At two sites in the Campania in Italy a round vat was divided in two apparently for this purpose (T4725). This is perhaps the *gemallar*, double (twin) container, that Columalla rejected preferring the more simple method entailing pouring and skimming (Columella 12.52.10). There are cases of three such collecting vats (T472302) and also of a whole series in a row connected one to the other. In some cases the expressed liquid flows to these large separating vats from the press-bed (T471312). At Barbariga in Yugoslavia there are two groups of vats of this type—one of five and one of eight. In other cases the separating vats are not connected directly to the press (T483). At Tipasa in Algeria there are seven square connected vats in a separate room, apparently without a channel or pipe connecting them to the press.

It has been suggested that an installation at Bettir 02 in North Syria is an oil separator working on this principle. It consists of a series of shallow stepped vats. The suggestion is that the lees would remain on the upper steps, and the oil flow to the bottom. It is not clear, however, how the lees or the oil would have been removed and the adjacent installation is probably a winery. It is not clear what purpose the stepped installation served.

The data suggests that overflow decantation was the usual method in Israel, Syria, Yugoslavia, Italy and North Africa.

11B3.3. Underflow Decantation. In underflow decantation the separated liquids were allowed to flow out in turn through an opening or spout at the bottom of the separating vessel or vat. Pottery separators using this principle are known from Cyprus and Crete in the Neolithic period and from Crete in the Early, Middle and Late Bronze Age, and Hellenistic period (T181). A pre-industrial separating vat from Methana on the Greek mainland also worked on the same principle (T0195). There is, however, one vessel from Vathypetro, Crete, from the Middle Bronze Age that used overflow decantation (T182). An interesting separating vat from Agrigento, Sicily (T485), which used underflow decantaticn was mistakenly reconstructed as an olive crusher.

A vessel with a spout at the bottom is not necessarily an olive separator, however, and could also serve other purposes, for example, as a small mobile crushing and extracting vessel for grapes or even for olives. It is very probable that vessels of this type found in Israel did not serve as oil separators.

11B3.4. Combined Overflow and Underflow Decantation. The fourth method is the most sophisticated, combining both overflow and underflow decantation. A vat is divided vertically into two unequal compartments that are connected at the bottom. The expressed liquid flows into the larger compartment and the opening joining the two compartments is left stoppered until the oil has separated. Then the lees are allowed to flow into the smaller compartment. Openings near the rim of both compartments allow the oil to flow out of one and the lees out of the other. There is only one clearly documented ancient example of this device at Madaure in Algeria (T47132), although there are hints that a similar separator existed at Ras el-Hammam Libya.

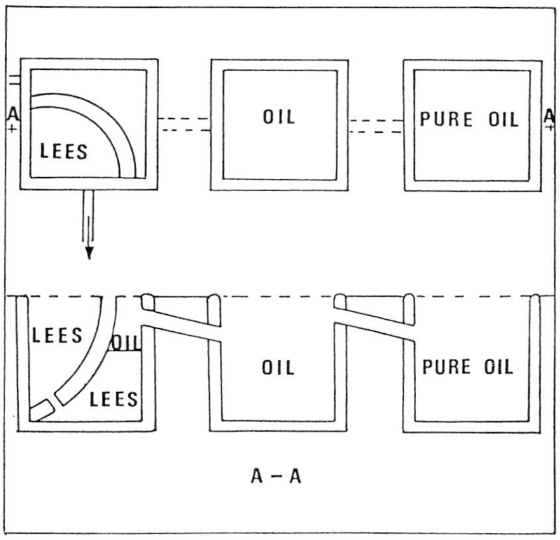

Figure 47. The Madaure oil separator (T47132) (based on Christofle 1930b).

This method for separating oil appears in several pre-industrial plants, each of which is slightly different. They are all in the western Mediterranean, two in Spain at Elche (T0191) and Huevar (T0192), on the island of Ibiza (T0193) and at Fez in Morroco (T0194).

The efficacy of this sophisticated method is proven by its survival in pre-industrial installations. However, it did not spread very far over the centuries.

Regional diversity is marked in oil separation. Overflow decantation was the method used in the

Photo 92. Pre-industrial oil separator (T0191): Elche, Spain (Photo: Carter Litchfield)

Levant, North Africa and southern Europe. Underflow decantation was that used in the Aegean, and the combined method in the western Mediterranean. The simple skimming method was apparently in use beside these more sophisticated methods throughout history.

11C. Spatial Patterns and Chronological Sequences

11C1. Technical Continuity, Regional Diversity and Cultural Identity

This study has shown that, to understand the significance of the typological differentiation of agricultural installations, the stress must be laid not on chronological sequences but on the interrelated effects of cultural continuity and regional diversity.

An example that demonstrates the close correlation between regional diversity and cultural continuity is central oil collection, which is found in Israel in oil presses from the Iron Age and was still in use in pre-industrial presses in the region until recently; yet during this period of over two and a half millennia it hardly spread further than the borders of Israel. Other striking examples of clearly defined regional diversity are the distribution of the various types of screw weights in Israel and of devices to anchor the beam in North Africa.

Cultural continuity and regional diversity are basic characteristics in all aspects of human culture and can be explained in various ways.

Regional diversity can be a result of objective geographic conditions. In this study some typological variations have been explained as having been caused by climatic or geological factors. It has been suggested that differences between methods of wine production in the northern and southern Mediterranean were a result of climatic factors and that the type of oil press used in the Golan region was effected by the Basalt rock formation of that region.

However, the present study has also shown that even in such a technical aspect of human culture as ancient agricultural installations, the primary reason for cultural continuity and regional diversity is the subjective human factor. The main cause for regional diversity and cultural continuity is clearly the isolation of one human group from its neighbour. This will be most marked where there are ethnic differences, particularly where these find expression in differences in language and religion and in political borders. These factors are usually also accompanied by local patriotism and pride in local traditions and techniques. Cultural and regional diversity is clearly also manifest in smaller cultural geographical areas within the larger ethnic and political units.

There are two main factors, however, that make regional diversity and cultural continuity more marked in agricultural installations than in other aspects of human culture. The first factor is the character of the installations themselves. They are usually very large and of unbreakable material, which results in their remaining in use for long periods and in their usually not being transported for great distances. A pottery vessel will rarely remain unbroken for more than a few years, whereas there are in Europe pre-industrial oil and wine presses that were in use for four or five hundred years, and in Israel there are rock-cut wineries that were perhaps in use for thousands of years. Whereas small objects such as pottery vessels, ornaments and weapons are often objects of trade and were therefore vehicles by which artistic motives and craft methods were transferred over great distances, agricultural installations were usually made in the immediate vicinity and rarely transported very far.

The second factor that made the cultural continuity and regional diversity in agricultural installations greater than that in other aspects of human activity was the character of the people who made and used them. Rural peasant societies are proverbially closed and conservative, although modern analogies may in this case be to some extent misleading. In ancient societies a much greater part of society was involved in and connected to agricultural activities. The cities of the Bronze and Iron Age in Israel and the Levant were basically agricultural communities, and the classical and Talmudic written sources show that the cultured elite were conversant with the technical details of the installations used. Nevertheless, the men who actually made and used the installations were not of this elite and probably often never left the immediate surroundings of their homes to see other installations, nor did they usually have many opportunities for other contacts with distant cultures.

Not only is regional diversity particularly marked in agricultural installations, but, because the stone parts are often large and sometimes even on a monumental scale, they have survived and can be recorded easily without the need for excavation. As a result it has been possible to draw up detailed distribution maps of various types of installation.

Geographical analysis of archaeological and other data is usually based on spatial units—districts/regions—which are determined on the basis of a combination of geographical factors—usually morphology, climate and geology—which sometimes include historical regional concepts. The data are then analysed quantitatively according to these pre-determined districts, in order to reach historical and other conclusions.

Detailed distribution maps of human cultural elements, such as the installations discussed in this study, allow for an additional method of spatial analysis. The distribution maps are in themselves 'cultural districts' and should be regarded as independent geocultural sources in their own right. These 'cultural districts' can be compared separately on the one hand to each of the various basic geographical maps—those of morphology, climate, geology, etc.—and on the other to maps of political and cultural regions based on ancient written sources. Such comparison can be used in various ways. The first stage is to explain the typological diversity and to determine if it is connected to objective geographical factors and/or to ethnic/cultural entities and, if the latter is the case, to define them in ethnic/cultural terms. If the technical characteristics of an ethnic culture have been determined, it is possible to continue to a second stage and use the distribution maps to define independently the geographical borders of the cultural spheres of influence and in some cases to follow changes in the extent of these over a period of time. Having determined by the same means the origins of a technical device, it is possible to continue to a third stage and to follow the direction of its diffusion to other regions and countries.

Several attempts at analysis of this type have been made in this study. Phoenician and Judean technical cultures were defined technically and spatially and it was shown that their extent was different in early and late antiquity—lower and eastern upper Galilee being Phoenician in character in early antiquity and Judean in late. A cultural subdivision in Judean cultural sphere of infuence was discerned in late antiquity. A separate Samarian culture was delineated. Similarly, specific Greek and Roman elements were defined and their diffusion traced.

The fact that the distribution areas of the various types of installation and of the technical cultures here defined coincide to such a great degree to the borders of the cultural/ethnic/political units raises the question of the relationship of the regional variation in types of installation to ethnicity and cultural identity.

In recent years scholars have focused research on a reassessment of ethnicity as manifested in archaeological finds. As opposed to the previous simple identification of regionally defined elements of material culture with ethnic groups, attempts have been made to analyse and define the nature of the style and of variation in material culture and the relationship of these to ethnicity, basing the results largely on analogies from ethnological data.

'Style' has been defined as 'formal variation in material culture that transmits information about personal and social identity' (Weissner 1983: 256; compare Binford 1962: 220) and is seen therefore as of active social or ethnic significance. Sacket (1982: 81-109) has termed this approach as 'iconological' and has suggested alternatively that variations in material culture are to be regarded as primarily of passive social or ethnic significance and termed them as 'isochrestic', literally 'equivalent in use', that is, a result of 'choosing specific lines of procedure from the nearly infinite arc of possibilities and sticking to them' (Sacket 1982: 72-73). Further debate (Sacket 1985; Weissner 1985; Shennan 1989) has led to a synthesis by which stylistic variation in elements of material culture, when proven to be ethnically connected, can be divided into two main categories according to the part they play in subjective cultural/ethnic identity. The first group are those manifesting an active ethnic significance, the second are those evincing only passive ethnic significance. Elements of a culturally active significance can be defined as those recognized both by those who belong to the human group concerned and by those who do not as symbolizing its ethnic and cultural identity. These elements, therefore, help to define and strengthen this identity and to distinguish members of the group or their property from those of other groups. The clearest elements evincing active cultural significance are those that, by definition, designate the group concerned, for example, actual symbols, be they political (flags, insignia) or religious (cross, crescent, menora or magen-david). There are, however, other elements that can be included in this group with almost the same degree of certitude. Articles of clothing often come into this category. Headwear, for example, is often the most recognizable distinguishing feature of human groups. Two of the innumerable examples that demonstrate clothing and particularly headwear as being cultural symbols are the manner in which ethnic groups were depicted in ancient Egyptian wall paintings and reliefs (e.g. the Sea Peoples, Egyptians and Libyans in the Medinat Habu reliefs [Nelson 1930: Pls. 19, 32, 37, etc.]) and, over three thousand years later, the significance of headwear in a city such as modern Jerusalem, as an element distinguishing the many national and religious groups that live side by side in this city.

Elements of passive cultural significance are to be defined as those that, although clearly associated with specific human groups, as shown by their geographical distribution, are nevertheless not regarded by the groups concerned or by their neighbours as symbolizing their cultural identity.

The question is whether it is possible to assign agricultural installations to one of these two categories. Attempts at answering this question can draw on two types of source: archaeological data and written documents.

The archaeological data, however, although rich, are difficult to interpret. There are several cases in which certain types of installations were clearly introduced from afar. Two examples are the unusual screw weights from Bet Loya and ᶜEin el-Judeida (Frankel, Patrich and Tsafrir 1990). Both of these are from monasteries and were probably introduced by foreign monks from their original homelands. In these cases it is even possible to suggest from which regions they were brought. There is no evidence, however, to suggest that the introduction of these devices from afar was regarded as of more than purely technical significance. In most monasteries in the country the methods employed were those used in the neighbouring villages, some of which were inhabited by Jews or Samaritans. The oil press from the monastery at Tabgha is of the same type as that used in the Jewish villages of the region. Similarly, the screw weight used in presses in the monasteries at Siyar el-Ghanam and at ᶜEizariya is the Samaria weight, typical of the region in which they are situated.

Analysis of the ethnic significance of archaeological evidence has usually focused on prehistoric periods. In these cases conclusions could be based on archaeological evidence alone, although aid was also sought from theoretical models based partly on anthropological data. In the case under discussion, however, written sources are available that provide clear evidence as to how people in antiquity regarded regional differences in agricultural techniques. There are actually references in both the Talmudic texts and in the writings of the Roman agronomists that show clearly that the writers were aware of different methods being used in different areas. Columella (12.52.7), while discussing the various types of olive-crushing installations, specifically states: 'The above machines are used however according to condition and local custom.' A passage in the Tosefta (*Ṭoh.* 10.12) implies the same: 'However in a place in which it was customary to place [the olives for pressing] in a *bd* they place them in a *bd* [where it was customary to place them], in a *bydyda* they placed them in a *bydyda* and [where it was customary to place them in a] *qtkw* [they placed them] in a *qtkw*' (see Appendix 1).

In neither passage, however, is an ethnic difference implied. Moreover, neither in the Talmudic nor in the Latin sources are the different types of presses given ethnic appellations. It is of significance that in the Mishnah (*B. Bat.* 4.5), when discussing legal problems connected to the sale of an oil press, the press referred to was clearly the Phoenician slotted-pier lever and weights press (see Appendix 1).

The written sources examined, although limited to one period, the first centuries of the present era, derive from two separate geographical areas and from texts rich in references to agricultural installations. These sources show clearly that, while people were conscious of regional differences, they did not regard them as of ethnic significance or associate them with cultural identity.

It is of course possible that, at other periods or under specific circumstances, the subjective approach to such installations was otherwise, as is perhaps hinted at by pre-industrial terminology such as 'pressoir à la Genoise' (Humbel 1976: fig. 39).

A corollary of the importance of the regional diversity in agricultural installations is that the ancient written sources must also be placed in their correct geographical context.

Cato grew up in Latium and, although he mentions other regions, the press he described is almost certainly the 'Latin press'. Pliny's detailed survey of the history of the press is also extremely important technically and chronologically but describes the situation in Italy and is not necessarily relevant for other regions. Columella's origin from Gades in Spain finds expression in his preferences and aids in identifying the *mola olearia*. Hero of Alexandria was an inventor and did not only describe installations but gave specifications for improved installations. However, he based his presses on installations that existed, and which he knew, these being clearly from the eastern Mediterranean. He described the lever and weights press, a device that never reached Roman Italy, and a lever and screw press equipped with a screw weight—a device that was as yet unknown in Italy at the period when he wrote. His description of the closed dovetail mortice was almost certainly based on a Phoenician prototype.

11C2. Integrated Cultures and Eclectic Cultures

In this study certain types of spatial pattern have been isolated. These can be easily defined. Their interpretation, however, poses considerable problems and any suggested explanations are inevitably hypothetical, cannot be proven objectively, and are therefore open to revision.

Both in Judaea and Phoenicia, comparatively few installation types were recorded, each of which appears in large numbers in clearly defined distribution areas. These installation types are typologically connected to one another. This pattern almost certainly shows a closed integrated culture, influenced little from outside, changes consisting mainly of internal development. In both cases when technical innovations, such as the screw, were introduced from outside they were incorporated in existing installations.

In the Jerusalem region a very different cultural pattern is manifest, which in many ways is the diametrical opposite of that found in Judaea and Phoenicia. In the Jerusalem area many types of installations were recorded, but very often only one example of each type appears, or very few. There are usually no typological connections between the different types. This eclectic culture is the result of an open society with myriad connections with other cultures. In the case of Jerusalem these were probably at least partly a result of people from other countries actually settling in the area. There are also very few signs of local development. One possible exception is the unique Ḥanita mortice with connecting channel from Tur, Beth Phage, which is perhaps a local innovation resulting from knowledge of different systems.

11C3. Central Sophistication and Peripheral Diversification

The two spatial patterns to be discussed in this section, central sophistication and peripheral diversification, are to some extent contradictions of one another.

In many cases much invention, imagination and originality was devoted to one specific part of the press, which entailed not only technical ingenuity but clearly also aesthetic values and sometimes almost monumental effects—what Mattingly (1988a) called 'Megalithic Madness'. This is the case in the Iron Age central vats from Israel, the North Syrian beam niches, the North African perforated piers and is to some extent true also of the beam weights of the region of Caesarea in Algeria.

In some cases this phenomenon finds expression in a spatial pattern whereby the most sophisticated sub-type is located near the centre of the distribution area of the main type. The most clearly defined example is that of the Tabgha screw press base with central basin and radial grooves (T73222) found around the Sea of Galilee in the centre of the distribution area of the ʿEin Nashut screw press base with open mortices and rounded corners (T732), the acme being perhaps the Weradim press base (T73223), with a central opening as well.

The spatial pattern that is in some ways the opposite of central sophistication is peripheral diversification. In this case in the main distribution area there is great uniformity and the exceptional sub-types are found in peripheral areas. The most clearly defined examples of this pattern is that of the Judaean grooved-pier screw press. The main type with square perforated mortice (T711) is found only in Judaea while the variant types are found only outside the distribution areas of the main type, partly in the immediate vicinity and partly in Galilee. The variants farther from the centre differ more from the main type than those nearer to it. The explanation for this pattern, which is very common in other facets of human culture—language, art, religion—is one of central orthodoxy and peripheral liberalism or heresy.

If, as has been suggested here, Judaea in late antiquity is the cultural nucleus of all Israel, then the great variability of press types in Samaria and Galilee when compared to the uniformity of those found in Judaea can also be regarded as an example of peripheral variability, while the defined spatial pattern of the Samaria screw weight would represent the forming of a new independent technical culture.

An example that perhaps combines both spatial patterns is that of the Bet Mirsham Iron Age central vat. The standard type is the cylindrical vat with groove and bore (T261111), which is the dominant type in the region to the north-west of Jerusalem. The variant with separating vat (T26112116) represents central sophistication. The other variant types, square in shape, without a bore, or with radial grooves, are examples of peripheral diversification being found in the west in the coastal plain and to the north in the Carmel region.

11C4. Slow Development and Fast Development

In the attempt to isolate spatial patterns that show slow or fast development or penetration of new methods into a region, both the definition of the patterns and their interpretation are more hypothetical. The discussion will be limited mainly to western Galilee, the region for which the most detailed data has been included in the catalogue.

It has been suggested above that slotted piers (T421) penetrated inland slowly into the more closed and less hospitable hills of western Galilee from west to east and that at each stage the piers were slightly more sophisticated. A similar pattern can be discerned in this region regarding screw weights, in that the simpler types do also not appear in the inland region where only the more sophisticated socketed Dinʿila weight is found. This suggests that the lever and screw weight penetrated into the eastern hills later than into the lowland to the west and south and, by the time that it did penetrate this region, the Dinʿila weight was that generally in use.

If the spatial pattern of slightly graded variation reflects slow development or penetration, a pattern of uniformity would reflect a fast penetration. This, however, is clearly not always the case. The great uniformity of types of beam weight and screw weight over long periods and wide areas clearly negates any such model. However, in some cases such uniformity could represent a fast introduction of a new method. To return to western Galilee—at the site of Dinʿila itself—all the late presses are of one type, perforated piers (T4233) with Dinʿila weights. This suggests that this site represents the final stage in the technical development of the region and that the oil presses were introduced over a short period at a stage when not only was the Dinʿila weight the main type but the piers in use had also changed and the slotted piers had been replaced by the perforated type.

11C5. Cultural Diffusion

In spite of the parallel development of separate closed integrated technical cultures, and although some techniques certainly developed independently in different regions, it is equally certain that other techniques originated and were diffused from one centre.

The early history of the lever and weights press remains uncertain. All that can be determined is that it was never part of the Roman cultural tradition, just as the lever and drum press was apparently limited only to the Roman sphere of influence.

The round rotary olive crusher was almost certainly diffused from one centre although, at the present stage of research, all that can be determined is that it originated in the eastern Mediterranean and was invented before the Hellenistic period.

It has been suggested above that the rectangular Semana beam weight with external dovetailed mortices reached the western Mediterranean from the Aegean and the slotted piers reached the western Mediterranean from the Levant.

Pliny has informed us that the screw was introduced to Rome from Greece at about the beginning of the present era or a little earlier. The earliest archaeological evidence for the use of the screw for pressing in Israel is from the second century CE, but, even if the later date for Hero of Alexandria is accepted, it is clear from his writings that the screw was in use in the Levant at an earlier date. The clarification of the earlier history of the screw must await further research in the northern Levant, Anatolia and the Aegean.

The model for the development of the socketed screw weights proposed above shows that, after the Dinʿila weight was already in use in the Levant, the Samaria weight went through several stages of development in Italy, and from there it reached other regions of the Mediterranean, in Israel becoming the main type used in the centre of the country. Thus, finally the Samaria and Dinʿila screw weights were in use side by side, one in Samaria and the other in Galilee. This demonstrates once again the slowness of diffusion of new techniques and the degree to which each technical culture maintained its independence.

Chart 5. Wineries – Dimensions

Site and installation no.	Site name	Period	Type	Treading Floor		Intermediate Vat/Cavity		Collecting Vat		
				Form	Area	Form	Area	Form	Area	Volume
Sample 10 km Square 16/26	Median				2.5 m^2				1 m^2	
000-1101-80-003(07)	Yatir, Naḥal West	Byz.	T121	Rectangle	2 m^2			Rectangle	0.30 m^2	
000-1101-90-003(07)	Yatir, Naḥal East	Byz.	T121	Rectangle	3.5 m^2			Rectangle	2.25 m^2	
000-1103-42-001(01)	Shivta, Ḥorevot-South	Byz.	T92	Square	33 m^2	Rectangle	0.5 m^2	Two-Round		6.7 m^3 6.7 m^3
000-1103-42-002(01)	Shivta, Ḥorevot-North	Byz.	T92	Square	35 m^2	?	?	Two-Round		5 m^3, 3.4 m^3
000-1103-42-003(01)	Shivta, Ḥorevot-Central	Byz.	T92	Square	25 m^2	Rectangle	2.0 m^2	Round		3.6 m^3
000-1105-86-001(01)	Halusa, Ḥorevott	Byz.	T92	Square	28.5 m^2	Rectangle	1.5 m^2	Two-Round		3.14 m^3, 3.14 m^3
000-1202-72-001(01)	ʿAvedat, Ḥorevot South-West	Byz.	T92	Square	33 m^2	Rectangle	0.5 m^2	Two-Round		6.5 m^3, 4.04 m^3
000-1202-82-001(01)	ʿAvedat, Ḥorevot South	Byz.	T92	Square	33 m^2	Rectangle	1.5 m^2	Round		8.8 m^3
000-1202-82-002(01)	ʿAvedat, Ḥorevot East	Byz.	T92	Square	16.3 m^2	Square	1.7 m^2	Round		4.2 m^3
000-1202-83-001(01)	ʿAvedat, Ḥorevot North	Byz.	T92	Square	33 m^2	Square	1.3 m^2	Three-Round		6.5 m^3, 4 m^3, ?
000-1311-91-001(01)	Marʿash, Kh.			Square	25 m^2			Rectangle	2 m^2	
000-1314-15-001(01)	Reḥovot-Dueran Kh..	Byz.	T92	Square	33 m^2	Rectangle	2.3 m^2	Two-Square		6.4 m^3, 6.4 m^3
000-1317-03-001(01)	Michal, T South		T121	Square	10.9 m^2			Rectangle	4 m^2	
000-1317-14-001(01)	Michal, T	Hel.	T9111	Rectangle	32 m^2	Square	2.2 m^2	Square	3.8 m^2	7.2 m^3
000-1317-14-001(02)	Michal, T	Hel.	T9111	Rectangle	20.25 m^2	Square	5.7 m^2	Square	5.2 m^2	
000-1317-14-001(03)	Michal, T	Iron Age	T9131	Four-Rectangle	6.6 m^2, 7.8 m^2			Eight-Rectangle		1.2 m^3, 2.5 m^3
000-1317-87-001(01)	ʿAzzun, Kh-Tabsur.		T932	Trapeze	16 m^2			Round		1.6 m^2
000-1409-28-001(01)	Benaya, Ḥ.		T961	Square	30.25 m^2	Square	1 m^2	Square		7.4 m^3
000-1413-98-001(01)	Ayalon Park Ḥanyon Hamispe		T961	Square	17 m^2	Square	0.5 m^2	Square		1 m^3
000-1413-98-002(01)	Ayalon Park, Ḥanyon Hamayanot		T121	Rectangle	24 m^2			Rectangle		3 m^3
000-1414-98-001(01)	Modiʿim		T4121	Square	9 m^2	Round	0.8 m^2	Rectangle	2 m^2	
000-1415-42-001(01)	Ben-Shemen-South		T9111	Square	25 m^2	Square	1.8 m^2	Square	2.25 m^2	
000-1415-53-001(01)	Ben-Shemen-North		T9111	Square	22 m^2	Square	1.64 m^2	Square	2.25 m^2	
000-1416-38-001(01)	Afeq, T-Antipatris	L.B.	T121	Rectangle	9.6 m^2			Round		1.12 m^3
000-1416-79-012(01)	Kafr Qasim-East		T9121	Square	15.4 m^2	Square	1.65 m^2	Round	1.33 m^2	
000-1416-79-016(01)`	Kafr Qasim-West		T9111	Square	28.1 m^2	Square	1.38 m^2	Rectangle	5 m^2	
000-1416-84-004(01)	Teena, Ḥ		T9111	Square	9.9 m^2	Square	0.28 m^2	Square	1.3 m^2	
000-1511-85-001(01)	Judur, Kh	Byz.	T961	Trapeze	18.5 m^2	Square	1.4 m^2	Round		8 m^3
000-1511-85-001(02)	Judur, Kh	Byz.	T961	Trapeze	12 m^2	Square	0.9 m^2			1.7 m^3
000-1513-00-001(01)	ʿArtuf		T9111	Rectangle	8.5 m^2	Rectangle	0.8 m^2	Square	1.0 m^2	1.0 m^3
000-1513-08-001(01)	ʿAqd, Kh-el-Ayalon Park		T9113	Square	16 m^2	Square	0.64 m^2	Rectangle		5.4 m^3
000-1513-12-001(01)	Ishwʿa		T9113	Irreg.	7.5 m^2			A. Square B. Rectangle		0.65 m^3 0.95 m^3
000-1514-27-001(01)	Shelat, H		T9111	Square	12.25 m^2	Square	0.5 m^2	Square		2.7 m^3
000-1516-53-001(01)	Deir Saman.Kh.			Round	14.2 m^2	Square	1 m^2	Rectangle	1 m^2	

Chart 5 (cont.)

Site and installation no.	Site name	Period	Type	Treading Floor		Intermediate Vat/Cavity		Collecting Vat		
				Form	Area	Form	Area	Form	Area	Volume
000-1518-13-001(01)	Sur Natan			Rectangle	c. 48 m^2			Rectangle	ca. 20 m^2	
000-1518-13-002(01)	Dardar, Ḥ			Rectangle	7.75 m^2			Rectangle	2.43 m^2	
000-1518-43-003(01)	Majdal el-		T9111	Square	8.4 m^2	Square	0.8 m^2	Square		4.2m2
000-1518-43-003(02)	Majdal el		T9112	Square	9.75 m^2			A Rectangle B Square	2.8 m^2 0.6 m^2	
000-1523-40-001(01)	Sumaq, Ḥ		T9612	Square	15.6 m^2	Round	0.20 m^2	Rectangle	2.24 m^2	
000-1523-86-001(01)	Jeleme, Kh. (ʿAsfana, Kh)		T963	Rectangle	23 m^2	Square	0.25 m^2	Square		7.2 m^3
000-1527-93-001(01)	Liman West		T951	Square (1 of 2)	16.3 m^2	Rectangle	0.25 m^2	Square	5.28 m^2	
000-1613-39-001(01)	Abu Zaʿrur		T9111	Square	18.9 m^2	Square	1.0 m^2	Square		7.4 m^3
000-1613-91-001(01)	Jerusalem, Masleva		T9111	Rectangle	38 m^2	Square	0.72 m^2	Square	4.0 m^2	
000-1616-08-001(01)	Burak, Kh.el-		T9112	Rectangle	31.6 m^2			A. Rectangle B. Square		4.2 m^3 1.95 m^3
000-1626-74-001(01)	Mana, Ḥ-East			A. Trapeze B. Rectangle Stage 2 Addition	12 m^2 23.25			Rectangle Stage 2 Deepened		4.35 m^3 6.8 m^3
000-1627-06-001(01)	Misrefot Yam, H-South		T95002	Square	17.2 m^2	Rectangle	0.30 m^2	Round	4.5 m^2	
000-1627-06-002(01)	Misrefot Yam, H		T962	Square	33.5 m^2	Square	0.17 m^2	Rectangle		10.8 m^3
000-1627-75-006(01)	Yaʿara, West		T951	Rectangle	14.8 m^2	Square	0.10 m^2	Square		107 m^3
000-1627-32-001(01)	Tur, el-Beth Phage			Square	36 m^2			A. Square B Square		1.4 m^2 5.9 m^3
000-1714-02-001(01)	ʿAtara, Kh		T9111	Square	34 m^2	Square	0.65 m^2	Square		2.31 m^3
000-1714-03-001(01)	Nasba, T.el	Iron Age	T111	Trapeze	7.3 m^2			Rectangle		0.6 m^3
000-1714-03-001(02)	Nasba, T.el-	Iron Age	T111	Trapeze	5 m^2			Round		0.5 m^3
000-1721-04-001(02)	Tiʿinnik, T. (Taanach)	EB-MB	T111	Rectangle	6.2 m^2			Rectangle		0.22 m^3
000-1726-05-003(01)	Habay, Kh.el-		T962	Square	16 m^2	Rectangle	0.4 m^2	Square		18 m^3
000-1726-27-003(01)	Kenisa, Ḥ-East			Trapeze	9.5 m^2			Oval		5 m^3
000-1914-20-001(01)	Qassab, T-el-		T9111	Square	49 m^2	Square	0.3 m^2	Square	4 m^2	
000-1921-17-001(02)	Bet Hashita		T961	Trapeze	45.6 m^2	Square	0.2 m^2	Rectangle	3 m^2	
000-1926-91-001(01)	Ramat Razim			Irreg. Circle	14 m^2			Square		2.8 m^3
000-2111-67-001(01)	Mareighat, el-		T9111	Rectangle	23 m^2	Square	1 m^2	Square	1.52 m^2	
02-0-0236-00-000(01)	Tipasa-Algeria		T9802	A Rectangle B Rectangle	10 m^2 3.9 m^2			Rectangle	6.6 m^2	
02-0-0436-00-006(01)	Azefoun 1-Algeria		T9802	A Rectangle B Rectangle	2.3 m^2 2.9 m^2			Rectangle	1.44 m^2	
05-0-3122-00-001(01)	Arminna-Egypt		T98103	Rectangle	4.6 m^2	Square	0.10 m^2	Rectangle		3.8 m^3
05-0-3222-00-001(01)	Tomas-Egypt		T98103	Rectangle	6.4 m^2	Square	0.10 m^2	Rectangle		4.5 m^3
05-0-3028-00-001(01)	Crocodopolis Theadelphia	Byz.	T98023	Rectangle	15 m^2			Round		6 m^3
09-0-3636-00-003(04)	Deit Mismis-Syria		T98	Rectangle	3.75 m^2			Bell		ca 0.8 m^3

Chart 5 (cont.)

Site and installation no.	Site name	Period	Type	Treading Floor		Intermediate Vat/Cavity		Collecting Vat		
				Form	Area	Form	Area	Form	Area	Volume
09-0-3636-00-050(01)	Behyo-Syria		T98	Rectangle	10.5 m^2			Bell		0.4 m^3
14-0-3344-00-002(01)	Tiritake-Crimea Russia	Hel	T98	Rectangle	12.07 m^2			Rectangle		5.58 m^3
14-0-3344-00-002(02)	Tiritake-Crimea Russia	Rom	T980002	A Trapeze B Trapeze	10 m^2 4 m^2			Rectangle Rectangle		4.2 m^3 1.34 m^3
14-3344-00-002(03)	Tiritake-Crimea Russia	Rom	T981003	A Rectangle B Trapeze C Square	6 m^2 1.9 m^2 3.6 m^2	Rectangle	0.30 m^2	Rectangle Rectangle Rectangle		1.5 m^3 2.5 m^3 1.9 m^3
18-0-2237-00-001(01)	Mycenae=Greece-	Hel	T98	Trapeze	4 m^2			Round		0.10 m^3
18-0-2338-00-002(01)	Athens-Acropolis Greece		T983	Trapeze	14 m^2			Amphora		0.05 m^3
18-0-1344-00-002(01)	Verige Bay Val Catena-Yugoslavia	T985	L-shaped	124 m^2 (!)			Rectangle		18 m^3	
20-0-1344-00-070(02)	Banjole-Yugoslavia		T95	Rectangle	4 m^2			Round		0.65 m^3
23-0-0643-00-001(01)	Losnich-Germany	Rom	T98	Rectangle	27 m^2			Rectangle		2.6 m^3
23-0-0649-00-002(01)	Piesport-Trier Germany	4th 5th Cent. CE	T981002	Two Rectangle	20 m^2	Two Quarter Circle	5 m^2	Two Rectangle	2.8 m^2	
23-0-0649-00-003(01)	Neumage-Dhron Germany	Rom	T98	Rectangle	8.8 m^2			Rectangle		1.4 m^3
23-0-0749-00-001(01)	Maring Noviand	Rom	T982	A Rectangle B Rectangle	8.35 m^2 6 m^2			Rectangle		1.33 m^3
27-0-1440-00-001(02)	Pompei-Villa of the Mysteries Italy	Rom	T985	Rectangle	24 m^2			Pithoi		
27-0-1440-00-001(03)	Pompei-Vineyard Italy	Rom	T985	L shaped	13.5 m^2	Rectangle	0.70 m^2	Rectangle		0.80 m^3
27-0-1440-00-006(01)	Boscoreale-Pisanella-Italy	Rom	T985	Rectangle	Rectangle	24 m^2		Rectangle (Pithoi)		12.9 m^3
28-0-0144000-001(01)	Allas Les Mines Dordogne-France		T980002	A Rectangle B Rectangle	8.1 m^2 4.3 m^2			A Rectangle B Round		3.4 m^3 0.25 m^3
28-0-0543-00-206(01)	Lambesc-Grand Verger-France	Rom	T98?	Rectangle	8.9 m^2			Rectangle		2.0 m^3
28-0-0643-00-080)	Roquebrussanne-Le Grand Loouil (Brun Site 80) France	Rom	AT98 B Possibly Oil Press	A Rectangle B Rectangle	17.5 m^2 40 m^2			A Rectangle B Two Rectangle		A 3.2 m^3 B Each 4.6 m^3
28-1-0044-00-002(01)	Cadillac-France	Rom?	T98		2.8 m^2			Rectangle		1.25 m^{-3}
50-0-3335-00-001(01)	Dhioros-Cyprus	Byz	T98	?	?			Round		2.6 m^3
77-0-2537-00-001(02)	Delos	Byz?	T98	Rectangle	11.25 m^2			Rectangle		6 m^3
78-0-2437-00-001(01)	Mikonos	Pre-Ind	T026	Rectangle	9 m^2			Rectangle	1.6 m^2	

Chart 6. Beam length and mechanical advantage of beam presses

Number	Name	Date	Type	B.L.	M.A.	
000-1409-16-001-04	Bet Mirsham (Bet Mirsim, T)	Iron Age II	T201130001	6 m	2.4	
000-1409-16-001-05	Bet Mirsham (Bet Mirsim, T)	Iron Age II	T201130001	5.5 m	4.23	
000-1409-16-001-06	Bet Mirsham (Bet Mirsim, T)	Iron Age II	T201130001	6 m	4.6	
000-1409-16-001-07	Bet Mirsham (Bet Mirsim, T)	Iron Age II	T201130001	6 m and 5 m	4 and 3.4	Room 2
000-1409-16-001-07	Bet Mirsham (Bet Mirsim, T)	Iron Age II	T2011320001	7 m and 5 m	4.7 and 2.5	Room 3
000-1409-16-001-08	Bet Mirsham (Bet Mirsim, T)	Iron Age II	T201170001	5.8 m.	4.46	
000-1411-01-001-02	Maresha, T.	Hel.	T40210002	5 m	3.33	
000-1411-01-001-08	Maresha, T	Hel.	T401210002	5.8 m	2.52	
000-1411-01-001-15	Maresha, T.	Hel.	T40121002	5.5 m	2.75	
000-1415-37-002	Tirat Yehuda	Hel.	T401146102	9 m	3.6	
000-1514-08-001	Midya-el	Pre-Ind.	T012161	9 m	2.25	
00-1614-76-001	Tira, Kh. el-		T401110008	4.70 m	1.8	
000-1627-63-001(1).	Zabadi, H.	Rom.	T402221203	West 7 m	3.5	
000-1627-63-001(1)	Zabadi, H.	Rom.	T402221203	East 9 m	4.5	
000-1628-41-001(01)	Umm el-ʿAmad	Hel.	T402321101	6 m	5	
000-1628-41-001(02)	Umm el-ʿAmad	Byz.	T402221207??	6.5 m	3.25	
000-1712-23-001(01)	Siyar el-Ghanam, Kh.	Byz.	T403130008	8 m	2.6	
000-1712-23-001(02)	Siyar-el-Ghanam, Kh.	Byz.	T403121108	7 m	2.3	
000-1726-30-001(01	Quseir Kh.el-(Rajmi, Kh).	Byz.	T402260047	7 m	3.3	
000-1726-30-001(02) Stage 1	Quseir Kh.el-(Rajmi, Kh).	Byz.	T402260047	8 m	4	
000-1726-30-001(03) Stage 2	Quseir Kh.el-(Rajmi, Kh).	Byz.	T403160047	8 m	4	
000-1726-30-001(04)	Quseir Kh.el-(Rajmi, Kh).	Byz.	T401260047	8 m	4	
000-1727-05-001(01)	A. Karkara, H.	Byz.	T402221203/8	9 m	4.5	
000-1727-05-001(01)	B. Karkara, H.	Byz.	T402221203/8	8 m	4	
000-1727-34-001(01)	Diniʿla, Kh.	Med.?	T403121108	7 m	4.3	
000-1727-34-001(02)	Diniʿla.Kh.	Byz.	T403121108	6.5 m	4	
000-1727-34-001(03)	Diniʿla.Kh.	Rom?	?	8 m	2.7	
000-2125-96-001(01)	Gamla (Salem, Tel-el)	Rom.	T401150002	6 m	2.4	
000-2316-63-001	Yajuz. Kh.		T401110002	6.5 m	2.6	
01-1-0535-00-001(01)0	Cotta-Morocco		T406191313	10 m	4	
03-0-0935-00-001(01)	Sbeitla-Sufetula-Tunisia		T403225304	9 m	6	
04-0-1331-00-001(01)	Amud, el-Libya	Rom.	T403298304	5.5 m	2.75	
04-0-1332-00-059(01)	Senam Terrʿgurt (Cowper Site 59)-Libya		T403227004	10 m	3.33	
04-0-1332-00-110(01)	(Oates Site 10)-Libya		T403227004	9 m	3	
04-0-1332-00-201(01)	Senam Rubdir-Libya		T403227004	9 m	3-av.	
08-0-3533-00-004	Kfar Hay-Lebanon	Pre-Ind.	T01114	5.6 m	4.3	
09-0-3636-00-012(01)	Kafr Nabo-G. Siman-Syria		T404321408	9 m	3.6	
20-0-1643-00-001(01)	Salone-Kapljuc-Yugoslavia	Late Rom.	T405125316	6.5 m	4.3	
27-0-1241-00-001	Capena-Monte Canino-Italy		T405191305	6 m	4	
29-1-0636-00-001(01)	Ubrique-Cadiz—Spain	Pre-Ind.	T01218050	4.5 m	3	
29-1-0637-00-001(01)	Villaneuva-Ariscal Seville-Spain	1574 CE	T01217057	12.9 m	2	
29-1-0637-00-002(01)	Huevar-Huelva-Spain	Pre-Ind.	T01218057	13.8 m	6.2	
30-1-0739-00-001(01)	Idanha-Portugal	Pre-Ind.	T0121610	6 m	2.4	
30-1-0740-00-001(01)	Enaxabarda-Portugal	Pre-Ind.	T0121610	6 m	2	
50-0-3335-00-004	Salamis-Cyprus	Byz.	R401121128	5.5 m	5.5	

Appendix 1

Oil and Wine Production in Ancient Hebrew Literature

A1.1. Terms in the Hebrew Bible

A1.1.1. Installations in the Hebrew Bible

There are three biblical terms for wine/oil-processing installations: *yqb* (appears 15 times), *gt* (appears 5 times), *pwrh* (appears twice). All three terms almost certainly refer to the simple treading installation/winery (Ts111, 121, 131).

A1.1.1.1. *yqb*. The root *yqb* is close to the Arabic *wqb*, a rock-cut depression, and to *nqb*, a hole or tunnel. The basic meaning of the term is clearly a rock-cut vat. The term *yqb* appears in connection to wine or must (Jer. 48.33; Job 24.11). Isaiah (5.2) describes hewing such an installation in the bed-rock; treading in a *yqb* is referred to twice (Isa. 16.10; Job 24.11); and the term appears once as a parallel to *gt* (Joel 4.13).

The term *yqb* appears several times coupled with *gwrn*, the threshing floor, the former representing the two liquid staple products, wine and oil, and the latter the others, the cereals. In one case wine and oil are referred to together in connection to *yqb* (Joel 2.24) and in the other cases both are apparently implied (Num. 18.27, 30; Deut. 15.14, 16.13; 2 Kgs 6.27; Hos. 9.2).

In later Mishnaic and Talmudic literature, the term *yqb* is no longer used to designate wineries.

A1.1.1.2. *gt, gnt*. The word *gt* consisted originally of three letters: *gnt*. The assimilation of the *n* led to gemination of the *t*. The *t*, however, was not one of the root letters of the word, which was either *ygn* (Gesenius 1907: 387) or *gnn* (Rainey 1966: n. 1).

In Ugaritic sources from the Late Bronze Age the term *gt* refers to a store or administrative centre of an agricultural economic unit, private or royal. The meaning was apparently broadened to refer to such a unit as a whole (Geltzer 1963: 4-7, 22; 1965: 118-25; Yanovskaiah 1963: 66-68, 88, nn. 118, 119). Geltzer suggests (1965: 122) that the original meaning of the term *gt* was wine press (winery) and that the BDL *gt* of Ugaritic documents were those who trod the grapes. In Ugaritic documents, however, the term *gt* does not appear as a winery, and the Accadian parallel of *gt*, *Dimtu,* means 'tower', but also has the wider meaning of 'agricultural unit' (Rainey 1966: 36 n. 2; Heltzer 1979, 1982: 49-79). This meaning connects to other Hebrew words that derive from the root *GNN-YGN*, to protect (Isa. 31.5; Zech. 9.15; 12.8), *mgn*, a shield, and *gn*, a garden, a protected field. The development of the word *gt* (GNT) was apparently 'protect/shield', a tower in the midst of an agricultural unit, and then 'the agricultural unit as a whole'. It is from the latter meaning that the many toponyms consisting of or incorporating the unit *gt* derive (Rainey 1966: 36 n. 2). It has been suggested that the reference to *gt* in Judg. 6.11 is to the fortified centre of such an economic unit (Rainey 1996: 36 n. 2). This is very probable, as this is the earliest appearance of the term in the Bible and it specifically refers to flight from an enemy. Rainey suggests that the other biblical references to *gt* are also not to wineries, but to similar agricultural administrative centres. However, of the four other references, three are concerned with treading in a *gt* (Lam. 1.15; Isa. 63.2; Neh. 13.15) and the fourth (Joel 4[3].13) also refers to producing wine or oil. As far as can be determined, these passages all date to the end of the First Temple period or later, and taking into account the Mishnaic meaning of the term *gt*, which is unequivocably a winery, it is evident that by this period the meaning of the term had narrowed to the winery that stood in the administrative centre rather than the centre itself.

A1.1.1.3. *pwrh*. The term *pwrh* appears only twice in the Hebrew Bible. The reference in Isaiah (63.3) implies a small winery, but that in Haggai (2.16) suggests that the meaning of the term is a liquid measure. In Accadian *puru* is a stone bowl (Von Soden 1959).

A1.1.1.4. *yqb*, *gt* and *pwrh*: conclusions. *yqb* is in most cases mentioned as containing oil or wine; in four cases clearly so (Joel 2.24; Num 18.27, 30) and three times by implication (Deut. 15.14; 16.13: 2 Kgs 6.27). Only rarely is the term mentioned in connection to treading (Isa. 16.10; Job 24.11). The term *gt*, however, is in most cases mentioned in connection with treading—three references out of five (Lam. 1.15; Isa. 63.2; Neh. 13.15). This led Rashi (*Yom.* 76a) to define the difference between the *yqb* and the *gt* as follows: 'every *yqb* in the bible refers to the vat in front of the *gt* where the wine is placed and not the grapes'. Borowski (1980: 115) suggests that the *yqb* is the rock-cut winery installation (T111), *gt* the built installation (T12), and *pwrh* the free-standing type (T131). The biblical text, however, does not allow for such clear-cut definitions. As for Rashi's explanation, it is clear that both terms are used for the whole installation including both treading floor and collecting vat: *yqb*, Isa. 5.2; *gt*, Joel 4.13. The three terms derived from different sources but almost certainly referred to the same simple wineries, although the term *gt* implied that the treading floor was the main part of the installation, while the term *yqb* implied that it was the collecting vat that was the main element. The comparatively complex specialized oil presses (Ts2) that are known from Israel in the Iron Age are not reflected in biblical terminology, which does not distinguish between installations for the production of wine and oil.

A1.1.2. Types of Oil in the Hebrew Bible: *šmn hṭwb, šmn hmr, šmn rwqḥ, šmn twrq, šmn rʿnn, šmn ktyt, šmn rḥṣ*

Something can be learned about how oil was produced and processed in the biblical period from the various types of oil that appear in the bible. *šmn rwqḥ* (Qoh. 10.1), is spiced oil, the same adjective being used for wine (Cant. 8.2). *šmn hmr* (Esth. 2.12) is oil mixed with myrrh. *šmn hṭwb* (Ps. 133.2; 2 Kgs 20.13), similarly to *yyn ṭwb*, is not simply good oil but spiced oil, as is demonstrated by parallels from other Semitic languages (Old Babylonian [Paul 1975: n. 12], Ugaritic [Gordon 1965: 1084: 1, 4, 6, 9] and Arabic [Sasson 1981: n. 9]).

The meaning of *šmn twrq* (Cant. 1.3) is more problematical. The term *twrq* is usually explained as being third person imperfect, feminine of *ryq/rwq*, 'to empty', and translated as 'poured forth', 'poured out' (Gesenius 1907: 937-38; Mandelkern 1969: 1085; Authorised Version; New English Bible—with note that Hebrew is uncertain). The *Biblica Hebraica* (Kittel 1973: 1201), however, suggests correcting the term to *tmrwq* (compare Esth. 2.3, 9, 12), meaning 'cosmetics'. *šmn rʿnn*, 'fresh oil' (Ps. 92.11) is probably the oil made of unripe olives, the 'virgin oil'—the *ʾnpyqnwn* of Talmudic literature (Stager 1985: 75).

The term *šmn ktyt* appears in the Hebrew Bible twice as oil to be used for the lamp in the tabernacle (Exod. 27.20; Lev. 24.2), twice as oil for the ritual offerings (Exod. 29.40; Num. 28.5) and once as oil paid by King Solomon to King Hiram of Tyre (1 Kgs 5.25).

Philologically, the root *ktt* is related to *ktš* and means to pound or beat, the term for mortar being *mktš*. In the Mishnaic discussion regarding the type of oil to be used in the temple ritual, Rabbi Yehuda apparently wanted to return to the methods used in the ritual of the First Temple. He explained '*ktyt* is it not *ktwš* (pounded?)', and prescribed that the ritual oil should be prepared by pounding the olives in a mortar and pressing by using only the direct weight of stones. *šmn ktyt* is without doubt the first oil as prescribed in the Mishnah, which was extracted by only pounding the olives without pressing.

It is of interest that the sacred oil of the Mandaeans was prepared by pounding sesame seeds in a mortar. Stol (1985) suggests the Accadian verb *halāṣum* refers to preparing oil by crushing in a mortar and *ṣāhitum* to that prepared using a mill. *ṣamnum halṣum*, prepared by the former method, was the finer oil.

šmn rḥṣ is a term appearing together with wine in the Samaria *ostraca*. The root *rḥṣ* is connected to washing. However, Sasson (1981) has shown that the grammatical construction of the term *šmn rḥṣ* does not allow for the meaning previously suggested, that this was oil used for cosmetic purposes, but implies the manner in which it was produced and thus also its quality. Stager (1983) has suggested that the term 'washed oil' showed that it was produced by adding hot water to the extracted liquid, as is the oil known in Arabic today as *zāt ṭafaḥ*. He also suggested that the term *šmn rḥṣ* was the northern Israelite equivalent of the southern Judaean *šmn ktyt* and that the latter was also prepared in the same manner. The use of hot water in the extracting of oil was, however, not universal in pre-industrial techniques, especially not in the Levant where the weather is usually still warm in the season of the olive harvest. The same was probably true in the past also, and it is very probable that *šmn ktyt* and *šmn rḥṣ* were produced by different methods, the former using a mortar only and the latter using hot water also.

A1.2. Installations in Talmudic Literature

A1.2.1. Installations Connected to Wine Only

A1.2.1.1. *gt*, winery (Hebrew). In later Hebrew literature the term *gt* is the term used for a winery and is very rarely used in connection to oil production. There are a few exceptions. In one case it is stated that the olives are brought up to a *gt* (*Exod. R.* 36.1). Another exception is the place name Gethsemane (*gt šmnym*, Mt. 26.36). This name, however, probably originated in earlier periods. The terms *byt hgytwt*, 'the house of the wineries', and *byt hgyt*, 'the house of the winery' (*t. Ter.* 3.7; *t. Ohol.* 18.13), are rare and refer apparently to wineries that were in buildings. The vintage season is also called *gt*, *mgt lgt*, from vintage to vintage (*t. Ṭoh.* 11.16).

Although at this period the use of the term *gt* was narrowed to exclude oil-producing installations, it was also broadened and applied to other rock-cut pits and depressions, a use not found in biblical Hebrew: for example, the pit in which the sin offering was burnt (*Zeb.* 14.1), the rock-hewn courtyard onto which burial caves opened (*t. Ohol.* 15.7) and other simple pits (*Kil.* 5.3; 5.4).

In Talmudic literature the terminology for the parts of the winery are not uniform. Often the treading floor is called *hgt hʿlywnh*, the upper winery, and the collecting vat *hgt hthtwnh*, the lower winery (*Maʿas.* 1.7; *Ter.* 8.9; *t. Maʿas.* 1.7; *t. Ṭoh.* 11.15; *t. Ter.* 7.19). Generally, however, the collecting vat is called simply *bwr*, pit/vat (*ʿAbod. Zar.* 5.10; *Ṭoh.* 10.6; *t. Ter.* 3.6; *t. Ṭoh.* 8.11; *t. Ṭoh.* 11.12). However, to distinguish the collecting vat of a winery from other pits, e.g. *bwr mym*, the water/pit-cistern, it was often called *bwršlgt* (one word), 'vat of the winery' (*t. Miq.* 3.6) or *bwršlyyn* (one word), 'vat of wine' (*ʿAbod. Zar.* 4.10; *Ter.* 10.5; *t. Ṭoh.* 9.14; *t. Ter.* 7.15). The treading floor was often simply *gt*, 'he left his treading floor (*gt*) and his vat (*bwr*)' (*t. Ṭoh.* 8.11), 'one treading floor (*gt*) to two vats (*bwrwt*); two treading floors (*gytwt*) to one vat (*bwr*)' (*t. Ter.* 3.7). This is a direct continuation of the biblical linguistic use of the term *gt*, by which it designates the whole winery but also only the treading floor.

Small treading appliances are also called *gt*. These can be of stone but also of wood and of pottery (*ʿAbod. Zar.* 4.8).

A trodden winery, *gt bʿwṭh* (*ʿAbod. Zar.* 4.8) was one in which the grapes had been trodden, but the must had not flowed to the vat—clearly the same as a stoppered winery, *gt pqwqh* (*ʿAbod. Zar.* 55b; *y. ʿAbod. Zar.* 4.8, 44a). Stoppering was possible when the treading floor and collecting vat were connected by a bore, *ṣynwr* (*Maʿas.* 1.7; *t. Maʿas.* 1.7). Covering the vat during fermentation is also referred to (*Ṭoh.* 10.7; *t. Ṭoh.* 11.13). Winery T94, for example, was equipped for both operations.

A1.2.1.2. *ʿṣrh, mʿṣrʾ, mʿṣrtʾ*, winery (Aramaic). The term *ʿṣr* means 'to press'. *ʿṣrh* is an Aramaic term used in the Babylonian Talmud for a winery (*B. Meṣ.* 60b; *B. Qam.* 27b). In the Targum *gt* (Isa. 63.2) and *yqb* (Isa. 16.10) are translated as *mʿṣrʾ*. *mʿsrtʾ* (*ʿAbod. Zar.* 70a, 74b) is an almost identical form. Jastrow (1926: 818) suggests it is a winery building, equivalent to *byt hgt*.

The term *mʿsrʾ* in Palestinian spoken Arabic is used for a winery (Dalman 1928–42: IV, 354) but also for an oil press (Dalman 1928–42: IV, 193, 214, 221, etc.).

A1.2.1.3. *zyyrʾ* (*mʿṣrʾ zyyrʾ*), press; *mkbš*, press. The term *zyyrʾ* appears once in connection with wine production (*ʿAbod. Zar.* 60A) and in three parallel passages which are apparently connected to laundering. (*Šab.* 123a; *y. Šab.* 17.2, 16b; *y. Beṣ.* 1.6, 60c). In each of the latter, three utensils are mentioned (the three vary so that four are mentioned in all) and in two cases *zyyrʾ* is explained *bzyyrʾ dw ʿṣr byh*, 'in the *zyyr* (in/with) which the pressing was done' (*y. Šab.*; *Beṣah*).

The passage in *ʿAbod. Zar.* 60a reads '[wine produced by a Gentile in a] *mʿṣrʾ zyyrʾ* was permitted by R. Papi.' The following discussion is concerned with direct and indirect action implying that in the *mʾṣrʾ zyyrʾ* the action was indirect. Rashi explains '*mʾṣr zyyrʾ*, a winery (*gt*), in which one does not tread [the grapes] but presses them with a beam'. Jastrow (1926: 393-94) translates 'press, perforated tub'. Jastrow's translation 'perforated tub' is apparently to explain the utensil in which the launderers pressed the textiles. However, launderers used presses already in Roman times (see p. 159) and *byh* can mean both 'in' and 'with'. Therefore the term *zyyrʾ* almost certainly means a press, but the context does not allow us to determine whether a specific type of press was meant, as Rashi implies, or whether it is a general term.

The term *mkbš* (*Šab.* 20.5; *y. Šab.* 17.1, 16A; *t. Šab.* 16(17).7) is also a clothes press, and in this case too the texts do not suffice to determine of what type it was.

A1.2.1.4. *ʿgwly* (*ʿygwly*) *hgt*, beam weights. The term *ʿgwly hgt* appears in three passages:

1. 'They (may) load (*ṭwʿnyn*) the beams of the oil press (*byt hbd*) and *ʿygwly hgt* (before the Sabbath)' (*Šab.* 1.9).

2. 'They (may) load beams of the oil press (*byt hbd*) and hang (*twlyn*) *ʿygwly hgt* (before the Sabbath)' (*t. Šab.* 1.21) (parallel of passage 1).

3. '(The area) between the *ʿgwlym* and the grape skins counts as public domain' (*Ṭoh.* 10.8).

The root *ʿgl* means 'round' and from the context in passages 1 and 2 it is clear that *ʿgwlym* are objects used during the pressing of grape skins and stalks after treading. The legal question that arose was whether or not it was permitted to set up the presses before the Sabbath, and let the pressing process continue during the Sabbath, albeit without human intervention.

Danby (1933: 101, 732) translates *ʿgwlym* as 'rollers'. However, he explains (1933: 732 n. 1) these as being 'beams for squeezing the grapes'. Krauss (1910–12: II, 611) has suggested that *ʿgwlym* were round stones that were laid on the grape skins. It is possible that the heavy rollers used in lower Galilee and the Carmel region to extract the must from the grapes (Ts861, 862) were afterwards placed on boards to squeeze out the remaining must. However, it is more likely that *ʿgwly hgt* were beam weights (Lieberman 1959: III, 19 n. 67; Hirschfeld 1983: 210). Passage 2 clearly refers to hanging the *ʿgwlym* and the term *ogl/agal* (plural *ugul*) is used for beam weights in modern Arabic (Dalman 1928–42: IV, 214-15). If the *ʿgwlym* are weights, the third passage is logical—the treading floor between the beam-end on which the weights were hung and the pile of grape skins being pressed was considered to be public domain as regards injunctions of ritual cleanliness, the rules being less stringent in the public domain than in private domain.

If the explanation given here for the term *ʿygwly hgt* is correct, both the oil press and the wine press referred to in passages 1 and 2 are lever and weights presses. The two presses are referred to completely differently, however. In the case of the oil press, only the beam is mentioned and in the case of the wine press only the weights. The explanation is perhaps that in the oil press the main operation was setting up the large beam, whereas in the small wine press the main operation was hanging the weights.

A1.2.1.5. *tpwḥ*, cup-mark in centre of treading floor; *lḥm*, press cake. The term *tpwḥ* (literally 'apple') appears in two parallel passages. The term *lḥm* (literally 'bread'/'loaf') appearing in one of these, and also in a similar meaning in a third passage.

1. 'they (may) take a trodden winery from the gentile even if he takes (grape skins) in his hands and puts them (*nwtn*, literally "gives to") on the *tpwḥ* (apple)' (*ʿAbod. Zar.* 4.8).

2. 'If the Gentile works (*ʿsh*, literally "does") with the Israelite in the winery and he lifts up the "loaf" (*lḥm*) toward the "apple" (*tpwḥ*) and takes the "loaf" from the "apple" even if the wine drips from his hands it is permitted' (*t. ʿAbod. Zar.* 7[8].3).

The discussion in both passages is concerned with whether wine is libation wine (*yyn nsk*) and therefore idolatrous and forbidden.

The word *lḥm*, loaf, appears in a similar meaning in relation to an olive press (*t. Men.* 9.7; see discussion below). The *lḥm*/loaf is, in these passages, clearly the mass of olive-pulp cake or grape-skin cake formed after most of the liquid has been expressed.

The term 'apple', *tpwḥ*, appears also in relation to the altar in the Temple (*Tam.* 2.2; *Ḥul.* 90b). Krauss (1910-12: II, 235) explained the term both as used in relation to the altar and in relation to wineries as a pile of ashes or grape skins shaped in both cases like an apple. Urmann (1974: 173 n. 6) and Hirschfeld (1981: 384 n. 12) have suggested that the *tpwḥ* is a depression which is in the form of the negative of an apple.

There are other cases in Talmudic literature in which a depression shaped like the negative of a fruit has been named after it—e.g. *ʿdšh*, lentil, the term used for a concave press bed (see below, p. 193).

The small hole found in the centre of many treading floors (T171) is almost certainly the Mishnaic *tpwḥ*.

A1.2.1.6. *mšṭyḥ* (*mšṭḥ*) *šl ʿlym*, auxililary treading floor; *mšṭyḥ* (*mšṭḥ*) *šl ʾdmh*, compartments. Halachic discussion in the Mishnah (*Ṭoh.* 10.4, 5) and Tosefta (*t. Ṭoh.* 11.8, 9) distinguishes between two types of floors (*mšṭyḥ*, *mšṭḥ*, literally 'spreading', 'spreading area'), one 'of earth' and one 'of leaves'. The Halachic question that arises is if the must that drips from the grapes lying on these floors makes the material it drops on susceptible to ritual uncleanness. Wine is one of the seven liquids (*Makš.* 6.4) that make objects susceptible to ritual uncleanness, but only when they drop on an object intentionally.

The discussion raises various possibilities, but in each case the ruling was more stringent as regards the 'floor of leaves' than as regards that of earth. Urmann (1974: 174 n. 10) has suggested that the difference was that on the 'floor of leaves' the first must was intentionally collected, whereas on the 'floor of earth' it was an incidental by-product. The auxiliary treading floors with a small collecting vat connected to the main treading floor (T803) clearly conform to the 'floor of leaves', whereas the large compartments found in the wine presses of the Negev (T802) fit the term 'floor of earth'. The small winery compartments connected to the main treading floor (T8012) possibly also served to extract first must and would then also fit the term 'floor of leaves'.

Technically, it is of interest that the grapes spread out in order to produce first must were placed on leaves to facilitate the flow of the must.

A1.2.2. Installations Connected to Oil Production

A1.2.2.1. The oil press of the Mishnah. The most detailed reference to an olive press in Talmudic literature is that in the Mishnah.

'If a man sold an olive press *byt-bd* he has sold also the *ym* and the *mml* and the *btwlwt* but has not sold the *ʿkyrym* and the *glgl* and the *qwrh*, but if he had said "It and all that is in it" all these are sold also. R. Eleazar says: "If a man sold an olive press he has sold the *qwrh* also"' (*B. Bat.* 4.5).

In chs. 4 and 5 of the tractate *Baba Batra* in the Mishnah there is a discussion on certain legal aspects of contracts of sale. The legal principle involved is that, when a man sells property, he has sold only the fixed components unless he specifically states that he is selling 'it and all that is in it', in which case he has sold also the moveable components. Thus the first three terms in the verse under discussion refer to fixed parts of the olive press and the last three to moveable parts.

In the chapter of the Babylonian Talmud in which the verse is discussed (*B. Bat.* 67b) there is an interpretation of the text, including Aramaic translations of the six terms and a list of additional parts of the oil press, some of which are sold and others not. In the Jerusalem Talmud (*y. B. Bat.* 4.5, 14c) and in the Tosefta (*t. B. Bat.* 3.3) this 'second list' appears together with a 'third list' of components not sold even if the vendor has stated that he is selling all that is in the oil press.

A1.2.2.1.1. *byt bd*, oil press; *qwrh*, press-beam. *byt bd* was the term used for an oil press throughout Talmudic literature. *bd* apparently means 'beam' and *byt bd* 'the house of the beam'. The term usually used for press-beam is *qwrh* and *bd* has been explained as meaning 'vat' or 'tank' (Jastrow 1926: 138). In the discussion on the Mishnaic oil press quoted above, however (*B. Bat.* 4.5), Rabbi Eleazar clearly understood *bd* to mean 'press-beam', and therefore stipulated that, in spite of the beam's being portable, it should nevertheless be sold with the press. The Babylonian Talmud (*B. Bat.* 67b) quotes the Mishnaic text and explains: 'Rabbi Eleazar says, he who sells the oil press (*byt bd*) sold the beam *qwrh*, for is the oil press not so called because of the beam.'

In the parallel section in the Jerusalem Talmud appears: 'If it has no beam it is not called *byt bd*' (*y. B. Bat.* 4.5, 14c). Here the reference is clearly to a direct-pressure screw press (T7) which has no press-beam. In practice, however, the term *byt bd* was used for all oil presses, while *gt* was used for all wineries, even those equipped with a beam press.

The term *bd* came to be used for the olive-picking season (see below, the Baraitha of the Lulabim, and *t. B. Meṣ.* 8.27) and also for the quantity of olives that was pressed in one batch (*Ṭoh.* 9.7; *t. Ṭoh.* 9.12; Latin: *factus*, Plinius 15.6.23).

The man who worked in the oil press was known as a *bdd*.

It is worthy of note that the term *bd* has survived in modern spoken Arabic in the region, variously, as the term for the press building as a whole (Dalman 1928–42: IV, 214), as that for the olive crusher (Dalman 1928–42: IV, 205), and as that for the press beam (Cresswell 1965: 37). It is also of great interest that Cyrillus in his Greek translation of section 19.2.19 in the Digesta, Justinian's famous legal corpus translates the Latin *praelum* in both its meanings, press beam and oil-press as βάδδηv clearly the Aramaic *bdyn* (Scheltma and Holwerde 1957: 1184).

A1.2.2.1.2. *ym*, collecting vat; *mml*, press-bed. The word *ym* means 'sea' and is used also for large vats or bowls (e.g. 'the molten sea', 1 Kgs 7.23-25). In the Babylonian Talmud (*B. Bat.* 67b) the term is translated as *ṭ[lp]ḥ*, a lentil, presumably meaning a vat shaped like the imprint of a lentil. The term *ym* does not appear in other references in Talmudic literature connected to oil presses.

The word *mml* is explained in the Babylonian Talmud (*B. Bat.* 67b) thus: 'R. Aba the son of Memel [says]: *mml* [means] *mprkt*'—apparently to be understood as a crushing implement.

The term *mml* appears also in two legal discussions, each of which appears in two versions.

1a. Oil (is liable for tithing) when it has flowed down into the collecting vat, but even after it has flowed down one may still take of the oil (un-tithed) from the olive pulp frail (*ʿql*) or from between the *mml* and between the boards (*hpṣym*) (*Maʿas.* 1.7).

1b. Oil (is liable for tithing) when it has flowed down into the collecting vat but even after it has flowed down one may still take of the oil (un-tithed) from the olive pulp frail (*ʿql*) and from the *mml* and from between the boards (*hpṣym*) (*t. Maʿas.* 1.7).

2a. When may one give a heave-offering of olives? When they are put in the press and R. Shimeon says when they are crushed. R. Yossi the son of Yehuda says when one brings the olives in a basket and puts them in the *mml* and walks on them back and forth. They said to him olives are not similar to grapes, grapes are soft and give of their wine, olives are hard and do not give of their oil (*t. Ter.* 3.13).

2b. R. Yossi the son of Yodeh says if one wants to one may bring the olives and put them under the *mml* and walk back and forth on them. They said to him olives and grapes are not similar (*y. Ter.* 3.4, 42b).

In most of the traditional commentaries, for example in that attributed to R. Gershom (*B. Bat.* 67b), and in the Rambam's commentary on the Mishnah, these terms are explained as the crushing basin (*ym*) and millstone (*mml*). This interpretation was based on the Aramaic translation of the terms in the Babylonian Talmud, was accepted by Goldman (1907: 42, 43) and this is the meaning of these terms in modern Hebrew.

Not all the early commentators, however, explained the terms in this way and Rav Hai Gaon explained '*mprkt*' a vessel of earthenware or stone made in order to contain the material that was crushed' (Lieberman 1959: I, 30).

Krauss (1910–12: II, 221-22; nn. 526, 527) was the first to suggest that the Rambam and other commentators were mistaken, and, basing his conclusions on the other references to *mml* (1a, 1b, 2a, 2b above), suggested that *mml* was the press-bed and *ym* the collecting vat. The question that arises when discussing any text dealing with olive oil production is whether it is concerned with the first stage of the process, the crushing, or with the second, the pressing. Verses 1a and 1b are without doubt connected to the pressing stage as shown by the reference to olive pulp frails, pressing boards and collecting vats. Within this context, almost the only part of the press the *mml* can be is the press-bed. Verses 2a and 2b are less clear and *prima facie* here it is the crushing stage that is referred to. This and the Aramaic translation of *mml* to *mprkt*' in the Babylonian Talmud led Lieberman (1955–88: I, 330) to suggest that '*Memel* was a general term for various parts of the oil press'. However, in verses 2a-b, R. Yossi apparently describes the production of olive oil in one stage only by treading the olives instead of the usual two stages of crushing and pressing. Treading olives was one of the primitive ways of producing olive oil as is evident in Mic. 6.15. Dalman (1928–42: IV, 207) doubted if the reference in Micah to treading olives should be taken literally. There is, however, much evidence that olive oil was extracted by treading (see above, p. 46).

It should be stressed also that, until recent times, use was made of primitive methods of production alongside more advanced and sophisticated techniques (Dalman 1928–42: IV, Pl. 47). Such methods are specifically referred to in other Talmudic texts such as those concerned with producing oil in the sabbatical year (see below, p. 192) and with the production of oil for use in the Temple (see below, p. 196).

If R. Yossi (2a-b above) has here been understood correctly, the *mml* he is referring to is the press-bed on which the olives would be trodden and the oil would then run into the adjacent collecting vat. There are other reasons to prefer Krauss's suggestion that the *mml* is the press-bed. In the 'second list', appearing in all the sources except the Mishnah, it is stated 'he sold the lower millstone (*rḥym*) but not the upper [millstone]' (see below, p. 190). The lower and upper millstone of the oil press are clearly the crushing basin and millstone. This text provides the correct terms for these installations, thus making it clear that the *ym* and *mml* are not the crushing basin and millstone. In addition, this text demonstrates that, if the *mml* was the millstone, it would have been one of the press components that was sold, and would not have been included among the components that are not sold, as it is in the Mishnah.

The Aramaic translation of *mml* in the Babylonian Talmud is presumably mistaken, or possibly we have not understood it correctly.

A1.2.2.1.3. *btwlwt*, slotted piers. One of the meanings of this word is 'maidens'. In the Babylonian Talmud (*B. Bat.* 67b) it is stated: 'R. Yohanan said: [*btwlwt* are] posts of cedar wood in which one puts the [press]-beam.' R. Yohanan lived in Eretz-Israel in the third century and his description clearly refers to the slotted-press piers, the function of which was to secure the press-beam. The writers of the early commentaries, such as that of the Rambam and that attributed to R. Gershom, who were familiar with similar wooden press piers in Europe, and explained the *btwlwt* accordingly, pointed out the similar names given them by the local peasants—'twins' (*Zwilling*, German; *jumelles*, *gemelles*, French) or sisters (*sorores*). Goldman (1907: 43 n. 4) was the first to recognize the similarity between the *btwlwt* and the *arbores* of Cato (18).

In recent years scholars based suggestions for identification of *btwlwt* on the finds of archaeological excavations in this country. However, it was not known that the slotted piers existed in Galilee. As a result, it was suggested that the *btwlwt* were the press-bed (Vilbosh 1947) or the plain press piers (Yeivin 1966: 59). However, with the discovery of the slotted piers in this country, there is no doubt that these are the *btwlwt* of the Mishnah.

A1.2.2.1.4. *glgl*, winch. The meaning of the word is already 'wheel' in biblical Hebrew. *glgl* is the term used for the wheel used to raise the bucket from the well (*ʿErub.* 10.14; *t. B. Bat.* 2.16), and is without doubt the wheel or winch that served to raise the press weight.

A1.2.2.1.5. *ʿkyrym* (*ʿkydym*), beam weights. This term appears in many forms, but it was already recognized in the commentary of the Gaonim to the tractate *Ṭohorot* (Epstein 1924: 27) that the same term appears in the tractate *Kelim* in the Mishnah (*Kel.* 12.8) and in the Tosefta (*t. Kel.* 7.12). The term appears in different manuscripts as *kyryn*, *kyrym*, *kydym*, *ʿbyrym*, *ʿbwdym*, *hbdym*, *ʿbydym*, *ʿbyrym*, *ʿkyryn* and many more forms, but the original form was it seems *ʿkydym* (Agmati 1961: 239a, 240).

In the Babylonian Talmud the term is translated into Aramaic as *kbšy*, a word connected to pressing. The early commentaries, the Rambam and the commentary attributed to Rabbi Gershon explained the term as pressing boards, the boards placed above the olive pulp frails, as did Krauss (1910–12: II, 223). Goldman (1907: 43) suggested that the

ʿkyrym/*ʿkydym* were the press weights. This explanation not only conforms to the character of the presses in this country in the Roman period, but is confirmed by the text in *Kelim*, where *ʿkyrym*/*ʿkydym* appears together with two other words *mṭwtlt*, a builder's plumb weight; *mšqlt*, a weight (for use on a balance), *ʿkyrym*/*ʿkydym* beng the third of the three. This is clearly a list of three weights arranged in order of size, the largest being the press weight.

A1.2.2.1.6. The oil press of the Mishnah: conclusions. The three fixed parts of the Mishnaic oil press are, therefore: the *ym*, collecting vat; the *mml*, press-bed and the *btwlwt*, slotted piers; and the three moveable parts are the *ʿkyrym*/*ʿkydym*, the press weights, the *glgl*, the drum for raising the weights, and the *qwrh*, the beam. These are the six parts of the lever and weights press of western Galilee—the Zabadi press. The fact that the crushing basin and millstone are not included also accords with the plans of oil presses in this region where these are situated in a separate room (Zabadi; Karkara). The six parts specified are not the parts of the oil press of Judaea in which the press-beam is not secured in piers and in which there is a central collecting vat instead of press-bed and collecting vat.

The centres of Jewish learning where the Mishnah was crystallized and edited were in western Lower Galilee—Usha, Bet Shearim, Ṣippori, and Shafar'am, which explains the Mishnaic oil press being of a type that was Phoenician in character.

A1.2.2.2. The Oil Press of the Mishnah: Additional Texts

A1.2.2.2.1. The second list. The 'second list' list appears in slightly varying versions in (1) the Babylonian Talmud (*B. Bat.* 67b), (2) the Jerusalem Talmud (*B. Bat.* 4.5, 14c), and (3) the Tosefta (*B. Bat.* 3.3).

1. Sold the *nsrym* and the *yqbym* and the *mprkwt* and the lower *ryḥym* but not the upper and in any case not the *ʿbyrym* the *sqyn* and the *mrswpyn*.
2. Sold the *yqbym* and the *ʾswrym* and the *mprkwt* and the lower *ryḥym* but not the upper and in any case not the *sqyn* and the *mrswpyn*.
3. Sold the *yṣydyn* (*yṣyryn*) and the *yq*bym and the *mprkwt* and the lower *ryḥym* but not the upper *ryḥym* and in any case not the the *sqyn* and the *mrswpyn*.

A1.2.2.2.1.1. *nsrym*, *ʾswrym*, *yṣydyn*, *ysyryn* = *syrym*, pots/vats. It is not clear what was the form of the original term. *nsrym* in the Babylonian Talmud is the only one of the three forms of the term that has a clear meaning—'sawn boards'—and was therefore explained as the pressing board or frame to enclose the olive pulp (R. Gershon, Rashi *B. Bat.* 67b). However, as the various forms are so different, the form in the Babylonian Talmud was probably a scribal correction to an understandable form, perhaps effected by the term *nsrym* appearing in the following sentence in the Mishnah in relation to a bath house (*B. Bat.* 4.6). R. Joseph HaLevi ibn Migash (Agmati 1961: 239b) gives the form *syrim*, 'pots'. As the part of the press under discussion was sold and was therefore a permanent part of the press, it was probably one of the various types of vats or tanks in the pressroom, known as a *syr*, pot.

A1.2.2.2.1.2. *yqbym*, vats. The Arukh (Kohut 1926: IV, 153-54) explains *yqbym* as the baskets in which the olives are brought from the field. It is, however, one of the permanent parts of the press and the Rashbam explains it as a *ym*, vat, as does Krauss (1910–12: II, 222 n. 523). This is very similar to its meaning in biblical Hebrew. Thus the *yqbym* are probably another type of vat or tank in the pressroom.

A1.2.2.2.1.3. *mprkwt*, upper and lower *ryḥym*, crushing devices. The upper and lower *ryḥym* are the terms used for the upper and lower stones of a corn mill in biblical and later Hebrew—in this case clearly referring to the crushing basin and crushing wheel (Ts3). As pointed out above, these were mistakenly thought to be termed *ym* and *mml*.

mprkwt derives from the root *prk*, to crush, crack, and is clearly also a crushing device. Being among the components sold, it was also a permanent part of the press, either another term for the crushing device as a whole or that for a particular type of it.

A1.2.2.2.1.4. (*ʿbyrym*) *sqyn*, *mrswpyn*, sacks, bags. The *ʿbyrym* appear in the Babylonian Talmud only and are probably the *ʿkyrim* that appear in the Mishnah.

The meaning of *sqym* is 'sacks' and *mrswpyn* had a similar meaning (Latin *marsupium*, Greek μαρσύπιον, pouch, bag, purse).

A1.2.2.2.2. The oil press of the Mishnah: additions—the third list. The first (Mishnah) and second list (Babylonian Talmud, Jerusalem Talmud, Tosefta) both distinguish between permanent parts of the press that are sold with it and moveable parts that are only sold when the proviso 'the press and all that is in it' appears in the contract. The third list that appears in the Tosefta (3) and Jerusalem Talmud (2) only enumerates components that are 'not sold in any case', in other words which are not parts of the press at all and must therefore in all cases be sold separately.

2.		*hyswʿyn hškwyyn*	*hḥrwtyn*
3.	*hbʾr* (*hbwr*) *hsyh*	*hysyʿyn*	*hdwtywt hmʿrwt*

The three items in the Jerusalem Talmud (2) are very unclear. However, they are apparently all three corrupt forms of terms that appear in the Tosefta where there are also two additional terms. The meaning of all six terms in the Tosefta is almost certain:

bʾr: well
bwr: a pit, in this case probably an underground water cistern, according to Rashi (*B. Bat.* 17a); round
syḥ (*hsškwyyn*): an undergound water cistern, according to Rashi (*B. Bat.* 17a) rectangular in shape
ysʿyyn (*yswʿyn*): an outhouse or extension to a building (*B. Bat.* 4.1)
dwt (*dwtywt*, *ḥrwtyn*): a building built underground (*B. Bat.* 4.2; *B. Bat.* 64a)
mʿrwt: caves

The second and third lists are clearly logical additions to the Mishnaic text and, although later in date, do not present a different stage technically.

A1.2.3. Texts Concerned both with Oil and Wine Production

A1.2.3.1. Wine and oil production in the Sabbatical year. Discussions as to which methods of wine and oil production were permitted in the Sabbatical year appear in four texts: (1) in the Mishnah (*Šeb.* 8.6), (2) Jerusalem Talmud (*y. Šeb.*, 8.6, 38b), (3) Tosefta (*t. Šeb.* 6.27), (4) Sifra Har Sinai ch. 1.

As regards wine, there is no difference between the texts and the injunction is clear: 'One may not tread grapes in a winery (*gt*) but only in an *ʿrybh*.' *ʿrybh* is a trough or tub. The term *ʿrybh* is used for a tanner's trough (*Kel.* 15.1; Epstein 1924: 37), a trough for mixing mortar (*Kel.* 20.2; Epstein 1924: 54), and a footbath (*Yad.* 4.1).

As regards olives, the text in the Mishnah (1) and Sifra (4) are identical: 'one does not process (*ʿwsym*, literally 'do') olives in a *bd* and a *qwṭb* but one pounded them and put in a *bwdydh*; R. Shimon ben Yohai says 'crush them in the olive press (*byt bd*) and put them in a *bwdydh*'.

In the Jerusalem Talmud the injunction is: 'One does not process the olives in a *bd* and in a *qwṭby* but our teachers permitted processing in a *qwṭby*. Rabbi Yohanan taught to those [the students] of Rabbi Yanai to crush in a crushing basin (*ryḥym*) as did Rabbi Shimeon (ben Yochai) and to process them in a *qwṭby* as did the Rabbis.'

In the Tosefta the injunction is: 'Olives of the Sabbatical year are not processed in a *qynby* (*qtby*), Rabbi Shimeon Ben Gamliel and his school enacted that they should be processed in a *qynby*, Rabbi Yehuda says "crush it in a crushing basin (*rḥym*) that was not used in other years", Rabbi Shimon says "one pounds and then separates (*mqph*) in a trough (*ʿrybh*)".'

	Authority (Source)	Crushing	Pressing
1a	Anon (Mishnah)	not in *bd*	and a *qwṭb*
1b		but pound (*kwtš*)	and put in *bwdydh*
1c	Shimeon Ben Yochai (Tosefta)	to pound (*kwtš*)	and separate (*mqphh*) in a trough (*ʿrybh*)
2	Shimeon Ben Yochai Mishnah	to crush (*twḥn*) in *byt bd*	and put in *bwdydh*
3a	Our teachers (Jerusalem Talmud)		put in *qwṭby*
3b	Shimon Ben Gamliel and his school (Tosefta)		in a *qynby*
4a	R. Yohanan (Jerusalem Talmud)	to crush in *ryḥym* as Shimon (Ben Yohai) see (2)	to put in *qwṭby* as our teachers-see (3a)
4b	R. Yehuda (Tosefta)	To crush in *ryḥym* not used in other years	

On the basis of the Mishnah and Jerusalem Talmud, the two stages in the production process can be distinguished in each injunction—crushing and pressing—and four changes can be traced in the laws, each progressively more lenient (1b, 2, 3a, 4a, in chart).

The version in the Tosefta is different in several ways. In one case (1c) it contradicts the tradition in the other sources attributing to Shimeon Ben Yohai a more stringent enactment than the anonymous one of the Mishnah. In the second (3b) it confirms the tradition in the other sources only attributing to Rabbi Shimeon Ben Gamliel the ordinance attributed to 'our teachers' in the Jerusalem Talmud (3a). In the third (4b) it adds a separate tradition in the name of Rabbi Yehuda.

The text refers to two methods of crushing olives: pounding (*kwtš*) in a mortar (1b and 1c) and using a crushing basin/oil mill (1a, 2, 4a, 4b). From R. Yohanan's statement it is clear that *bd* (1), *byt bd* (2) and *ryḥym* (4a) all refer to the same type of installation and that the *bd* and *byt bd* in this case are general terms for the whole oil press building (contra Jastrow 1926: 138), but implying specifically the crushing basin.

For the second stage three methods are mentioned. The first apparently involved skimming the oil off the lees in a trough. In this case the verb *qph* (1c) is used for separating oil. Its more usual meaning is skimming off floating dregs. The other two methods involve the use of two different installations, a *bwdydh* and a *qwṭb/qwṭby qynby*, presumably three different forms of the same term.

A1.2.3.1.1. *bwdydh*. From the context it is clear that this is a simple pressing device. In *Genesis Rabbah* (31.2), 'make oil within their walls' (Job 24.11) is explained 'they were making small *bddywt*', a description of producing oil in small installations in the olive yard. In local spoken Arabic *bdwdy* is the term used for small rock-cut mortars (Dalman 1928–42: IV, 236). Therefore it is usual to understand this term as referring to small simple pressing installations Ts1 (Krauss 1910–12: II, 223).

It is not clear whether the word is a diminutive of *bd* or derives from *bdyd*, a small depression dug round a tree for irrigation (*t. Šeb.* 1: 7; *t. M. Qaṭ.* 1:2; *M. Qaṭ.* 4b).

A1.2.3.1.2. *qwṭb*, *qwṭby*, (*qynby*, *qtkw*). *qwṭb*, *qwṭby*, *qynby* appear in the various texts relating to producing oil in the Sabbatical year and *qtkw* appears in another text in the Tosefta (*t. Ṭoh.* 10.12):

> In those places where it was usual to place [the olives] in a *bd* they should place in a *bd* [where it was usual to place them] in a *bydydh* [they should place them] in a *bdydh* and [where a] *qtkw* [was usual] in a *qtkw*.

Lieberman (1959: II, 569) has shown these four terms all to be one: *qwṭb*, *qwṭby*. Krauss (1910–12: II, 600 n. 598) understood from the text in the Mishnah (1a in chart across) that *bd* and *qwṭb* were two parts of one installation and suggested the *qwṭb* to be the weight that hung from the *bd*, understanding *bd* as a press-beam. However, these two terms clearly refer to two different stages in processing the olives.

In later Hebrew *qwṭb* is a pole and in Arabic it is both the pole star and a round beam (Jastrow 1926: 1326, Goldman 1907: 46). This has led scholars to explain the term either as the beam itself (Arukh-Kohut 1926: VII, 77), or as a beam press (Jastrow 1926: 1326).

Rashi (in commentary to *Šeb.* 8.6) suggests that *bd* is a large oil press, *qwṭb* a small one and *bwdydh* a very small one. However, as already pointed out, comparison to the parallel texts (Nos. 1a and 2 in chart, p. 191) shows that, although *bd* or *byt bd* usually means the press as a whole, in the passage that Rashi was commenting on, the term refers to the crushing installation in the oil press.

The text in the Tosefta (*Ṭoh.* 10.12; p. 191 above) suggests that *bd* and *qwṭb* are different types of presses used in different regions. There is not sufficient evidence, however, on which to base any suggestion as to which specific type *qwṭb* refers to.

A1.2.3.2. The Baraitha of the Lulabim: *lwlbyn*, screw; *dpyn*, pressing board; *ʿdšyn*, press-bed; *ʿqlym*, frails. The Talmudic 'Baraitha of the Lulabim' appears in five slightly varying versions twice in the Babylonian Talmud, twice in the Tosefta and once in the Jerusalem Talmud.

Version A has been presented complete in Hebrew and English as it is the longest. The variations in the other versions appear in the notes, numbers 1-12, referring to the sections of the text, and letters A-E referring to the versions A: *Nid.* 65a-b; B: *ʿAbod. Zar.* 75a; C: *t. ʿAbod. Zar.* 8(9).3; D: *t. Ṭoh.* 11:16; E: *y. ʿAbod. Zar.* 5.14, 45b. The section in the Mishnah parallel to these (*Ṭoh.* 10.8) is very different and is quoted in full separately below.

(1) הרי שהיו גתין ובית בדין טמאות
(2) ובקש לעשותן בטהרה (3) כיצד הוא עשה (4) הדפין והלולבין
והעדשין (5) מדיחו (6) העקלים של נצרים
(7) ושל בצבוץ של שיפא ושל גמי מיישנן (8) וכמה מיישנן? - י״ב
חודש (9) רשב״ג אומר: מניחן מגת לגת ומבד
לבד (10) היינו ת״ק! איכה ביניינו חרפי ואפלי (11) רבי יוסי
אומר: הרוצה לטהר מיד מגעילן ברותחין
או חולטן במי זיתים (12) רשב״ג אומר משום ר׳ יוסי מניחן תחת
צינור שמימיו מקלחין או במעיין שמימיו רודפין: וכמה? עונה.

1. (If a person's wine presses and oil presses were unclean) 2. (and he wished to purify them) 3. (what does he do?) 4. (The *dpyn* and the *lwlbyn* and the *ʾdšyn*) 5. (he washes them) 6. (the *ʿqlym* of *nṣrym*) 7. (and of *bṣbwṣ* of *šypʾ* and of *gmy* he puts them aside) 8. (and for how long does he put them aside?—for 12 months) 9. (Raban Shimeon Ben Gamliel says: He leaves them from vintage to vintage from pressing season to pressing season) 10. (and his ruling is accepted. There is a difference between early fruit and late fruit) 11. (Rabbi Jose says: he who wishes to cleanse them forthwith pours boiling water over them or scalds them with olive lees [literally 'olive water']) 12. (Raban Shimeon ben Gamliel citing R. Jose ruled: he puts them beneath a pipe of running water or a flowing spring. For how long? A season?).

1. C.D. מי (instead of הרי) בית is missing
B.C.D. טמאין instead of טמאות
E. היתה גיתו טמאיה
2. B.C. מבקש לטהרן
E. מבקש לטהרה
3. B.C.D.E. missing
4. B. הדפין והעדשין והלולבין
C. הדפין הלולבין והעדשות
D. דפין (missing) הלולבין והעדשים
E. הלולבין הדפין והעדשה
5. C. מנגבן והן טהורות
D. מנגבין והן טהורין
E. מנגבן וטהורין
6. B. והעקלין של נצרין
C. העקלין של נצרין
D. העקלין של נוצרין
E. העקל של נצרים
7. B. ושל בצבוץ מנגבן ושל שיפה ושל גמי מישנן
C. ושל בצבוץ צריך לנגב ושל חשיפה ושל גמי צריך לישן
D. ושל בצבץ צריך לנגב ושל השיפה ושל גמי צריך לישן
E. ושל בטבוט ושל עץ של שיפה של חוטין של גמי מיישנין
8. B. כמה מיישנן missing
C. ומה ישונן שנים עשר חדש
D. כמה הוא מישנן שנים עשר חודש
E. כמה מיישנן ,כל שני עשר חודש missing
9. E. מבד לבד מגת לגת
10. C.D.E. missing
11. B. לטהרן
C. לטהרן מיד חולטן במי רותחין או מולגן במי זיתים
D. לטהרן מיד חולטן ברותחין או מולגן במי זיתים
E. missing except for אם רצה לטהרן מיד which was included in section 12
12. B. רשב״ג משום רבי יוסי אומר
C.D.E. או בנהר שמימיו מהלכין מלא (E כדי) עונה
E. משום רבי יוסי missing
C.D. ...בנהר שמימיו מהלכין בצינור שמימיו מקלחין

The parallel text in the Mishnah reads: 'The vessels of the olive press and the wine press (*gt*, winery) and the frail (*ʿql*) if they are of wood need only to be dried and they are clean, if they are of *gmy* they must be put aside for 12 months or scalded in hot water. Rabbi Jose says it suffices to put them in the stream of a river' (*Ṭoh.* 10.8).

It is of interest that Columella (12.52.22) also speaks of cleaning the frails in very hot water and also if possible in running water or a pond.

The Mishnah gives the Halachic ruling in the tersest form. Of the other five versions, those in the Tosefta C, D are almost identical and almost certainly represent the source of this Baraitha. In the two versions in the Bablyonian Talmud (A, B) there are explanatory and stylistic embellishmants. In that in *Niddah* (A), however, some of the original niceties have been lost. In the version in the Jersualem Talmud there are some fundamental changes which are not purely textual but apparently reflect a different technical context.

The text refers to *lwlbyn*, *ʿdšyn* and *dpyn,* which are almost certainly components of a press. The usual meaning of *lwlbyn* in both biblical and later Hebrew is palm branch, and the term is referred to in particular as regards the feast of Succoth (Tabernacles). Rashi (*ʿAbod. Zar.* 75A) suggested therefore that in the present context these were brooms made of palm branches used in the press room. The Arukh (Kohut 1926: V, 26-27) explains the term as 'the two posts supporting the beams of the press'. The term *lulav* in Arabic, however, means screw and the term was in use in the local Arabic dialect in the central regions of Israel for the screw of the oil press (Dalman 1928–42: IV, 218, 219). Dalman (1928–42: IV, 234) suggested this to also be the meaning in the text under discussion. As no beam is mentioned, it is probably one of the direct-pressure screw presses that is referred to here. The *ʿdšh*, literally a lentil, here probably implies a press-bed, and the *dpyn* (boards) probably the pressing boards/platen.

According to versions B, C, D, it was sufficient to dry (wipe?) the frails (*ʿql*) of *nṣryn* (wickerwork) and *bṣbwṣ* (hemp [Jastrow 1926: 183] or horse-tail, *equisetum* [Löw 1928: IV, 70]), whereas those of *šyfa* (plaited wood shavings [Jastrow 1926: 1565]; reed grass, *typha*? [Löw 1928: I, 578-82]) and *gmy* (papyrus [Löw 1928: I, 559-72]; reed grass [Jastrow 1926: 252]) are to be left for a year or to be scalded in hot water or olive lees or in running water. This ruling conforms to that of the Mishnah.

In a passage that appears in *ʿAbod. Zar.* 75a-75b only, Rabbi Yehuda discusses the cleansing of other containers used in wine and oil production: straining bags (*rwwqy*), of hair (*mzyʾ*), of wool (*ʿmrʾ*), of flax and wicker baskets (*dkwly*) and strainers (*ḥlʾtʾ*, Rashi) of plaited palm fibre, twigs and flax. He adds that if there are knots in them they are to be untied.

The sages mentioned in the text, Shimeon ben Gamliel (II) and Jose (ben Halafta), were active in the middle of the second century CE. However, they are not quoted as using the actual terms for the press that appear in the text and therefore their mention does not necessarily provide a date for the use of the press referred to. Its mention in the Tosefta, however, almost certainly shows it to originate in the Roman period, which, if the explanation for the terms presented here is correct, is additional evidence for the use of the direct-screw press at this period.

The text in the Jerusalem Talmud (E) differs from the other four versions in several important points that are apparently connected. (1) only the wine press (*gt*, winery) is mentioned; (2) the frail (*ʾql*) is in the singular and not in the plural; (3) two items were added to the list of substances of which the frails were made—wood and rope (*ḥwtyn*). This suggests that the text in the Jerusalem Talmud is referring to the single fixed-screw press (Ts81–83) used only in wineries and in which the usual frails cannot be used and, instead, the trodden grape skins and stalks are enclosed either by boards or within a rope wound to form a cylinder.

A1.2.4. Installations in Other Texts

A1.2.4.1. *ʿwqh*, collecting vat. In all the texts but one where this term appears (*Maʿas.* 1.7; *t. Maʿas.* 1.7; *y. Maʿas.* 2.6, 50a; *Sifre Num.* 121) it is stated that oil flows into an *ʿwqh*. In the one exception (*t. Ṭoh.* 10.2) the word *yqt* is apparently to be read as *ʿql*, a pressing frail. *ʿwqh* is thus clearly a collecting vat.

A1.2.4.2. *ʿqrb*, beam clamp; *ʾnqly*, wall hook. 'The *ʿqrab* of the oil press is susceptible (to uncleanness) and the *ʾnqly* in the walls are not susceptible' (*Kel.* 12.3). *ʾnqly* derives from the ανκαλε, a hook. *ʿqrab* means literally 'a scorpion'. The *ʾnqly* are not susceptible to uncleanliness because they are attached to a stone. The fact that the *ʿqrab* was susceptible suggests it was attached to the wooden beam and probably served to fasten the drum that raised the weight in a lever and weights press.

A1.2.4.3. *ʾwllh*, batch? press-bed? In the Babylonian Talmud (*B. Meṣ.* 105a) it is stated that one puts 'three *kwr* (a measure of capacity) in the *ʾwllh*'; or puts 'one *kwr* in the *ʾwllh*'.

Rashi explained the *wllh* to be the batch to be pressed, as did Krauss (1910–12: II, 59 n. 30). (compare *bd* and *factus* above, p. 188). Brand (1962: 315) suggested that the *ʾwllh* was another term to designate the press-bed.

A1.2.4.4. The seat attached to the press-beam. Mishnah *Kel.* 20.3 reads: 'if they fixed a seat to a press-beam, the seat is susceptible [to uncleanness] but the beam does not count as a connective with it: if the beam-end was used as a seat it is not susceptible to uncleanness because they say to him [that sits upon it] "Get up and let us do our work." '

In certain cases implements used for work were susceptible to uncleanness and if they were not so used they were not susceptible.

In this case the explanation of the injunction is apparently that the drum to raise the weights was at the end of the beam. If a seat could be used to operate the drum, it was susceptible, but if it was placed at the end of the beam and hindered the operation of the drum, it was not susceptible (Brand 1962: 310).

A1.2.5. The Oil for the Temple Offering

The relevant texts are: *Men.* 8.3-5; 85b-86b; *t. Men.* 9.5-8; *Sifra-Amor* 13.

The discussion deals with the various factors that affect the quality of oil and can be divided into five subjects:

1. The places from which the best oil was brought.
2. Unsuitable oils.
3. The three types of olives, classified according to the treatment received before crushing.
4. The three types of oil graded according to pressing method.
5. The nine available varieties of oil resulting from extracting the three types of oil from each of the three types of olives.

The four texts vary. In some cases they are similar. In some cases they complement one another, while in others, particularly as regards the fourth subject, there are marked differences which reflect differences of opinion between the sages and a development in the accepted Halacha.

A1.2.5.1. The places from which the best oil was brought. The Mishnah states: 'Tekoa comes first in its quality of oil.' Abba Saul says: 'Second to it is Regeb beyond Jordan.' The Tosefta adds: 'Eleazar ben Yaakob says third is Gush Halav in Galilee.' In the Babylonian Talmud there is a narrative about a man from Gush Halav who sold a hundred myriads of oil to the people of Laodicea in Syria. This narrative is remarkably similar to a narrative about the sale of oil by John of Gischalla (Gush Halav) to the Jews of Syria (Josephus, *War* 2.21.2).

A1.2.5.2. Unsuitable oils. The Mishnah brings a list of fields from which offerings are not be be brought: 'not from *byt hzblym*, manured fields, *byt hšlḥym*, irrigated fields, or from those (having other crops) which were sown between them' (known today as intercropping).

This section provides valuable data as to the agricultural methods of the time even if they were here regarded as exceptional and therefore made the olives unfit for offering. The parallel section in the Tosefta speaks of bringing 'special olives'.

The following section enumerates the types of oil that are unfit; the Mishnah: *ʾnpyqnwn* (Latin: *omphacium*, Greek: ὀμφάκιον; ὄμφαξ, unripe grape), bitter, green, summer oil made of unripe olives (see discussion above, pp. 42-43). In the Tosefta it is explained that this is unfit because it is only 'oil sap' (*srp*). The Mishnah: 'not shrivelled olives (*grgrym*) that were soaked in water (*nšrw*)'; the parallel in the Tosefta: 'soaked (*šrwy*)'. The Mishnah: 'not pickled (*kbwšym*) or stewed (*šlwqym*)'; in the Tosefta appears stewed (*šlwq*), cooked (*mbwšl*), of the lees (*šmrym*) and ill-smelling. *šwlqy zytym*, olive stewers/seethers and their main tool of trade *ywrh šl šwlqy zytym*, the olive seethers boiler (*Kel.* 5.5), are mentioned often in Talmudic texts (Ayali 1984: 71) and apparently produced oil by boiling olives, probably those of poor quality, such as windfalls.

A1.2.5.3. Three types of olive. The discussion on the types of olive graded according to how they were treated before pressing and that on the types of oil graded according to the method of pressing are combined. In each case the type of olive is defined and then the way each of the three types of oil were produced is explained. Therefore the section on the types of oil was repeated in an identical form three times. The opening sentence in the Mishnah explains this: 'There are three [grades of] olives and of these three [times] three grades of oil.'

The subject appears in the Mishnah, Sifra and in a Baraitha in the Babylonian Talmud. The texts of the Mishnah and Sifra are fuller, while that in the Talmud is shorter but more similar to that in the Sifra.

Mishnah: 'The first olive ripens (*grgr*) on the olive tree'; the Sifra adds: 'and is brought into the olive press (*bd*)'. The version in the Baraitha is identical to the Sifra, only instead of *grgr* appears *glgl*. The meaning of *grgr* is explained as 'to ripen' or 'to shrivel', which is in effect to over-ripen. *glgl* can be explained as rounded, fully ripened (Rashi, *Men.* 86a; Jastrow 1926: 264).

Mishnah: 'The second olive is ripened (*grgr*) on top of the roof'; the Sifra adds: 'and is brought into the olive press'; in the Baraitha appears only 'and so with the second olive'.

Mishnah: 'The third olive is stored (*ʿwṭnw*) in the house until it is over-ripe (*ylqh*) and taken up and dried on the roof.' In Sifra: 'The third olive is stored in the press room (*byt bd*) until it is over-ripe and then taken up and placed in a pile (*tmrh*) on top of the roof and brought into the house.' In the Talmud this section reads: 'and taken up on top of the roof and made into a sort of pile (*tmrh*) until the liquid flows and then brought into the press room.' The usual meaning of *tmr* is 'date' and *tmrh* has been translated as 'in the same manner as dates' (Soncino Talmud *Qod.* 1.521), but the root *tmr* also means to rise straight up and the context suggests that here the term means 'a pile'.

Rashi (*Men.* 86a; Soncino Talmud, *Qod.* 1.519-20) explains the three types of olive as representing three stages in picking: in order of ripening, those at the top of the tree first, those at the roof-top height second, and those left third. This, however, is not the way olives ripen and clearly is a misinterpretation of the text, which refers to three different ways of treating the olives after picking but before pressing.

Both Cato (64) and Columella (12.52.18-19) argue that leaving the olives in a store-room before picking does not increase the yield of oil. At first glance the text under discussion would suggest otherwise. However, the situation that the text reflects is one in which the olives are picked faster than the presses can cope with, resulting in the need for storing them in the interim period. Columella specifically says: 'we have already given instructions that each day's picking should be immediately placed under the millstone and the press. However, since the large quantities of berries sometimes defeats the efforts of the men who work the presses there ought to be a store-room to which the fruit may be conveyed.' (Columella 12.52.1-3). Storing olives on the roof awaiting their turn at the oil press is a sight still seen in villages in Israel. The first and best olive is that taken directly to the press after picking; the second is stored for a short time on the roof; the third for a longer time in the store-room, thus necessitating drying them and moving them again before pressing.

A1.2.5.4. The three types of oil. This subject appears in all four sources, Mishnah, Sifra, the Baraitha in the Babylonian Talmud and in the Tosefta. In all four texts the differing opinion of Rabbi Yehuda ben Ilai (mid second century CE) is also brought. In all the sources but the Mishnah Rabbi Yehuda's opinion is brought separately. In the Mishnah, however, as already pointed out in the Babylonian Talmud (*Men.* 86a) in those cases where the compiler (Yehudah Ha-Nasi) accepted R. Yehuda's view, his opinion is brought without R. Yehuda's name being mentioned, whereas where the compiler did not accept R. Yehuda's opinion, R. Yehuda's differing view is quoted mentioning his name, and placed after the precept prescribed in the Mishnah which represented the view of Yehuda Ha-Nasi, the compiler.

In the version in the Baraitha, to produce the first oil, 'the olives were crushed in a crushing basin (*ryḥym*) and placed in baskets'. To produce the second oil, olives were then pressed under a beam (*qwrh*). To produce the third oil, the beam was dismantled, the olive pulp crushed a second time and once again pressed under the beam. In the Sifra the beam was already used to produce the first oil and the pulp was crushed and pressed twice more to produce the second and third oil.

In the Baraitha appears: 'R. Yehuda says "The olives were not crushed in a crushing basin (*rḥym*) but in a mortar, they were not pressed under a beam but with stones and they were not put into baskets but around the basket."' In the Tosefta the last clause states: 'they did not put the loaf (*lḥm*, cake of crushed olive, see above, p. 187) in the basket, but the basket in the loaf'—the meaning being apparently that R. Yehuda was of the view that the stones that pressed the pulp were to be placed in a basket on the olive pulp. Another statement of Rabbi Yehuda that appears both in the Baraitha and the Sifra explains his approach. R. Yehuda explains the meaning of the biblical term *šmn ktyt* (see above, p. 186) by asking a rhetorical question: 'beaten (*ktyt*): is beaten (*ktyt*) not pounded (*ktwš*)?' *ktwš* is from the same root as *mktšt*, mortar; *šmn ktyt* was the term used in the Bible for the finest oil for the temple offering and Rabbi Yehuda was clearly of the opinion that the oil for the temple when it is rebuilt should be made in the same way, using a mortar to pound the olives and pressing by using stones placed directly on the olive pulp without a beam.

The Mishnah states as follows: '[The olives] are pounded and put in the basket. Rabbi Yehuda says around the basket. That is the first [oil]. Placed under the beam. Rabbi Yehudah says with stones that is the second [oil] crushed once more and again placed under the beam that is the third [oil].'

The method prescribed by the Mishnah was a compromise between that described in the Baraitha in the Babylonian Talmud and the method suggested by Rabbi Yehuda ben Ilai. In the method described in the Baraitha, the olives were first crushed in a crushing basin and then the pulp was left without pressing and the oil that dripped out was the first oil. The pulp was then pressed to produce the second oil and then the pulp was crushed a second time and pressed a second time to produce the third oil. Rabbi Yehuda was of the opinion that the oil for the temple should be from olives pounded in a mortar and pressed using weights but no beam. The compromise in the Mishna accepted Rabbi Yehuda's view that the first oil should be from olives pounded in a mortar; however, the second and third oils were as prescribed in the Baraitha apparently including crushing in a crushing basin, as the term *ṭḥn* is used.

In classical literature there are several references to three grades of oil and, although these are without doubt primarily the products of consecutive pressings (Columella 12.52.11), there is also some confusion as to terminology. Diocletian's price list refers to three grades of oil *olei flos*, *olei sequentis* and *olei cibari* (Lauffer 1971: 102-103) and Pliny (20.6.23) explains *flos* as the first oil to flow from the press. In the Greek version of the price list, however, the equivalent of *flos* is ὀμφάκιον, *omphakion*, virgin oil made from unripe olives, while Columella (12.52.11) refers to the first oil produced without actual pressing as *lixivium*.

A1.2.5.5. The nine grades of oil in the Mishnah. In the next paragraph it is explained that three types of oil from three types of olive produced nine types of oil—the first of the first was the finest, the second of the first was equivalent to the first of the second and the third of the first was equivalent to the second of the second and the first of the third and so on, producing five grades of oil. The finest oil was for the temple candelabrum, *mnwrh*, the other grades for the other temple offerings

A1		
A2 =	B1	
A3 =	B2 =	C1
	B3 =	C2
		C3

A1.2.6. *mwḥl*, *my zytym*, Olive Lees

Olive lees were usually termed *mwḥl* (*Makš*. 6.5; *Ṭoh*. 9.3; *t. Ṭoh*. 10.2, 3; *Šab*. 144b), but the term *my zytym*, olive water, was also used (see versions A, C and D of the 'Baraitha of the Lulabim', above).

The juice that exuded from the olives when left standing and the watery lees left after the oil had been separated were both called *mwḥl*. The main Halachic discussion was whether the lees should be regarded as one of the liquids that makes the material it drops on susceptible to ritual uncleanness (see above). Among the arguments brought were that there were always some drops of oil left in the lees and so the lees should be regarded as oil.

A1.2.7. Variations in Terminology in Talmudic Literature

The texts and terms discussed above demonstrate that in Talmudic literature a variety of terms was used both for the plants as a whole and for the various components. In the accompanying table the terms have been arranged according to the main relevant texts or groups of texts.

It has already been shown that some of these texts can be related to specific types of installation. The Mishnaic oil press (Text 1) refers to the Zabadi press, the Phoenecian lever and weights press; the press of the 'Baraitha of the Lulabim' in the Babylonian Talmud and Tosefta (Text 3) is a direct-screw press (Ts7); and that in the Jerusalem Talmud (Text 4) the single fixed-screw wine press (Ts81–83).

Terms mentioned in other texts could clearly refer to some of the many other types of installation which the archaeological evidence shows to have been in use. Modern ethnological evidence makes it clear, however, that different terms are often used for exactly the same installations even in neighbouring villages (Dalman 1925–42, IV, 204-206). It is therefore extremely difficult to determine the exact meaning of many of the Talmudic terms.

A1.2.7.1. The complete oil press. The usual term for an olive press is clearly *byt bd* (Texts 1, 3, 5) or simply *bd* (Texts 5, 10). This in some cases does not include the crushing equipment (Text 1); in others it clearly does and possibly sometimes specifically refers to such equipment (Text 5).

The term *qwṭb* (Text 5; *qtkw*, Text 10) is also an oil press, the relevant texts showing clearly that it is not an alternative term for *byt bd*. Rashi suggested that it refers to a smaller version but more probably a specific type is meant.

bwdydh (Text 5; *bydydh*, Text 10) is clearly a small simple press perhaps without beam.

*zyyr*ᵓ and *mkbš* are other terms for press, but again what type they refer to is not clear.

A1.2.7.2. Press-bed. In the Mishnaic oil press (Text 1) the press bed is called a *mml*, a term that appears in other texts also (Text 7). The *ᶜdšyn* (singular *ᶜdšh*) of Texts 3 and 4 is perhaps a screw base of the ᶜEin Nashut type, while it is possible that *ᵓwllh* (Text 9) is another term for press-bed.

A1.2.7.3. Collecting vat. The term used for a collecting vat in the Mishnaic oil press (Text 1) *ym*, sea, does not appear in this meaning in other texts, the more common term being *ʿwqh* (Text 7), while the general term *bwr*, pit, was also used (*Ṭoh.* 9.3). The *syrym* and *yqbym* of Text 2 are probably separating or storing vats or vessels.

A1.2.7.4. Press-beam. The term *qwrh* for press-beam is common (Texts 1, 6, 8), the term *bd* appearing only as a part of the name of the oil press, which raises some doubts as to whether this is the source of the term.

A1.2.7.5. Winch, slotted piers, screw. Only one term appears in the texts for each of these components and each of these only in one text (winch, *glgl*, and slotted piers, *btwlwt*, Text 1; screw, *lwlb*, Texts 3, 4). There is no doubt, however, as to the meaning of the terms.

A1.2.7.6. Beam weight. The beam weight of the Mishnaic oil press appears also in the tractate *Kelim* (12.8; *t. Kel.* 7.12) in a great number of variants—the original term was probably *ʿqydym*. Another term was used for the weights of a lever and weights wine press, *ʿgwly hgt* (Text 10).

A1.2.7.7. Crushing equipment. The usual term for the rotary olive crusher was upper and lower *rḥyym* (Texts 2, 5, 6). However, *mprykwt* (Text 2) also refers to crushing equipment and it is not clear to what type. A *mktšt* is a mortar (Text 6).

A1.2.7.8. Frails. The common term for pressing frails is *ʿql* (Texts 3, 4, 7). However, in Text 6 the term *slym* (baskets) clearly refers to a frail. The *sqyn* and *mrswpyn* of Text 2 refer perhaps to bags and baskets for other purposes.

A1.2.7.9. Wooden components. Various terms in the texts clearly refer to wooden components of the press, the pressing board (platen) and press frames. The same term is not repeated in two different texts, however, and the exact meaning of *psym* (strips, Text 7), *dpyn* (boards, Texts 3, 4) and *nsryn* (sawn boards, if this is not a textual error, Text 2) cannot be determined.

Chart 7. Terms used for installations in Talmudic Literature

	1	2	3	4	5	6	7	8	9	10
	Mishnaic oil press	2nd list and 3rd list	Beraitha of the Lulabim: Tosefta and Babylonian Talmud	Beraitha of the Lulabim: Jerusalem Talmud	Oil in the Sabbatical year	Oil for the temple	Tithes	Sabbath observance		Ritual cleanliness
	B. Bat. 4.5	*B. Bat*. 67b *y. B. Bat*. 4.5, 14c *t. B. Bat*. 3.3	*Nid*. 65a-b *ʿAbod. Zar*. 75a *t. ʿAbod. Zar*. 8(9).3 *t. Ṭoh*. 11.16	*y. ʿAbod. Zar*. 5.14, 45b	*Šeb*. 8.6 *y. Šeb*. 8.16, 38b *t. Šeb*. 6.27 *Sifra Har Sinai* 1	*Men*. 8.3-5 *Men*. 85b-86b *t. Men*. 9.5-8 *Sifra Amor* 13	*Maʿas*. 1.7 *t. Maʿas. Rish* 1.17 *t. Ter*. 3.13 *y. Ter*. 3.42b	*Šab*. 1.9 *t. Šab*. 1.21 (*Ṭoh*. 10.8)	*B. Meṣ* 105a	*t. Ṭoh*. 10.12
Complete installation oil press/ winery	*byt bd* (oil)		*byt bd* (oil) *gt* (wine)	*gt* (wine)	*bd/byt bd?* *bwdydh* *qwṭb?*					*bd: qtkw bydydh*
Press-bed	*mml*		*ʿdšyn*	*ʿdšyn*	*qwṭb?*		*mml*		*ʾllh?*	
Collecting vat	*ym*	*syrym?* *yqbym?*					*ʿwqh*			
Wooden components Presssing board Pressing frame		*nsrym?*	*dpyn*	*dpyn*			*psym*			
Pressing frails		*sqyn* *mrsqpyn*	*ʿqlyn*	*ʿql* of wood or rope		*slym*	*ʿql*			
Crushing basin and crushing wheel		*rḥyym* upper and lower *mprykwt*			*bd/byt bd?* *ryḥyym*	*mktst* (mortar) *ryḥym*				
Beam	*qwrh*					*qwrh*		*qwrh* (oil)		
Weights	*ʿkyrym* (*ʿkyrym*)							*ʿygwly hgt* (wine)		
Winch	*glgl*									
Slotted piers	*btwtwt*									
Screw			*lwlbyn*	*lwlbyn*						

Appendix 2

Types of Wine in Ancient Hebrew Literature

A2.1. Terms for Wine in the Hebrew Bible and Related Epigraphic Sources

A2.1.1.1. *yyn*. The usual term for wine in biblical and later Hebrew is *yyn,* one of the roots found both in Indo-European and Semitic language. The root was originally almost certainly *wyn*, as it remained in Latin and western European languages (Brown 1969: 147-51; Kutcher 1961: 34).

The term *yyn* appears not only in the Bible but is also attested in inscriptions that originate in Judaea—e.g. from Arad (Aharoni 1981: letters 1, 2, 3, 4, 8, 10, etc.), from Lachish (Ussishkin 1978: 83 inscription XXV) and from a decanter apparently from the Hebron region (Avigad 1972).

Further north the form was *yn*. This was the form in northern dialects of Hebrew: Samaria Ostraca (Reisner, Fisher and Lyon 1924: 227; Aharoni 1968: 315-27), in Phoenician-Shiqmona jar inscriptions (Cross 1968) and in Ugaritic (Gordon 1965: 410, 1093).

A2.1.1.2. *tyrwš*. *tyrwš* appears in the recurring biblical formula of *tyrwš* and *dgn*, the two staple crops wine and wheat (Gen. 27.28; Num. 18.12; 2 Kgs 18.32; etc.), and in the standard triad that includes also *yṣhr*, oil (Deut. 28.51; 33.28; 2 Chron. 32: 28; Neh. 13.5; etc.). *tyrwṣ* is understood to mean new wine, must, but here clearly represents wine.

A2.1.1.3. *ʿsys*. *ʿsys* is apparently a general term for fruit juice (Cant. 8.2) but referring also to new wine (Joel 1.5).

A2.1.1.4. *škr*. *škr* means literally 'intoxicating liquor'. In later Hebrew it refers to beer, date wine, etc., often with a descriptive noun in construct state. In the Bible the term often appears as a poetic parallel to or in association with *yyn* (e.g. Num. 6.3; Isa. 29.9; 56.12; Prov. 20.1; 31.6). In the Midrash various views are presented as to whether the two biblical terms were synonymous. Yossi Hagalili states *yyn* is *škr* and *škr* is *yyn*, while Eleazer Hagapar suggests that one is mixed wine, *mzwg*, and one pure, *ḥy* (living), although he wavers as to which is which (*Num. R.* 10.8 to Num. 6.3).

A2.1.1.5. *ḥmr*. The term *ḥmr* appears certainly twice in the Bible (Deut. 32.14; Ps. 75.9) and possibly a third time (Isa. 27.2). In the LXX, however, the latter is translated as 'pleasant', implying a *Vorlage* of *ḥmd*, resh and daleth being easily confused (compare to Amos 5.11). The term *ḥmr* appears also in a document from Ugarit (Gordon 1965: text 23), in two jar inscriptions from Shiqmona (Cross 1968) and on an *ostracon* from Arad (Aharoni 1981: number 2.5). Aharoni explains the meaning in the Arad *ostracon* as a liquid measure (compare Ezek. 45.11, 14). In Deuteronomy (32.14), however, the term is clearly a synonym for wine and the other contexts imply it to be a type of wine. The meaning of *ḥmr* as seething, bubbling, as water in a flowing river (Ps. 46.4) suggests the meaning here to be '(well)-fermented vintage wine' (Mandelkern 1969: 404; Cross 1968: n. 15; Lemaire 1977: 162, 229). *ḥmr* is the standard term for wine in Aramaic.

A2.1.1.6. *ḥmṣ*. The term *ḥmṣ* (Num. 6.3; Ps. 69.22) does not mean only vinegar but also sour wine—piquette—as is also illustrated by the epigraphic evidence in an Arad *ostracon* (Aharoni 1981: number 2.7; Lemaire 1977: 162, 229).

A2.1.1.7. *mz*. *mz* is a term that does not appear in the Bible but is attested for in two jar inscriptions, one from Hazor Stratum Va (eighth century BC), the other from Lachish Stratum II (early sixth century). By comparison to terms in other semitic languages, *mz* has been shown to mean 'extract' or 'essence' (Lemaire 1980).

A2.1.1.8. *smdr*. *smdr* is a biblical term (Cant. 2.13, 15; 7.13) which also appears in a jar inscription from Hazor (Yadin *et al.* 1959: inscription 7, Pl. 95.4), clearly there having a different meaning. The biblical meaning is almost certainly 'the flower of the vine', whereas the jar inscription refers to a type of wine. It has been suggested that this was wine made from seedless grapes—sultanas (Altbauer 1971)—but it is more probable that *smdr* was either wine flavoured with the flower of the wild vine, οἶνος οἰνάνθινος, *vinum oenanthium*, the *ʾlwṭyt*, *ʾlwntyt*, *ʾlwntyn* of Talmudic literature (see below) or perhaps wine made from unripe grapes, ὀμφάκιον, *omphacion* (Ahituv 1974–75).

A2.1.1.9. *msk, mzg*. There are several biblical terms that refer to wines mixed with other substances. The terms *msk* (Isa. 5.22; Prov. 9.2; 9.5; Ps. 75.9; 102: 10) and *mzg* both mean to mix wine and are clearly connected to one another and both to the Greek μίσγω and Latin *misceo*. In biblical terminology both *mzg* and *msk* mean to mix different wines, whereas in later Hebrew the former meant 'watering down wine' and the latter 'mixing wines' (Brown 1969: 153-54).

A2.1.2.0. *yyn ʾgnt*. The term *yyn ʾgnt*, the wine of the vat/bowl, appears in Ostracon 1 from Arad (Aharoni 1981), and the juxtaposition of *mzg* and *ʾgn* in Cant 7.3 suggests that *gn* was a technical term for a bowl in which wines were mixed, and *yyn ʾgnt* such a term for wine so mixed.

A2.1.2.1. ***yyn hrqḥ, yyn ṭwb***. *yyn hrqḥ* (Cant 8.2) is spiced wine. *yyn ṭwb* (Cant 7.10) does not, as would appear, mean simply 'good wine' but more specifically 'spiced or resinated wine' as is demonstrated by parallels from Old Babylonian (Paul 1975: n. 123, Ugaritic (Gordon 1965: 1084: 1, 4, 6, 9) and Arabic (Sasson 1981: n. 9).

A2.1.2.2. ***yn yšn***. *yn yšn*, old wine, a term that appears several times in the Samaria *ostraca*, is clearly also a technical term, probably meaning wine one year old (see *B. Bat*. 6.3 and p. 200).

A2.1.3. Wines Designated by Provenance

Designation of wine by provenance is perhaps the characteristic method in recent centuries. Biblical examples are *yyn ḥlbwn*, Helbon wine (Ezek. 27.183), *yyn hlbnwn*, wine from Lebanon (Hos. 14.8).

Epigraphic material enlarges the list considerably: *gt krml*, Gat Carmel, jar inscriptions from Shiqmona (Cross 1968); *gbʿwn*, Gibeon, stamped jar handles from el-Jib, biblical Gibeon (Pritchard 1959, Avigad 1959); and *mṣh*, Moṣah which appears on an encised jar handle from the Iron Age (Avigad 1972: 5-9) and on stamped handles from the Persian period (Avigad 1958).

As regards two different terms that both appear in inscriptions, there are differences of opinion as to whether they refer to the provenance of the wine or to its type. *yyn kḥl* refers either to a place of that name, several of which are known (Avigad 1972: 1-5), or to the dark colour of the wine (Demsky 1972). Similarly *yyn ʿšn* refers to the place Ashan (Lachish inscription XXV: Ussishkin 1978: 83-74) or to smoked wine (Demsky 1979).

One interesting jar inscription from the Persian period that clearly refers to the provenance of wine is one from Ashdod. It reads *krm zbdyh*, 'the vineyard of Zibdiyah' (Dothan 1972: 12; Hestrin 1972: number 165).

A2.1.4.1. The *lmlk*, seal impressions. Another group of epigraphic finds that possibly refers to the provenance of wine are the *lmlk* stamps. Many hundreds of these stamped handles have been found, almost all in Judaea. Each impression includes one of two symbols, either a four-winged scarab or a double-winged sun disc. In all cases above the symbol appears the word *lmlk* 'to (or of) the king', and below one of four names, three of these known towns: *ḥbrn* (Hebron), *swkh* (Socoh), *z(y)p* (Ziph), the fourth apparently also a town but as yet unidentified—*mmšt*. Recent research has shown these jars to date to the period of Hezekiah (Ussishkin 1977) and it would appear almost certain that their production was connected to preparations for war against Assyria, but the exact significance of the inscriptions remains a matter in dispute.

The term *bt* is a biblical liquid measure (1 Kgs 7.26; 2 Chron. 2.9); the term *bt lmlk*, the royal *bat* measure, has been recorded on two stamped handles from Tell en-Nasbeh and Lachish (Avigad 1953); and the term *lmlk* appears in the Shiqmona inscription in a context that clearly implies it meaning a royal measure (Cross 1968). Aharoni (1968: 340-46) suggested the *lmlk* to represent a royal measure and the towns administrative centres. Cross (1969) concurred with the view that the *lmlk* signified a standard measure, but suggested that the places were royal vineyards in line with the other inscriptions and records that designated wine by its provenance.

The fact that a *lmlk* stamp was found on a large *pithos* at Beer Sheba made its representing a standard measure doubtful (Aharoni 1973: 76-77). Ussishkin (1978: 77-80) has since shown that the volume of the jars varied considerably. The capacity of six complete jars was 43.00, 44.25, 44.75, 45.33, 46.67 and 51.80 litres. The variation between the largest and smallest was 8.8 litres, which confirmed doubts as to whether the *lmlk* referred to a standard royal measure. In recent researches it has been suggested therefore that the *lmlk* stamp refers to royal property. Rainey (1982) presents this view and sees the four names as those of royal vineyards, and Naʿaman (1986), while agreeing that the *lmlk* stamp designates royal property, sees the places as administrative centres.

Other jar handle seal impressions are often found associated with the *lmlk* stamps, some with private names, others concentric circles scratched on them.

A2.1.4.2. Other seal impressions and jar inscriptions. An additional group of jar inscriptions that should be mentioned here, although they also do not refer directly to a type of product, are three that refer apparently to the profession of those who used the jar. A jar from Dan dated to the ninth to eighth century BCE bears the inscription *lṭb(ḥ)yʾ*, 'belonging to the butchers, or cooks' (Avigad 1968–69:). A jar from Ein Gev from the ninth century BCE bears the inscription *lšqyʾ*, 'belonging to the cup-bearers' (Mazar *et al.*, 1964: 27). A jar from Hazor (Yadin *et al.* 1959: 19 Pl. CLXX, CLXIX) was reinterpreted by Naveh (1981) as reading *lmkbdm*, 'belonging to the food server'.

Many stamped jar handles and jar inscriptions are also found in contexts from the Persian and Hellenistic periods. These in most cases, however, refer neither to the type of product nor its provenance. They include many stamps with private names, e.g. those from Ramat Raḥel (Aharoni 1954: 145-46). Others of particular interest are those inscribed with the term *yhd* (Juda), *yrslm* (Jerusalem) and also stamps with names together with the term *pḥw*. These are usually thought to be the names of governors of Jerusalem, although there are those who understand *pḥw* to mean 'potters' (Avigad 1957, 1960, 1976: 35; Cross 1968; Lapp 1963).

A2.2.1. Types of Wine in Talmudic Literature

Although there are not specific instructions as to how to make wine in Talmudic literature, such as those in the works of the Roman agronomists, Cato, Varro, etc., we find there innumerable references to wine, primarily in connection to religious precepts (*Halacha*) but also in parables and anecdotes in the Midrash and Hagadah.

The main precepts and religious injunctions involving wine were those concerned with wines suitable and unsuitable for the temple offering (mainly *Men*. 8 and parallels; compare Pliny 14.23.119), the prohibition of the use of libation wine, *yyn nsk* (wine suspected of having been prepared for use in pagan practice) (mainly *ʿAbod. Zar.* 4, 5 and parallels) and of exposed (*mgwlh*) wine, wine that is forbidden because it might have been contaminated by having been drunk by a poisonous snake (mainly *B. Bat.* 97b; *ʿAbod. Zar.* 70a) and legal questions concerned with purchase and sale (*B. Bat.* 5, 6 and parallels).

In these passages many types of wine are mentioned, the Hebrew terms appear mainly in the Mishnah and Tosefta but sometimes also in Baraithot in the Talmudim. These are then often explained sometimes in Hebrew and sometimes in Aramaic in the Talmudim and Midrashim.

As in other cultures, the names and appellations of wines were connected to various characteristics, while some derived from Greek and Latin (Löw 1928: I, 56-63; Paul 1975; Gur 1960). In most cases no attempt has been made here to distinguish between an adjective used to describe a wine and an actual type of wine—an exception is in the case of wines named according to provenance.

A2.2.1.1. Wine named according to age

A2.2.1.1.1. New wine

yyn ḥdš, Hebrew (*Giṭ.* 67b; *Qid.* 20a; *t. B. Meṣ.* 9: 12)
ḥmr ḥdtʾ, Aramaic (*Giṭ.* 67b; *ʿAbod. Zar.* 66a)

A2.2.1.1.2. Old wine

yyn yšn, Hebrew (*Men.* 8.6; *B. Bat.* 6.3; *Yom.* 18a)
ḥmrʾ ʿtykʾ, Aramaic (*Giṭ.* 70a; *ʿAbod. Zar.* 66a)
ḥmrʾ ʿtyk, Aramaic (*Lam. R.* 2.16; 23b)
ʿtyk ʿtyky, Aramaic (*Pes.* 42b, as an explanation of *yyn yšn*)

A2.2.1.1.3. Vintage wine

yyn mywšn/myyšn, Hebrew (*B. Bat.* 6.3; *B. Bat.* 98b; *t. B. Bat.* 6.5; *Ned.* 9.8)

In the Mishnah old wine (*yyn yšn*) is defined as that of the previous year and vintage wine (*yyn mywšn*) as three years old (*B. Bat.* 6.3). The vendor of old wine was held responsible for its quality until the feast (of tabernacles), i.e. for a year (*t. B. Bat.* 6.5). Wine before it was 40 days old was not considered yet fit to be brought as an offering to the temple (*t. Men.* 9.12).

There was also controversy, however, as to whether old wine was fit to be brought as a temple offering, Rabbi (Yehuda Ha-Nasi, the compiler of the Mishnah) forbidding it and the Sages allowing it (*Men.* 8.6).

The qualities of old wine were lauded in many passages:

> R. Jose ben Yehuda of Kefar Ha-Babli said: He that learns from the young to what is he like? To one that eats unripe grapes and drinks wine from his winery. And he that learns from the aged to what is he like? To one that eats ripe grapes and drinks old wine. Rabbi said: Look not at the jar but on what is in it; there may be a new jar that is full of old wine and an old one in which is not even new wine (*Ab.* 7.20).

'Old wine is good for the stomach' (*Ned.* 9.8). In a discussion as to how a slave is to be treated, Abaye stipulates 'thou (shouldst not) drink old wine and he new wine' (*Qid.* 20a). Because of its potency, old wine was forbidden to the high-priest while preparing for the Day of Atonement (*Yom.* 18a).

A2.2.1.2. Wines named by colour

Pliny (14.11.80) speaks of four colours of wine: white, brown, blood-red and black; similar distinctions are found in Talmudic literature.

A2.2.1.2.1. Red wine.

yyn ʾdwm, Hebrew (*Nid.* 9.11; *Zeb.* 78b; *y. Šab.* 8.1, 11a)
ḥmrʾ swmqʾ, Aramaic (*Šab.* 129a).

Different coloured grapes produced different coloured wines: 'one vine has red wine, another black' (*Nid.* 9.11). Red wine was considered fine wine: 'R. Hisda said there are sixty kinds of wine, the best of all is red fragrant wine (*swmqʾ ryḥtnʾ*)' (*Giṭ.* 70a).

A2.2.1.2.2. Dark wines

yyn šḥwr, Hebrew, black wine (*Nid.* 9.11); *yyn kwšy* (literally 'Negro wine'), Hebrew (*t. Men.* 9.9; *B. Bat.* 97b)
yyn ḥrdly (literally 'mustard wine'), Hebrew (*Šab.* 62b; *Gen. R.* 98.10, 182a)

yyn kwšy is one of the wines that appears in the list of wines unfit for the temple ritual in the Tosefta and the Babylonian Talmud (*B. Bat.* 97b; *t. Men.* 9.9) but is missing from that in the Mishnah (*Men.* 9.6). In both sources the proviso is added that if brought to the temple it is permitted. The name *ḥrdly*, mustard wine, is usually taken to refer to the colour of the wine, according to Rashi black and translated in the Soncino Talmud as dark. In both the texts it appears together with *yyn gwrdly*, clearly a play on words (see below under white wines).

A2.2.1.2.3. White wines

yyn lbn, white wine, Hebrew (*Yom.* 18a; *Zeb.* 78b; *t. Miq.* 7.4)
ḥmr ḥywryn, white wine, Aramaic (*B. Bat.* 97b; *Ker.* 6a)
ḥmr ḥywryn ʿtyk, old white wine, Aramaic (*Ker.* 6a)
ṭylʾ hwwrt, white tillia (a type of wine), Aramaic (*Giṭ.* 70a)
bwrq, bwryq (t. Men. 9.9; *B. Bat.* 97b)
my bʾrg (*ʿAbod. Zar.* 30a)
yyn gwrdly (*Šab.* 62b; *Gen. R.* 98.10, 182a)
(*yyn*) *hḥlb*, literally '(wine) of milk' (*Gen. R.* 98.9, 181b)

White wine (*yyn lbn*) was considered particularly intoxicating and so forbidden for the high-priest while preparing for the Day of Atonement (*Yom.* 18a).

bwrq comes from the same root as lightning and is explained by Jastrow as a 'white effervescent wine—not fully fermented'. It is also one of the wines unfit for temple ritual that appears in the Tosefta and Babylonian Talmud but is missing from the parallel list in the Mishnah. It also, if brought, was allowed (*t. Men.* 9.9; *B. Bat.* 97b).

my bʾrg (*ʿAbod. Zar.* 30a) appears as an explanation for *yyn mtwk*, sweet wine (see below). It has been explained as a colloquial form of *bwrq* (Jastrow 1926: 136). Rashi suggests that the term derives from Persian, and Krauss (1910–12: II, 241) accepts this view. White tillia (*ṭylʾ hwwrt*) was described by R. Hisda as the worst of all the 60 types of wine (*Giṭ.* 70a). *Gwdrly* is a 'bad wine' (*Gen. R.* 98.10, 182a). In the passage in the tractate *Šabbat* (Babylonian Talmud) two types of wine represent two women, about whom the speaker in the text asked his neighbour in connection with erotic adventures: *yyn ḥrdly*, a dark lady, and *yyn gwrdly*, a fair one (Rashi). Jastrow (1926: 227) suggests that *gwrdly* hints at more specific sexual implications, presumably because the root *grd*, to scrape, is also used for the friction of the sexual act (Jastrow 1926: 265).

In *Gen. R.* 98.9, 'the milk' is brought in juxtaposition to 'the red', clearly referring to wines of different colours.

A2.2.1.2.4. Clear wine
yyn ṣlwl, Hebrew (*t. Ter.* 4.3; *Šab.* 109a, 139b)
ḥmr nqydʾ, Aramaic (*Giṭ.* 69a)

A2.2.1.3. Wines named by quality, taste and aroma
A2.2.1.3.1. *yyn yph*, fine wine, Hebrew
(*t. B. Bat.* 6.7, 8, 9)

A2.2.1.3.2. Aromatic wine
ḥmrʾ rytḥnʾ, aromatic wine, Aramaic (*Šab.* 110a)
swmqʾ rytḥnʾ, aromatic red wine, Aramaic (*Giṭ.* 70a)

A2.2.1.3.3. *yyn qšh*, strong wine (literally 'hard wine'), Hebrew
(*B. Bat.* 10a, applying to wine in general; *y. Sanh.* 10.2, 28d, applying to Ammonite wine)

A2.2.1.3.4. *ṭylʾ hrypʾ*, acrid tillia
(*ʿAbod. Zar.* 30a)

A2.2.1.3.5. *yyn ḥd*, sharp wine, Hebrew
(*ʿAbod. Zar.* 30a; *y. ʿAbod. Zar.* 2.3, 41a; *y. Ter.* 8.5, 45c)

yyn ḥd is one of the three wines (the other two wines *yyn mr* and *yyn mtwk*, are discussed in the following paragraphs) that appear in Hebrew in a Beraitha in the name of R. Joshua ben Levi as being neither forbidden after being exposed, presumably because a poisonous snake would not drink them, nor susceptible to being libation wine. There also appear three different interpretations in Aramaic as to the meaning of the Hebrew names of the three wines. Two of these are in the version in the Babylonian Talmud, one (a) apparently anonymous and the second (b) in the name of Rav Hama, who states that these wines were all improved wines (*ʿylwlyʾ*) and explains accordingly. The third Aramaic explanation of the Hebrew (c) is in the two versions in the Jerusalem Talmud and is in the name of Rabbi Simon. *ḥd* is explained in the first interpretation (a) as *ṭylʾ ḥryph dmṣr zyqy*, 'acrid tillia that makes the wine-skins burst', in the second (b) as *ḥmr plplyn*, pepper wine (see below), and in the third (c) as *qondytin* (see below), a spiced wine.

A2.2.1.3.6. (*yyn*) *mr*, bitter wine, Hebrew
(*ʿAbod. Zar.* 30a; *y. ʿAbod. Zar.* 2.3, 41a; *y. Ter.* 8: 5, 45c).

This the second of the three wines listed by R. Joshua ben Levi (see *ḥd*, previous paragraph) and is explained as (a) *yrqnʾ*, wine flavoured with herbs(?) (see below), and (b) and (c) as *ʾpsyntyn/psytygwn/psytṭwwn*, absinth wine (see below).

A2.2.1.3.7. Sweet wine
yyn mtwq, sweet wine, Hebrew (*Men.* 8.6; *ʿAbod. Zar.* 30a; *y. ʿAbod. Zar.* 2.3, 41a; *y. Ter.* 8.5, 45c)
ḥmrʾ ḥlyʾ/ḥwlyʾ, sweet wine, Aramaic (*ʿAbod. Zar.* 30a; *Men.* 87a)

Sweet wine, (*yyn*) *mtwq*, is the third wine in R. Joshua ben Levi's list and is explained (a) as *ḥlyʾ*, a simple translation, (b) as *my bʾrg*, apparently a white sparkling wine (see above) and (c) as *hmrʾ mbšl*, boiled wine (see below).

Sweet wine, *yyn mtwq*, is also in the list of wines unfit for temple ritual in the Mishnah (*Men.* 8.6) but missing from the parallel list in the Tosefta. If sweet wine was actually brought to the temple it was still forbidden.

In the Talmud (*Men.* 87a) an apparent contradiction is pointed out: *hlysṭywn* (see below), also a sweet wine, if brought to the temple is allowed, whereas ordinary sweet wine (*yyn mtwq*) is not. R. Ashi explains 'if the sweetness is by reason of the sun it is not offensive but if the sweetness is in the fruit it is offensive'.

A2.2.1.3.8. *yyn qwss*, pungent wine, Hebrew
(*B. Bat.* 6.2; *t. B. Bat.* 6.6; *B. Bat.* 97b; *t. Men.* 9.9)

yyn qwss is one of the wines unfit for the temple ritual and forbidden even if brought. It appears in the list in the Tosefta but not in the Mishnah. If a vendor sold a hundred jars of wine, up to ten could be pungent without need to pay indemnity.

A2.2.1.4. Wines named according to provenance
In the Mishnah (*Men.* 8.6) a list of places is recorded from which wine was brought to the temple:

qrwtym/qrwḥym, *ḥṭwlym/ḥtwlym*, *byt rhm*, *byt lbn* and *kpr sgnh*. It is possible that the text has been misunderstood and that *qrwtym* should read *prwtym*, a term that apparently derives from the Greek πρώτη, first (Yadin and Naveh 1989: 35). Similarly from the Talmud we learn that *yyn qrws*, wine jelly, came from *snyr* (Mt Hermon) (*Suk.* 12a; *Soṭ.* 48b). There is no evidence, however, that these places gave their name to types of wine, although, there were types of wine named after districts and foreign countries.

A2.2.1.4.1. *yyn hšrwny*, Sharon wine
(*Nid.* 2.7; *t. Nid.* 3.11; *Nid.* 21A; *Song R.* 7.3, 36c)

A2.2.1.4.2. *yyn krmly*, Carmel wine
(*t. Nid.* 3.11; *Nid.* 21a)

Sharon wine, when mixed one part with two parts water, is brought in the Mishnah as a standard for assessing colour (*Nid.* 2.7). In the parallel passage in the Tosefta it is stated that Carmel wine is similar and in that in the Babylonian Talmud it is stipulated that Sharon and Carmel wine are to be regarded in the same manner.

A2.2.1.4.3. *yyn ʿmwny*—Ammonite wine
(*y. Sanh.* 10.2, 28d; *Sanh.* 106a)

Ammonite wine is described as a strong wine (*qšh*, literally 'hard'), the aroma of which spreads far.

A2.2.1.4.4. *yyn hʾyṭlqy* (*bʾtlqy*?), Italian wine (in an Italian measure?)
(*Sanh.* 8.2; *t. Pes.* 1.28; *ʿErub.* 64b; *y. Šab.* 8.1, 11a; *y. ʿAbod. Zar.* 1.9, 40a)

There is some controversy whether these passages refer to Italian wine or wine measured according to an Italian measure (Lieberman 1955–88: IV, 499).

A2.2.1.4.5. *yyn pwrgyyt*ʾ, Phrygian wine/Plugata wine
(*Šab.* 147b; *Lev. R.* 5.3, *plwgt*)

Steinzaltz (1969: 651) explains *pwrgyyt*ʾ as Phrygian; however, Jastrow (1926: 1177) shows that the form *plwgt* (*Lev. R.* 5.3) is found in another text also (*Num. R.* 9.24, 29c) and there it is clearly a place near Tiberias.

A2.2.1.5. Names of wine related to the production process
A2.2.1.5.1. (*yyn*) *ʿylwy*ʾ, improved wine
(*ʿAbod. Zar.* 30a)

Rav Hama used the term *ʿylwy*ʾ in reference to *ḥmr wplplyn*, pepper wine, *ʾpsyntyn*, absinth wine, and *my bʾrg*. Compare with Pliny (14.18.98) who used the term *ficticius* (artificial wine).

A2.2.1.5.2. (*yyn*) *mʿšn*, smoked wine
(*Men.* 8.6)

This is one of the wines unfit for temple ritual even if brought. It appears in the Mishnah but not in the Tosefta. It has been suggested that the term appears on an Iron Age ostracon from Lachish (Demsky 1979; see above). Pliny mentions smoked wine (14.3.16).

A2.2.1.5.3. *yyn ṣymwqym*, raisin wine
(*t. Men.* 9.9; *B. Bat.* 97b)

Raisin wine was unfit for temple ritual but, if brought, accepted. It appears in the Tosefta but not in the Mishnah. Columella (12.39) quotes Mago the Punic agronomist on how to make *passum* raisin wine.

A2.2.1.5.4. Boiled wine
yyn mbšl, boiled wine, Hebrew (*Men.* 8.6; *t. ʿAbod. Zar.* 4(5).12; *y. Šab.* 8.1, 11a; *Ter.* 2.6; *t. Ter.* 4.4).
*ḥmr mbšl*ʾ, boiled wine, Aramaic (*ʿErub.* 29a; *y. Ter.* 8.5, 45c; *y. ʿAbod. Zar.* 2.3, 41a)

Boiled wine is one of those that appear in the Mishnah but not in the Tosefta as unfit for temple ritual and, if brought, not accepted. Rabbi Simon explains *yyn mtwq*, sweet wine, as boiled wine. In a discussion regarding the preparation of an *ʿEruv*, 'wine to eat' is mentioned which is then explained as boiled wine, further explained by Rashi as wine into which bread is dipped (*ʿErub.* 29a). (An *ʿEruv* in this case is a deposit of food left by a man in a place before the Sabbath, the place then counting as his abode and thus allowing for freer movement on the Sabbath; see Danby 1933: 793.)

Boiled wine is clearly the sweet concentrated boiled-down must of Roman sources known as *defrutum*, *sapa* or *carenum* (Pliny 14.9.80; Columella 12.20.2; Palladius 11.18; see above p. 42).

A2.2.1.5.5. *yyn šmrym*, lees-wine
(*t. Men.* 9.10; *B. Bat.* 97a, b)

Lees-wine is unfit for temple use, it appears in the list in the Tosefta, but is lacking in that in the Mishnah. If brought, it was not accepted.

This is apparently the *vinum faecatum* described by Cato (153) made by pressing the juice out of the wine lees.

A2.2.1.5.6. *tmd*, after-wine.
(*ltmd*, verb meaning 'to make after-wine')
(*ʿOr.* 1.8; *Maʿas.* 8.6; *Miq.* 7.2; *B. Bat.* 97a; *Pes.* 42b).

The passage in *ʿOrlah* (1.8) reads 'defective grapes, grape-pips, grape skins or *tmd* made from them'. That in *Maʿaserot* speaks of adding water and in *Miqwaʿot* (7.2) a distinction is made between fermented and unfermented *tmd*. These passages show *tmd* to be after-wine made by adding water to the grape skins and stalks after treading and then fermenting the liquid. After-wine was known as *lora* in Latin and δεθεριος, *deuterius* (seconds) in Greek, and various methods for its production are described by Pliny (14.12.863); Cato (25) and Columella (12.40).

There are, however, Talmudic discussions that imply that the passage in *Maʿaserot* speaks of adding water to lees and not to pressed grape skins (*B. Bat.* 97a) and a distinction is made between (*tmd*) *drwwq*ʾ (after-wine of the straining bag) and (*tmd*) *dpwrṣny* (after-wine of the kernels) (*Pes.* 42b). The latter is in effect lees-wine. It is to be noted that Pliny regarded lees-wine as a variety of after-wine (Pliny 14.12.86).

It has been suggested that the term *tmd* derived from the Latin *temetum* (Krauss 1910–12: II, 239; Dalman 1928–42: IV, 271), but this is not accepted by all scholars (Fraenkel 1898: 293). Others are of the opinion that the word appears already in the Bible and suggest that *tmyd* in Obad. 1.16, which is translated as 'wine' in the LXX should be read as *tmd* (Sukenik 1942: 22; Nestle 1903: 345). This, however, is also disputed and the *Vorlage* reflected in the Greek translation is explained instead as *ḥmr* (Kittel 1973: 929).

It is of interest that a Hebrew stamp with the word *tmd* engraved on it from the Hellenistic or Roman period has been found (Sukenik 1942).

A2.2.1.5.7. *yyn qrwš*, coagulated wine
(*t. Suk.* 3.15; *Suk.* 12a; *Soṭ.* 48b)

In all three texts *yyn qrwš* is compared to *ʿgwly dbylh*, 'round cakes of dried figs'. In the two texts in the Talmud it is stated that the wine came from *snyr*, Mt Hermon, which is covered in snow every year suggesting that it was coagulated because of the cold. In the Tosefta, however, it is told that every 60 or 70 years young priests descended into the pit, *šyt*, beside the altar into which the libation wine drained and brought out *yyn qrwš*. In this case it was clearly wine that had largely evaporated. Pliny (14.6.55) describes wine that was nearly 200 years old in a similar manner.

A2.2.1.5.8. *yyn mrtp*, wine from the wine cellar
(*t. Men.* 9.9; *B. Bat.* 97b)

yyn mrtp, apparently wine that had been stored for a long time in a wine cellar, was considered unfit for ritual use in the temple but, if brought, allowed. It appears in the list in the Tosefta.

A2.2.1.6. Diluted and undiluted wine
A2.2.1.6.1. Diluted wine
yyn hmzyg, *yyn hmzwg*, *yyn mzwgyn*, Hebrew (*t. Men.* 9.9; *t. Nid.* 3.11; *B. Bat.* 97b; *y. Šeq.* 3.2, 43a; *y. Šab.* 8: 1, 11a)

A2.2.1.6.2. Undiluted wine
yyn ḥy/ḥyyn, Hebrew (literally 'living wine') (*t. Ber.* 4.3; *y. Šab.* 8.1, 11a; *Sanh.* 70a)
ḥmrʾ ḥyyʾ, Aramaic (*Giṭ.* 67b, 69b)

yyn mzg in Talmudic Hebrew is diluted wine, although in biblical Hebrew *mzg* is almost certainly spiced wine.

Mixed wine was considered unfit for use in the temple but was considered the normal way to drink wine to such an extent that some contended the normal blessing on wine was only to be said on diluted wine. 'They do not say the benediction over the wine until water has been added to it, so R. Eleazer. But the sages say they may say the benediction' (*Ber.* 7.5). In the Tosefta (*Ber.* 4.3) a different benediction is even specified for undiluted wine. There is a rabbinical dispute as to how much to dilute wine, one part of wine to two parts of water or one part of wine to three parts of water (*Šab.* 75b-77a). In another passage mixtures consisting of a half, a third or a quarter of wine are referred to (*B. Meṣ.* 60a).

A2.2.1.6.3. *ḥmrʾ mrqʾ*, yellowish wine? pale wine, diluted wine?
(*Giṭ.* 67b, 69b; *ʿErub.* 29b; *Yeb.* 80a)

The name apparently derives from *mrwkʾ*, saffron-coloured (Jastrow 1926: 839, 847). It appears as the contradiction of undiluted wine, *ḥmrʾ hyyʾ*, in two texts (*Giṭ.* 67b, 69b) and has therefore also been understood as diluted wine.

A2.2.1.7. Wines with additives
A2.2.1.7.1. *yyn mbwsm*, spiced wine
(*B. Bat.* 6.3; *t. B. Bat.* 6.5)

In both texts it is stipulated that a vendor of spiced wine was responsible for its quality until Pentecost. The Tosefta goes into greater detail: for ordinary wine the vendor was responsible only until Passover, whereas for vintage wine, *yyn mywšn*, his responsibility was longer, until New Year.

A2.2.1.7.2. *ḥmr bsym*, *bsymʾ*, well-seasoned/sweet/over-fermented (sour) wine
(*Ber.* 56A; *B. Meṣ.* 60a; *B. Meṣ.* 69a; *y. Maʿas. Š.* 4.11, 55c; *Lam. R.* 1.12)

The word *bsym* apparently derives from the same root as *mbwsm*, but it would appear to refer to a quality and not a type of wine. In some cases it is clearly an approbatory term (*Ber.* 56a; *B. Bat.* 60a, 69a) meaning 'sweet' or 'fine', but in others it is a derogatory term meaning over-fermented, sour (*y. Maʿas. Š.* 4.11, 55c; *Lam. R.* 1.12).

A2.2.1.7.3. *yyn mṭwbl* (*mtwbl*), spiced wine
(*y. Šeq.* 3.2, 47c)

This is another term for spiced wine.

A2.2.1.7.4. *yyn qprysyn*, caper wine
(*Ker.* 6a)

yyn qprysyn is mentioned in the Mishnah as an ingredient of frankincense. Rashi explains *yyn qprysyn* as Cyprus wine and the island is known by that name in modern Hebrew. However, the correct explanation is almost certainly caper wine. The caper bush, κάππαρις in Greek, is known as *ṣlp* in Hebrew, but the flower was known as *qprs* (*Maʿas.* 6.4) from the Greek. *yyn qprysyn* was apparently grape wine flavoured with flowers of the caper.

A2.2.1.7.5. *ḥmr wplplyn*, wine with pepper
(*ʿAbod. Zar.* 30a)

Rav Hama explains sharp wine (d) as *ḥmr wplplyn*, wine flavoured with pepper.

A2.2.1.7.6. *yrnqʾ*, *yrqwnʾ* (Munich Manuscript), vegetable wine
(*ʿAbod. Zar.* 30a)

Bitter wine is explained as *yrqwnʾ* (*yrq*, green/vegetables), vegetable wine, wine flavoured by vegetables. Pliny speaks of wines flavoured by vegetables (14.19.105-106).

A2.2.1.7.7. *ʾsprgws*, asparagus beverage
(*Ber.* 51a; *Pes.* 110b; *Kid.* 70a)

The term *ʾsprgws* derives from the Greek ἀσπαραγος, *asparagus*, a word that also exists in Latin. A long discussion on the medicinal benefits of this drink (*Ber.* 51a) shows clearly that there was both asparagus wine, grape wine flavoured with asparagus, and a drink apparently made of asparagus alone. Pliny mentions asparagus wine (14.19.105).

A2.2.1.7.8. *qwryyṭy*, date wine?
(*Ber.* 50b; *ʿAbod. Zar.* 14b; *y. Šab.* 14.3, 14c *qrwrṭyn*)

From *ʿAbodah Zarah* it is clear that *qwryyṭy* is a type of date known also as the Nikolaus date. In the other sources it is a beverage. In *Berakot* a use of undiluted wine is for *qwryyṭy*, suggesting either that *qwryyṭy* was a date wine that was mixed with grape wine or more probably that grape wine was used in producing the drink.

A2.2.1.7.9. *yyn tpwḥym*, apple wine
(*Ter.* 11.2, 3; *t. ʿAbod. Zar.* 4[5].12; *t. Ber.* 4.2)

In the passage in *ʿAbodah Zarah* it is permitted to consume apple wine produced by Gentiles except that sold in the market when there is suspicion that wine is added, thus suggesting that the wine was made of apples alone, grape wine being sometimes added. In the second passage it is stipulated that the benediction on apple wine is like that on *mwryys* (see below), a sauce to which wine was also sometimes added.

Palladius (11.16) mentions apple wine, but it is not clear from the context whether it is based on grape wine or not.

A2.2.1.8. Wines the names of which derive from Greek or Latin
A2.2.1.8.1. *ʾynwmylyn*, *yyn ymylyn*, honey wine
(*Šab.* 20.2; *Ter.* 11.1; *y. Šab.* 14.2, 14c; *y. Šab.* 20.2, 17c; *ʿAbod. Zar.* 30a; *Šab.* 139b; *Šab.* 140a)

The name derives from the Greek οἰνόμελι, honey wine. The form *yynimylyn* is a reversal to the Hebrew *yyn* from the

Greek, *ʾynw* οἰνος (Kutcher 1961). The Latin name for honey wine is different in form but also derives from Greek *melitites* (Pliny 14.11.85).

According to the Talmudic sources, *ʾynwmylyn* is made of 'wine, honey and peppers' (*ʿAbod. Zar.* 30a; *y. Šab.* 14.2, 14c; *Šab.* 140a). Pliny (14.11.85) speaks of mixing 30 congia of must to six of honey and a cup of salt, and then boiling it. Pliny also speaks of wine made by 'adding pepper and honey' (14.19.108). The fact that it was permitted to make *ʾynwmylyn* on the Sabbath (*Šab.* 140a) suggests that the pepper and honey were added to the wine immediately before the meal.

A2.2.1.8.2. *ʾlwnṭyt, ʾlwntyt*, an aromatic wine

(*t. Dem.* 1.24; *t. ʿAbod. Zar.* 4(53.12; *t. Ter.* 9.15; *ʿAbod. Zar.* 30a; *Šab.* 140a, 147b; *y. Šab.* 19.3, 17a)

ʾlwntyt derived apparently from the Greek οἰνανθε, οἰνανθινος, *oenanthe, oenanthinos*, the flower of the wild vine. Pliny describes how the *oenanthe*, the dried flowers of the wild vine, is prepared (12.61.132) and (*vinum*) *oenanthinum* is produced (14.18.98) 'by steeping two pounds of the flowers of the wild vine in a jar of must'.

The Talmudic description as to how this wine is made does not conform to the original Greek name, suggesting that they no longer knew the true meaning of the word. 'What is *ʾynwmylyn* and what is *ʾlwnṭyt*? *ʾynwmylyn* is wine, honey and peppers, *ʾlwnṭyt* is old wine, clear water and *prsmwn* (balsam)' (*Šab.* 140a; similar in *ʿAbod. Zar.* 30a). The fact that, as opposed to *ʾynwmylyn*, the making of *ʾlwnṭyt* was forbidden on the Sabbath (*Šab.* 140a) suggests that this type of wine was mixed some time before drinking and allowed to mature.

A2.2.1.8.3. *hylysṭwn, ʾlysṭwn, ʾlywsṭwn*, sweet wine made of grapes left in sun

(*Men.* 8.6; *t. Men.* 9.9; *Men.* 87a; *B. Bat.* 97b)

The name derives from the Greek ἡλιάζω (*helias*, of the sun), ἡλιαστύς (*heliazu*, to bake in the sun; note that the eta appears as aleph and he).

hylysṭwn is the only wine unfit for temple use that appears both in the Mishnah and the Tosefta. If brought to the temple, it was accepted. In the discussion in the Babylonian Talmud (*Men.* 87a) it was noted that sweet wine was not allowed when brought to the temple, while *hylysṭwn*, which was also a sweet wine, was. The explanation brought was 'If the sweetness is by reason of the sun it is not nauseous, but if the sweetness is in the fruit itself it is nauseous.' This passage shows clearly that *hylysṭwn* was made of grapes left to dry in the sun. Columella (14.27) explains that sweet wine (*vinum dulce*) was made by leaving the grapes spread out in the sun for three days and then treading them at noon of the fourth. Pliny (14.10.77) describes a similar wine that he accredits to the Greeks.

A2.2.1.8.4. *qwndytwn*, spiced/peppered wine

(*y. Ter.* 8.5, 45c; *y. ʿAbod. Zar.* 2.3, 41a; *y. Beṣ.* 1.9, 60d; Pd *Pes. K.* 102b)

In the case of *qwndytwn* the Greek name χονδιτον was retained. It is also unchanged in Latin, *conditum*. R. Simon defines sharp wine (d) as *qwndytwn* (*y. Ter.* 8.5, 45c; *y. ʿAbod. Zar.* 2.3, 41a) while Rav Hama explains *ḥd* as pepper wine (*ʿAbod. Zar.* 30a). The Midrash (*Pes. K.* 102b) states that '*qwndytwn* has in it wine, honey and peppers'. There is also discussion as to whether grinding the spices for *qwndytwn* (*mysḥwk*) is allowed on the Sabbath (*y. Beṣ.* 1.9, 60d) which also suggests that the spices (pepper?) were added immediately before drinking. Pliny states that (*vinum*) *condita* and *piperata* (peppered) are alternative names for the same wine (14.19.108) made by 'adding pepper and honey'. In this case then, the Roman and Talmudic recipes are identical, both showing pepper to be the main spice in this wine.

A2.2.1.8.5. *ʾpsyntyn, psyntṭwn, psyntṭwwn, pstygwn*, absinth (wormwood wine)

(*ʿAbod. Zar.* 30a; *y. Ter.* 8.5, 45c; *y. ʿAbod. Zar.* 2.3, 41a; *Tanḥuma Vayešeb* 8)

The name derives from the Greek ἀψίνθιον (*apsinthion*), wormwood—the Latin term is the same, *absinthium*. The Hebrew term for wormwood is *lʿnh*, but no name of a wine derived from it.

Both R. Hama (*ʿAbod. Zar.* 30a) and R. Simon (*y. ʿAbod. Zar.* 2.3, 41a; *y. Ter.* 9.5, 45c) explain bitter wine (*yynmr*) as *ʾpsntyntyn*, and in the Midrash *Tanḥuma* Potiphar asked Joseph for *psntṭwn* instead of *qwndytwn*.

A2.2.1.8.6. *ʾynmrynwn*, myrtle wine?

(*Song R.* 4.141)

The term appears in the Midrash as an explanation for myrhh (*mrh*) (*Cant.* 4.14) The Musafia (Jastrow 1926: 53) explains this as deriving from Greek οἰνος μυρινης (*oinos murines*), myrtle wine. Dalman (1928–42: IV, 375) accepts this explanation, suggesting *enesmirninon* as an intermediate stage. Jastrow (1926: 52-53) has pointed out that the context does not point to a wine but to an unguent and suggests the source to be Greek ἐλαιον μυρρινον (*elayon murrinon*), myrtle oil.

Columella (12.38) devotes a long chapter to describing different types of myrtle wine—one being grape wine flavoured with myrtle berries, another made of honey and myrtle berries and the third of myrtle berries alone.

A2.2.1.8.7. *qrynʾ*, an Asiatic sweet wine

(*ʿAbod. Zar.* 30a; *y. ʿAbod. Zar.* 2.41a)

Explained as deriving from the Greek χαρινον, χαρυνον (Jastrow 1926: 1420) or καρινον, χαρυνον, *carenum*, a sweet wine boiled down one-third (Palladius 11.18), although Jastrow also points to the similarity to the Assyrian *karanuv, kurunnu*. R. Abahu explains that *qrynʾ* is a sweet wine (*ḥmrʾ ḥlyʾ*) from Asia, which corresponds well to its identification with *carenum*.

A2.2.1.9. Other types of wine

A2.2.1.9.1. *tylʾ, tlyyʾ* (Rashi writes *tylyʾ*), a bitter white wine

(*ʿAbod. Zar.* 28a; *ʿAbod. Zar.* 30a; *Giṭ.* 70a)

In *ʿAbodah Zarah* sharp wine (d) is explained as 'acrid *tylʾ* which bursts the wine skins'. In *Giṭṭin* Mar Uqbah states that if a man drinks white *tyl*ʾ he will suffer from debility and R. Hisda asserts that the worst of all the 60 wines known is white *tyl*ʾ. The fact that in one case the wine is described as acrid (*ḥryb*ʾ), in another as white (*hywr*ʾ) and in a third case appears without adjective, suggests that it was a bitter white wine, but of course this could mean that there were different types of *tyl*ʾ wines and therefore the need for the adjectives.

A2.2.1.9.2. *ḥmr*ʾ *d*ʾ*qrym*, partly fermented wine
(*ʿAbod. Zar.* 30a)

*ḥmr*ʾ *d*ʾ*qrym* is explained by Rashi as wine that had only partly fermented.

A2.2.2. Products Containing Wine

A2.2.2.1. *mwryys*, salted fish sauce sometimes containing wine
(*Ter.* 11.1; *t. Ber.* 4.2; *t. Dem.* 1.24; *t. ʿAbod. Zar.* 4(5).11, 13; *Ḥul.* 6a; *ʿAbod. Zar.* 34b; *ʿAbod. Zar.* 38b; *Pes.* 109a; *y. Ter.* 8.3, 45b; *y. Šab.* 1.4, 3c)

A2.2.2.2. *ḥylq ḥyqh*, fish sauce made of the *ḥylq* fish
(*ʿAbod. Zar.* 34b; *y. Šab.* 1.4, 3c)

mwryys is clearly connected to the Latin *muria, muries* and *ḥylq*/*ḥyqh* to the Latin *hallex* (*alec, allec, hallee*). *muria* and *hallex* were two of the four main Latin terms for fish sauces. The other two, which are apparently not represented as such in Talmudic literature are *liquamen* and *garum* (but see below, ʾ*nygrwn* and ʾ*ksygrwn*).

Although it would appear from the written sources that *liquamen* is a general or perhaps official term for all fish sources, and in one case is even applied to a salted pear drink (Palladius 3.25.12), while *muria* is primarily brine, it is clear from the Roman *tituli picti* on amphora (Curtis 1984–86: 227-31) that these were four distinct types of fish sauces (Corcoran 1962; Curtis 1983).

According to the numbers of *tituli picti* studied by Curtis (1984–86), the two types that are apparently absent from Talmudic sources were the most popular (*garum* 164, *liquamen* 69, *muria* 34, *hallex* 11).

Although the distinction between these four products is not completely certain, the manner in which they were made is clear from ancient sources (Pliny 31.43–44.93-96; *Geop.* 20.46) and from comparison to fish sauces made until recently: *nuoc-mam* from Indochina and *garos* or *rajihe* in Turkey (Grimal and Monod 1952). The intestines and other unusable parts of the fish were placed in brine and fermented in the sun, sometimes for as long as two or three months. Both Pliny and the *Geoponica* explain that *garum* is the liquid produced and the *hallex* the solid sediment that remains and as such a poorer product. However, Pliny also speaks of *hallex* especially made of small fish which is then often a luxury product. Pliny adds 'another kind [of *garum*? of *alex*?] is devoted to superstitious abstinence (*castimonarum*) and Jewish rites and is made of fish without scales' (clearly only with scales is meant). The latter has been connected to the *tituli picti*: *gar*(*um*) *cast*(*um*) or *mor*(*ia*) *cast*(*a*) (Curtis 1984–86: 223-24: Frey 1933: 372-74).

The *Geoponica* (20.46) in the chapter discussing fish sauces speaks also of adding wine to fish in a proportion of two to one. It is not clear, however, if the wine was added to salty fish sauce or if the reference is to a completely different product. *mwryys,* although derived from *muria,* is almost certainly the Talmudic equivalent of *garum*/*liquamen*/*muria*, salty fish sauces. The Talmudic discussion is usually connected to the addition of wine (*Ter.* 11.1; *t. Dem.* 1.24). It is clear, however, that wine was not always added. *mwryys* made by an expert (*mwryys* ʾ*wmn*) was presumed not to have wine in it (*t. ʿAbod. Zar.* 4(5).11). However, Abimie, the son of R. Abbahu (*ʿAbod. Zar.* 34b), explains that the first and second 'time' (extraction?) have no fat and therefore need no wine, whereas the third and subsequent times wine is added. This description makes it clear that the *mwryys* was extracted from the fish several times. In another passage (*ʿAbod. Zar.* 38d) it is explained that the wine is added to overcome the bad smell of the fish.

In the same chapter an incident is mentioned involving a shipload of *mwryys* arriving at the port of Akko and the question arises as to whether it was not from Tyre. It is of interest that there is evidence for the trade in fish sauce from Tyre where the grave of a merchant of *garum* was uncovered (Rey Coquais 1977: 154).

As regards *ḥylq,* the picture is more complicated. A type of fish named *ḥylq* is mentioned in the Mishnah (*ʿAbod. Zar.* 2.6) and is explained in the Babylonian Talmud (*ʿAbod. Zar.* 39a) as being the *swltnyt.*

The *ḥylq* mentioned in other passages, together with *mwryys,* is clearly a fish sauce made of the *ḥylq* fish. In the case of this sauce too, that made by an expert is above suspicion (*ʿAbod. Zar.* 34b). It is not quite clear, however, whether the suspicion is as to whether wine had been added or as to whether the fish is really *ḥylq* fish.

In this case it is very probable that the Latin *hellex* (*alec allec, hallec*) derived from the Levant. Jastrow (1926: 457) has suggested that 'the variations [in the Latin] indicate foreign origin'. The Hebrew *ḥylq* and Latin *allec* (etc.) was probably a more solid form of fish product, a paste rather than a sauce. As opposed to Pliny's presentation of the facts, it was probably first made from the small *ḥylq* fish and only later from the sediment of other fish pastes. It is of interest that the *Geoponica* (20.46) also mentions a fish named *alec.* Pliny's *allex*? *castimoniarium* could possibly be connected to the *ḥylq* ʾ*wmn* (*ḥylq* produced by an expert) permitted by the Rabbis (*ʿAbod. Zar.* 34b).

A2.2.2.3. ʾ*nygrwn*, ʾ*ngwrnwm*, ʾ*ngwryn*, sauce of wine and garum
(*Yom.* 76a; *Ber.* 35b-36a; *Šeb.* 23a; *t. Šeb.* 6.3; *t. Ter.* 9.10; *y. Beṣ.* 2.10, 61d; *y. Šeb.* 8.2, 38a)

The name ʾ*nygrwn* derives from the Greek οἰνογαρον, Latin, *enogarum*, garum and wine.

A2.2.2.4. ʾ*ksygrwn*, ʾ*nsygrwn*, *snygrwn*, ʾ*grṭwn*, ʾ*rgstwrwn*, sauce of vinegar and garum
(*Yom.* 86a; *Ber.* 35b; *t. Šeb.* 6.3; *t. Ter.* 9.10; *y. Šeb.* 8.2, 38a; *Eccles. R.* 1.18)

The name ʾ*ksygrwn* clearly derives from the Greek ὀξυγαρον, Latin *oxygarum*, vinegar and garum.

In Latin literature *oenogarum* was a sauce consisting primarily of garum and wine, to which other substances were often added. Apicius in his treatise on cooking speaks of *oenogaro simplici*, simple *oenogarum* (3.4.5 [77]; 3.20.1 [115]) but also gives various detailed recipes for *oenogarum* to be added to truffles (1.17.1 [35]; 7.16.3 [318]), fish (4.2.31 [158]; 10.3.II [465], 12 [466]); mixed entré (4.5.1 [175]); marrows (4.5.3 [177]; foie gras (7.3.1 [260]) and snails (7.18.1i [323]). The additional substances included pepper, *ligusticum* (lovage; Flower and Rosenbaum 1958: 121), coriander, rue, honey and oil.

Oxygarum was a sauce consisting of garum and vinegar, which sometimes was also mixed with other substances. Apicius (1.20.1, 2 [39-403]) gives two different recipes for *oxygarum digestibilem* which includes, as well as vinegar and garum, pepper, silis, gallici, cardamon, cumin, nard, mint, petrosilia and *ligusticum* (lovage).

ʾnygrwn and *ʾksygrwn* appear in the majority of the Talmudic sources together. In one passage which is repeated several times they appear as examples of substances that are to be eaten and not drunk (*y. Šeb.* 8.2, 38a; *t. Ter.* 9.10; *t. Šeb.* 6.3). The substances are explained by Rabba the son of Shmuel thus: '*ʾnygrwn* is beetroot juice, and *ʾksygrwn* is made of all other cooked vegetable juices' (*Ber.* 35b; *Yom.* 76a; *Šeb.* 23a). From another passage (*y. Beṣ.* 2.10, 61d) it is clear that ground peppers were an ingredient of *ʾnygrwn*, from others that wine was also added to *ʾnygrwn* (*Šeb.* 23a; *Yom.* 76a) and from the discussion in *Berakot* (35a-36a) that oil was also an integral part of *ʾnygrwn,* although usually not the main ingredient.

In the discussion of *ʾnygrwn* and *ʾksygrwn* in Talmudic literature there is mention of oil, pepper, wine and various vegetables. These and other substances appear in Apicius's recipes also, but clearly as additives, and simple *oenegarum,* lacking these additives, is also mentioned. In the Talmudic references the basic ingredient, fish sauce (garum) is never mentioned. It is difficult, however, to determine whether garum was taken for granted or whether these terms had come to mean vegetable sauces.

A2.2.3. Jar Inscriptions and Other Epigraphic Evidence from the Roman Period

There is apparently little epigraphic evidence of local origin from the Roman period that refers either to types of wine or their provenance. As opposed to the rich corpus of Latin *tituli picti* of the Roman world, which often designate both these characteristics, the local jar inscription consists largely of personal names, presumably those of the merchants (e.g. Hestrin 1972: numbers 244-47).

One interesting exception is that of the large limestone stamp (45 × 34 cm; thickness 15 cm; published by Sukenik 1942) on which are engrâved two separate inscriptions both in the negative. On the bottom the name, *hwšʿ*, Hosha, and on the narrow side the term *tmd, temed*, after-wine (see above, pp. 42, 202).

The epigraphic material from Masada provides an important corpus of material that illustrates the difference between the local inscriptions and those from abroad (Yadin and Naveh 1989; Cotton and Geiger 1989). The largest group of jar inscriptions is that of personal names (numbers 462-515). Another group refers to the contents of the jars (numbers 516-53); however wine does not appear, only other substances, mainly dried figs but also fish, meat and balsam. One group of inscriptions that is concerned with wine specifies almost only whether the contents of the jars are for tithes or whether they are fit or unfit for holy purposes (numbers 441-61). Of particular interest are inscriptions from jars in the storehouse specifying apparently the fitness of whole rows of jars. Two terms are used, *prwṭym* and *krwṭym*. These are clearly connected to the reference in the Mishnah (*Men.* 8.6) '*qerutim* and *haṭṭulim alpha* for wine', while *prwtym* apparently derives from the Greek πρώτη, first (Yadin and Naveh 1989: 35).

The Greek inscriptions, which are clearly also of local origin and often appear together on the same jar with a Hebrew/Aramaic inscription, are also nearly all personal names (numbers 867-927). A small group (numbers 854-66) bear the inscription καλον κεραμιον (beautiful jar), referring to the jar itself, not the contents.

The Latin *tiluli picti* are of a completely different character. These were all of foreign origin. They often designate the contents of the jar, which is usually wine but includes also honey (number 800), garum (number 826) and apples (number 822). The type of wine was often specified: Amineum (numbers 796, 797), Tarentium (number 818), Massic (number 819) and a group of jars of Philonianum wine from the estate of L. Laenius specified as having been sent to the wine cellars of King Herod (numbers 804-16). This group and another (numbers 795-99), the Caesar series, have also the Roman date according to the name of the consul or consuls.

A2.2.4. Conclusions

Brown (1969) has most beautifully demonstrated that one can speak of a Mediterranean vocabulary of the vine. It is clear from the short summary of terms used for wines in Talmudic literature that the types of wines referred to and their names are also part of a Mediterranean culture of the wine. Those terms that are obviously not Semitic derive usually from the Greek. In spite of Roman rule, ancient Israel in the Roman and Byzantine periods was in many ways part of a Hellenistic culture. The methods used are also very similar: adding honey, spices, leaving the grapes in the sun, boiling down the must, etc. In some cases, apparently, the names in the Hebrew and Aramaic no longer mean what would have been expected from their Greek origins. There are also some techniques attested for in the works of the Latin authors for which there is no evidence in the Talmudic texts. One of these is adding salt or salt water. This was apparently primarily a Greek technique but clearly known and practised in Roman Italy also (Pliny 14.10.77-79; 14.19.100; 14.21; 14.24; Cato 62; Columella 12.37).

However, we do not find the epigraphic evidence in Israel similar to the rich corpus of jar inscriptions found in the western Mediterranean. This suggests that the wine produced was largely consumed locally. The one exception is that of Gaza wine, for which there is written evidence. We also know today both the wineries where it was produced and the amphorae in which it was transported (Mayerson 1985; wineries T92; 'Gaza Amphora': Zemer 1977: Type 183; Peacocke and Williams 1986: Class 49; Riley 1979: Amphora Type 3; Sciallano and Sibella 1991: Amphora Late Roman 4).

Bibliography

Entries preceded by an asterisk are those relevant to the subject of the book but that have reached the notice of the author too late to be referred to in the text.

Abel, M.
1923 'Chronique II: Une ville romaine à Djifna', *Revue Biblique* 32: 111-14.

*Achaya, K.T.
1993 *Ghani: The Traditional Oilmill of India* (Kemblesville, PA: Olearius).

*Acquaviva, R.
1995 *Palmenti e Frantoi in Sicilia* (Siracusa: Zan Garastampa).

Adam, J.P.
1984 *La construction romaine: Matériaux et technique* (Paris: Picard).

Adams, W.Y.
1966 'The Vintage of Nubia', *Kush* 14: 262-83.

Adriani, A.
1939 *Le gobelet en argent des amours vendangeurs du Musée d'Alexandrie.* (Société royale archéologique d'Alexandre, cahier 1; Alexandria: Société de publications égyptiennes.

Addyman, P.V.
1962 *The Archaeology of the Sbeitla Area: An Interim Report* (Brathay Exploration Group Annual Report and Account of Expedition in 1962).

Agmati, Zachariah ben Judah
1961 *A Digest of Commentaries on the Tractate Babha Kamma, Babha Mesia and Babha Bhathera of the Babylonian Talmud* (London: British Museum).

Aharoni, Y.
1953 'Bet Sittah', *Israel Exploration Journal* 3 (Notes and News): 266.
1954 'Excavations at Ramat Raḥel, Preliminary Report', *Israel Exploration Journal* 6: 137-56.
1964 *Excavations at Ramat Raḥel Seasons 1961, 1962* (Rome: Centro di Studi Semitici, Istituto di Studi del Vicino Oriente, Universita).
1968 *The Land of the Bible* (London: Burns).
1973 'The Hebrew Inscriptions', in Y. Aharoni (ed.), *Beer Sheba I* (Tel Aviv: Tel Aviv University): 71-78.
1981 *Arad Inscriptions* (Jerusalem: Israel Exploration Society).

Ahituv, S.
1974-75 'The Meaning of Semadar', *Lešonenu* 89: 37-40 (Hebrew).

Ahlström, G.W.
1978 'Wine Presses and Cup-Marks of the Jenin Megiddo Survey', *Bulletin of the American Schools of Oriental Research* 231: 19-49.

Akerraz, A., and M. Lenoir
1981–82 'Les huileries de Volubilis', *Bulletin d'archéologie marocaine* 14: 69-99.

Alami Sounni, A.
1981–82 'Etude mécanique d'un pressoir de Volubilis', *Bulletin d'archéologie marocaine* 14: 121-51.

Albright, W.F.
1931 'The Excavations of Tell Beit Mirsim. I. The Pottery of the First Three Campaigns', *Annual of the American Schools of Oriental Research* 12.
1933 'The Excavations of Tell Beit Mirsim. IA. The Bronze Age of the Fourth Campaign', *Annual of the American Schools of Oriental Research* 13.
1938 'The Excavations of Tell Beit Mirsim: II', *Annual of the American Schools of Oriental Research* 17.
1941–43 'The Excavations of Tell Beit Mirsim: III', *Annual of the American Schools of Oriental Research* 21–22.

Alexandre-Bidon, D. (ed.)
1990 *Le pressoir mystique* (Actes du Colloque de Recloses; Paris: Cerf).

Almes, G., et les Amis du Vieux Lambex
1983 'L'huilerie de l'établissement gallo-romain du grand verger à Lambex', *Bulletin archéologique de Provence* 11: 1-7.

Alpago-Novello, L.
1957 'Resti di Centuriazione Romana Nella Val Belluna', *Atti della accademia nazionale dei Lincei* 8.12 (fasc. 1-2): 249-66.

Altbauer, M.
1971 'More on "Semader" on a Jar from Hazor', *Eretz Israel* 10: 64-66 (Hebrew).

Amiran, R.
1956 'The Millstones and the Potters' Wheel', *Eretz Israel* 4: 46-49 (Hebrew).
1960 'The Pottery of the Middle Bronze Age I in Palestine', *Israel Exploration Journal* 10: 204-25.
1970 *Ancient Pottery of the Holy Land* (New Brunswick, NJ: Rutgers University).

Amit, D.
1990 'ḥakhlile ʿanavim miyayin (Eyes Bluer than Wine)', *Gushpanka* May 1990: 9-11 (Hebrew)

Amouretti, M.C.
1981 'Des agronomes latins aux agronomes provençaux les moulins à huile', *Provence historique* 31 (fasc. 124): 84-100.
1986 *Le pain et l'huile dans la Grèce antique* (Paris: Annales littéraires de l'Université de Besancon, 67; Paris: Centre de recherche d'histoire ancienne).

*Amouretti, M.C., and J.-P. Brun (eds.)
1993 *Oil and Wine Production in the Mediterranean Area* (Bulletin de correspondence hellénique supp., 26; Paris: Ecole Française d'Athènes).

*Amouretti, M.C., J.P. Brun and D. Eitam (eds.)
1991 *Oil and Wine Production in the Mediteranean Area from the Bronze Age to the End of the XVIth Century* (Conference proceedings pre-publication; Aix en Provence: Centre Camille Julian [CNRS] Centre archéologique).

Amouretti, M.C., G. Comet, C. Ney and J.L. Paillet
1984 'A propos du pressoir à huile de l'archéologie industrielle à l'histoire', *Mélanges de l'école française de Rome, Antiquité* 96: 379-421.

Andel, T.H. Van, and C.N. Runnels
1987 *Beyond the Acropolis: A Rural Greek Past* (Stanford, CA: Stanford University).

Andersen, F.G.
1985 *Shiloh. II. The Remains from the Hellenistic to the Mamluk Periods. The Danish Expedition at Tall Sailun Palestine in 1926, 1929, 1932 and 1963* (Copenhagen: National Museum of Denmark).

Anderson, J.G.C.
1903 *A Journey of Exploration in Pontus* (Brussels: Studia Pontica).

Andreussi, M.
n.d. *Vicus Matrine* (Forma Italiae Regio, 7.4; Rome: De Luca).

Angelini, L.
1967 *Bergamo e la Bergamasca* (Bergamo).

Apicius
1965 *Apicius: De Re Coquinaria (L'art culinaire)* (trans. J. Andre; Paris: Klicksieck).

Argoud, G., O. Callot and B. Helly
1980 *Salamine de Chypre*. XI. *Une residence byzantine, l'huilerie* (Paris: de Boccard).

Armani, V.
1962 'Le pressoir à huile "U Frangu"', *Corse Historique* 5–6: 71-74.

Arrighi, G.
1962 'Notizia di un frantoio Pompeiano per Ulive', *Klearchos* 13–14: 83-86.

Aviam, M.
1986–87 'Olive Growing and Viticulture in Upper Galilee in Ancient Times', *Israel Land and People* 4(22): 197-210 (Hebrew).
1990a 'Tel Yodfat Oil Press', *Hadashot Arkheologiot* 95: 18 (Hebrew).
1990b 'Horvath Hesheq: A Unique Church in Upper Galilee. Preliminary Report', in Bottini, Di Signi and Alliati 1990: 351-78.

Avigad, N.
1953 'Another Bat le-Melekh Inscription', *Israel Exploration Journal* 3: 121-22.
1957 'A New Class of Yehud Stamps', *Israel Exploration Journal* 7: 146-53.
1958 'New Light on the MSH Seal Impressions', *Israel Exploration Journal* 8: 113-19.
1959 'Some Notes on the Hebrew Inscriptions from Gibeon', *Israel Exploration Journal* 9: 130-33.
1960 'YEHUD or HA'IR', *Bulletin of the American Schools of Oriental Research* 158: 23-27.
1968–69 'An Inscribed Bowl from Dan', *Palestine Exploration Quarterly* 100–101: 42-44.
1972 'Two Hebrew Inscriptions on Wine Jars', *Israel Exploration Journal* 22: 1-9.
1976 *Bullae and Seals from a Post-Exilic Judean Archive* (Qedem, 4; Jerusalem: Hebrew University).

Avitsur, S.
1965 *The Palestinian Plough* (Tel Aviv: Ha-Sadeh) (Hebrew).
1976 *Man and his Work: Historical Atlas of Tools and Workshops in the Holy Land* (Israel Exploration Society; Jerusalem: Carta) (Hebrew).
1986 *Bet Ha-Bad Ha-Mesorati* (The Traditional Oil Press) (Man and his Work Library, 6; Tel Aviv: Ha-Aretz Museum) (Hebrew).

Avi-Yonah, M.
1977 *The Holy Land* (Grand Rapids: Baker Book House).

Avi-Yonah, M., and E. Stern (eds.)
1978 *Encyclopedia of Archaeological Excavations* (4 vols.; Jerusalem: Massada).

Avner, R.
1991 'Jerusalem Pisgat Ze'ev Deir Ghazali', *Hadashot Arkheologiot* 97: 60-62 (Hebrew).

Avni, G., and S. Gudovitz
1987 'The Oil Press and the Subterranean Storage Complexes at Achuzat Chazan', in Heltzer and Eitam 1987: 6* 12* (Hebrew).

Ayali, M.
1984 *A Nomenclature of Workers and Artisans in the Talmudic and Midrashic Literature* (Tel Aviv: Hakibbutz Hameuchad) (Hebrew).

Ayalon, E.
1982 'Ha-Aretz Museum Tel Aviv', *Hadashot Arkhiologiot* 80-81: 17 (Hebrew).
1984 'Two Wine Presses from the Roman Period at the Ha'aretz Museum Centre', *Tel Aviv* 11: 173-74.
1988 'mitqanei haqlaut ve-melakha be-arbel (The Agricultural and Craft Installations at Arbel)', in Z. Ilan and A. Ezdarekhet (eds.), *Arbel* (Tel Aviv: The Section for Homeland Studies of the Kibbutz Movement): 159-87 (Hebrew).
1990a 'H. Migdal', *Hadashot Arkhiologiot* 95: 44-45 (Hebrew).
1990b 'The Hassuna Oil Plant in Lod', *Cathedra* 56: 176-81 (Hebrew).

Ayalon, E., H. Gilboa and Y. Harpaz
1987 'An Arab Public Building and Crusader Remains at Tel Qasile, Seasons Twelve and Thirteen', *Israel People and Land* 5–6: 9-22.

Bagatti, P.B.
1947 *I monumenti di Emmaus El-Qubeibeh dei Dintorni* (Publications of the Studium Biblicum Franciscanum, 4; Jerusalem: Tip dei PP Franciscani).

Baird, J.A. (commentator)
1979 *Wine and the Artist* (New York: Dover).

Baramki, D.C.
1933 'A Byzantine Bath at Qalandia', *Quarterly of the Department of Antiquities in Palestine* 3: 105-109.
1934–35 'Recent Discoveries of Byzantine Remains in Palestine', *Quarterly of the Department of Antiquities in Palestine* 4: 118-21.

Barbery, J., and J.P. Delhoume
1982 'La voie romaine de Piedmont Sufetula Masclianae (Djebel Mrhila, Tunisie Centrale)', *Antiquités africaines* 18: 27-43.

Barnett, R.D.
1957 *A Catalogue of the Nimrud Ivories in the British Museum* (London: British Museum).
1982 *Ancient Ivories in the Middle East* (Qedem, 14; Jerusalem: Hebrew University).

Baroja, J.C.
1956 'Sobre maquinarias de tradicion antiqua y mediaval', *Revistas dialectologia tradiciones populares (Madrid)* 12: 114-75.

Bartocchi, F.N.
1963 'La villa romana di Camarelle', *Klearchos* 5: 135-52.

Baruch, U.
1986 'The Late Holocene Vegetational History of Lake Kinneret (Sea of Galilee), Israel', *Paleorient* 12.2: 37-47.

Basch, L.
1981 'Carthage and Rome: Tenons and Mortises', *The Mariner's Mirror* 67: 245-50.

Ballu, A.
1925 *Rapport sur les travaux de fouilles et consolidations entrepris par le service des monuments historiques pendant l'exercise 1924* (Alger: Gouvernement Général de L'Algérie).
1927 *Rapport sur les travaux de fouilles et de consolidations effectués en 1926 par le service des monuments historique de l'Algérie* (Alger: Gouvernement Général de L'Algérie).

Bel, A.
1917 'La fabrication de l'huile d'olive à Fez et dans la region', *Bulletin de la Société de Géographie d'Alger et de l'Afrique de Nord*: 121-37.

*Bellard, C.G., P. Guerin, E.D. Cusi and G.P. Jorda
1993 'El vino en los inicios de la cultura Iberica: Nuevas excavaciones en l'Alt de Benamaquia, Denia', *Revista de Arqueologia* 13 (142): 16-27.

Ben David, C.
1989 'Olive Cultivation and Oil Production in Roman and Byzantine Golan' (Unpublished MA thesis, Bar Ilan University) (Hebrew).

Bellet, M.E.
1976 'Les huileries gallo-romaines de Provence', *Archeologia* 92: 53-59.

Benoit, M.F.
1936 'Notes et documents d'archéologie arlésienne XVI: Pressoirs d'olive à levier et contrepoids en Provence et en Afrique', *Mémoires d'Institut historique de Provence* 13: 106-26.
1960 'Circonscription d'Aix en Provence (région sud)', *Gallia* 18: 286-327.
1968 'Résultats historiques des fouilles d'Entremot (1946–1967)', *Gallia* 26: 1-31.

Ben-Tor, A., M. Avisar, R. Bonfil, I. Zeretsky and Y. Portugali
1987 'A Regional Study of Tel Yoqneʿam and its Vicinity', *Qadmoniot* 20 (77-78): 1-17 (Hebrew).

Ben-Tor, A., and Y. Portugali
1987 *Tell Qiri: A Village in the Jezreel Valley. Report of the Archeological Excavations 1975–1977* (Qedem, 24; Jerusalem: Hebrew University).

Benvenisti, M.
1970 *The Crusaders in the Holy Land* (Jerusalem: Israel Universities Press).

Ben Yaʿakov, Y.
1978 *ha-gefen ve-ha-gat be-har-hevron u-be-meqorot* (The Vine and the Wine Press on Mount Hebron and in the Written Sources) (Kefar 'Etsyon Field School) (Hebrew).

Berthier, A.
1967 'Etablissements agricoles antiques à Oued-Athmenia', *Bulletin d'archéologie algérienne* 1 (1962–65): 7-20.
1980 'Un habitat punique à Constantine', *Antiquités africaines* 16: 13-26.

Beyer, H.W., and H. Lietzman
1930 *Jüdische Katakombe der Villa Torlonia in Rom* (Jüdische Denkmäler, 1; Studien zur spätantiken Kunstgeschichte im Auftrage des deutschen archäologischen Instituts, 4; Berlin: W. de Gruyter).

Bickmore, D.P. (ed.)
1958 *Concise Oxford Atlas* (London: Oxford University Press).

Billiard, R.
1913 *La vigne dans l'antiquité* (Lyon: Lardanchet).

Binford, L.R.
1962 'Archaeology as Anthropology', *American Antiquity* 28: 217-25.

Biran, A.
1980 'Two Discoveries at Tel Dan', *Israel Exploration Journal* 30: 89-98.

Bizard, L.
1906 'Fouilles de Délos', *Bulletin de correspondance hellénique* 30: 483-672.

Blackman, A.M.
1953 *Archaeological Survey of Egypt: The Rock Tombs of Meir* (London: Egypt Exploration Fund): V.

Blankenhagen, P.V., M.A. Cotton and J.B. Ward-Perkins
1965 'Two Roman Villas at Francolise, Prov. Caserta. Interim Report on Excavations 1962–1964', *Papers of the British School at Rome* 33 (NS 20): 55-69.

Blanco Freijeiro, A.
1978 *Mosaicos romanos de Merida* (Corpus de Mosaicos Romanos de Espana, Fasc. 1; Madrid: Instituto español de arqueologia 'Rodrigo Caro' CSIC).

Blazquez Martinez, J.M.
1980 *Producion y comercio del aceite en la antiquedad (I congreso internacional, Madrid, 1980)* (Madrid: Universidad Complutense).

Blazquez Martinez, J.M., and J.R. Remesal Rodriquez
1983 *Producion y comercio del aceite en la antiquedad (II congreso internacional, Seville 1982)* (Madrid: Universidad Complutense).

Bliss, F.J., and R.A.S. Macalister
1902 *Excavations in Palestine during the Years 1898–1900* (London: Palestine Exploration Fund).

Blitzer, H.
1991 'Olive Cultivation and Oil Production in Minoan Crete', in Amouretti, Brun and Eitam 1991: 91-100.

Blümner, H.
1921 'Seife', *Paulys Realencyclopädie der classischen Altertumwissenschaft* 2.3: 1112-14.

Boak, A.E.R., and E.E. Peterson
1931 *Karanis: Topographical and Architectural Report of Excavations during the Seasons 1924–28* (Ann Arbor: University of Michigan).

Boardman, J.
1958–59 'Excavations at Pindakos in Chios', *Annual of the British School at Athens* 53–54: 295-309.
1976 'The Olive in the Mediterranean: Its Culture and Use', *Phil. Trans. R. Soc. London B* 275: 187-96.

Bojanovski, I.
1969 'Anticka uljara na mogorjelu i rekostrukcija Njenog Torkulara', *Goisnyak zavoda zu zastitu spomenika kulture Bosne i Hercegovine* 12: 27-54.

Bonello, V., V. Borg, M. De Azevedo, A. Ciascca, E. Coleiro, A. Davico, G. Garbini, S. Moscati, F.A. Pennacchietti, B. Pugliese and V. Scrinari
1964 *Missione archeologica italiana à Malta: Rapporto preliminare della campagna 1963* (Rome: Istituto di Studi del Vicino Oriente).

Borowski, O.
1980 *Agriculture in Iron Age Israel* (Ann Arbor: University Microfilms International).
1982 'A Note on the Iron Age Cult Installation at Tell Dan', *Israel Exploration Journal* 32: 58.

Bosanquet, R.C.
1901–02 'Excavations at Praesos I', *Annual of the British School at Rome* 8: 231-81.

Bottini, G.C., L. DiSigni and E. Alliati (eds.)
1990 *Christian Archaeology in the Holy Land: New Discoveries. Essays in Honour of Virgilio C. Corbo* (Studium Biblicum Franciscanum, Collectio Maior, 36; Jerusalem: Franciscan Printing Press).

Boucher, J.P.
1956 'Le temple rond de Tébessa Khalia', *Libyca* 4: 7-30.

Bouvier, M.M.
1990 'Cuves vinaires en Vaucluse', in M.R. Chevallier (ed.), *Archéologie de la vigne et du vin* (Caesarodunum, 24; Actes de Colloque 28–29 Mai 1988; Paris: De Bocard): 57-70.

Bouzon, H.
1970 'Le pressoir 'Casse-Cou' du pays de Loudun', *Société d'études folkloriques du centre ouest revue de recherches ethnographique* 4: 356-67.

Bowyer, P.R.
n.d. *A History of Cidermaking in England* (Department of Architecture, Brighton Polytechnic).

Brace, H.W.
1960 *History of Seed Crushing in Great Britain* (London: Land Books).

Brand, Y.
1962 'Bet Ha-Bad (The Oil Press)', *Sinai* 6: 303-25 (Hebrew).

Branigan, K.
1970 *The Foundation of Palatial Crete: A Survey of Crete in the Early Bronze Age* (London: Routledge & Kegan Paul).

Brogan, O.
1964 'The Roman Remains in the Wadi el-Amud', *Libya Antiqua* 1: 47-56.

Brøndsted, J.
1928 'La basilique des cinq martyrs à Kapljuc', in E. Dyggve and J. Brøndsted, *Recherches à Salone*, I (Copenhagen: J.H. Schultz Imprimerie de l'Université): 33-176.

Brown, J.P.
1969 'The Mediterranean Vocabulary of the Vine', *Vetus Testamentum* 19: 146-70.

Brown, F., S.R. Driver and C.A. Briggs
1907 *A Hebrew and English Lexicon of the Old Testament* (Oxford: Clarendon Press).

Brun, J.P.
1983 'Recherches récentes sur l'oléiculture antique en Provence: Les donnés archéologiques et leurs interpretation', in Miège 1983: 35-51.
1986 *L'oléiculture antique en Provence: Les huileries du department du Var* (Paris: CNRS).

Brun, J.P., M. Gerard and M. Pasqualani
1981 'La villa gallo-romaine de St. Michal à la Garde', *Histoire et archéologie* 57: 69-70.

Brun, J.P., and M. Pasqualani
1981 'Les huileries de l'Ormeau à Taradeau', *Histoire et archéologie* 57: 75-76.
1989 'La villa gallo-romaine de Saint Michel à la Garde (Var): Un domaine oléicole au haut-empire', *Gallia* 46: 103-62.

Bruneau, P., and P. Fraisse
1981 'Un pressoir à vin à Délos', *Bulletin de correspondance hellénique* 55: 128-53.
1984 'Pressoirs Deliens', *Bulletin de Correspondence hellénique* 58: 713-30.

Bull, R., J.A. Callaway, E.F. Campbell, J.F. Ross and G.E. Wright
1965 'The Fifth Campaign at Balatah (Shechem)', *Bulletin of the American Schools of Oriental Research* 180: 7-41.

Bunimovitz, S.
1987 'Minoan-Mycenaean Olive Oil Production and Trade', in Heltzer and Eitam 1987: 11-15.

Butler, H.C.
1903 *Architecture and Other Arts: Publication of the American Archaeological Expedition 1899–1900*, II (New York: The Century).

Cadogan, G.
1986 'Maroni II', *Report of the Department of Antiquities Cyprus* 1986: 40-44.

Callot, O.
1982 'Remarques sur les huileries de Khan Khalde (Liban)', in *Archéologie au Levant, Recueil à la mémoire de Roger Saideh* (Lyon: Collection de la Maison de l'Orient Méditerranée, 12): 420-28.
1984 *Huileries antiques de Syrie du Nord* (Paris: P. Geuthner).
1987 'Les huileries du Bronze Récent à Ougarit: Premiers élément pour une étude', in M. Yon, *Le centre de la ville Ras Shamra Ougarit III, 38e 44e campaignes (1978–1984)* (Paris: Editions Recherche sur les Civilisations): 197-212.

Callot, O., M. Reddé and J.P. Vallat
1986 'Un contrepoids ou pierre d'ancrage de pressoir (companie septentorale)', *Mélanges de l'école française de Rome. Antiquité* 98: 129-40.

Camps-Fabrer, H.
1953 *L'olivier et l'huile dans l'Afrique romaine* (Alger: Imprimerie Officielle).
1957 'Une nouvelle huilerie romaine dans la region de Perigotville', *Libyca* 5: 43-46.
1983 'L'olivier et son importance économique dans l'Afrique du Nord Antique', in Miège 1983: 53-78

Canivet, P., and M.T. Fortuna
1968 'Recherches sur le site de Nikertai', *Annales archéologiques arabes syriennes* 17: 7-54.

Capart, J.
1925 *Thebes: La gloire d'un grand passé* (Brussels: Vromat).

Caprino, C.
1944–45 'Guidonia villa rustica con "torcularium" ', *Notizie degli scavi di antichita* 1944–45: 39-51.

Carandini, A., and T. Tatton-Brown
1980 'Excavations at the Roman Villa of "Sette Finestre" ', in K. Painter (ed.), *Roman Villas in Italy* (London: British Museum): 9-22.

Carmi, I., and D. Segal
1994–95 ^{14}C Dates of Olive Stones from an Underwater Site at Kfar Samir', *Mitekufat Haeven, Journal of the Israel Prehistoric Society* 26: 146.

Caro Baraja, J.
1963 'Remarques sur la vie agraire en Andalousie', *Etudes rurales* 10: 81-101.

Carreras, M.
1945–46 'Los hallaggos arqueologicos de Porporas (Reus)', *Boletin arqueologico de la sociedad arqueologica tarraconense* 45.4: 82-91.
1948 'Las instalaciones agricol-industriales ibero romanas de Porporas', *Boletin arqueologico de la sociadad arqueologica tarraconense* 48.4: 65-70.

Caruana, A.A.
1888 'Remains of an Ancient Greek Building Discovered in Malta', *American Journal of Archaeology* 4 (1st series): 450-54.

Casanova, A.
1966 'Technologie et communautés rurales: Notes sur les pressoirs préindustriels de Corse', *Corse historique* 21–22: 37-61.
1968 'Typologie et diffusion des pressoirs préindustriels dans les communautés rurales de Corse', *Corse historique* 31–32: 39-56.
1990 *Paysans et machines à la fin du XVIII siècle: Essai d'ethnologie historique* (Annales Littéraires de l'Université de Besançon, 415; Paris: Belles Lettres).

Catani, E.
1976 'I frantoi della fattoria bizantina di El-Beida', *Quadreni di archeologia della Libia* 8: 435-48.

Catling, H.W.
1972 'An Early Byzantine Pottery Factory at Dhiorios in Cyprus', *Levant* 4: 2-17.
1976–77 'The Knossos Area 1974–1976', *Archaeological Reports for 1976–1977*: 3-23.

Cato
1979 'Marcus Porcius Cato: De Agri Cultura', in W.D. Hooper (trans.) *Cato and Varro* (London: Loeb): 2-157.

Cesnola, L.P. di,
1877 *Cyprus: Its Ancient Cities, Tombs and Temples* (London: Murray).

Cerruti, J.
1981 'Down the Ancient Appian Way', *National Geographic Magazine* 159.6: 717-47.

Chaillan, M.
1930–31 'Fond de pressoir à huile sur pierre à Rainure découvert à l'Oppidum d'Entremont', *Bulletin du comité des travaux historiques et scientifiques* 1930–31: 451-53.

Chevallier, R (ed.)
1990 *Archéologie de la vigne et du vin: Actes du colloque 1988* (Caesarodunam, 24; Paris: Université de Tours, De Bocard).

Chouquer, G., F. Favory and J.P. Vallat
1987 *Structures agraires en Italie centro meridionale: Cadastres et paysages rural* (Rome: Ecole française de Rome, 100).

Christofle, M.
1930a *Rapport sur les travaux et consolidations affectués en 1927, 1928 et 1929 par le service des monuments historiques de l'Algérie* (Alger: Governement Général de L'Algérie).

1930b *Essai de restitution d'un moulin à huile de l'époque romaine à Madaure (Constantin)* (Alger: Ancienne maison Bastide-Jourdan, Jules Carbonel).

Chronique
1955 'Chronique des fouilles et découvertes archéologiques en Grèce en 1954', *Bulletin de correspondance hellénique* 79: 205-376.
1980 'Chronique des fouilles et découvertes Archéologiques en Grèce en 1979', *Bulletin de correspondance hellénique* 104: 581-688.

Clastrier, M.S.
1910 'Pierre molaire à rainure du Grand Arbois (Bouches-du-Rhone)', *Bulletin de la société préhistorique de France* 7: 61-211.

Clemont-Ganneau, C.S.
1896–99 *Archaeological Researches in Palestine during the Years 1873–1874* (2 vols.; London: Palestine Exploration Fund).

Cohen, A.
1989 'A Soap Factory in Ottoman Jerusalem', *Cathedra* 52: 120-24 (Hebrew).

Cohen, R.
1987 'Excavations at Moa 1981–1985', *Qadmoniyot* 20 (77-78): 26-31 (Hebrew).

Columella
1969–79 *Columella, Lucius Junius Moderatus: De Re Rustica* (trans. H.B. Ash, E.S. Forster and E. Heffner; 3 vols.; London: Loeb).

Conder, C.R.
1889 *The Survey of Eastern Palestine*, I (London: Palestine Exploration Fund).

Conder, C.R., and H.H. Kitchener
1881–83 *The Survey of Western Palestine* (3 vols.; London: Palestine Exploration Fund).

Coon, C.S.
1931 *Tribes of the Rif* (Harvard African Studies, 9; Cambridge: Peabody Museum of Harvard University).

Corbo, P.V.
1955 *Gli scavi di Kh. Siyar el-Ghanam (Campo dei Pastori) e i monasteri dei Dintorni* (Publications of the Studium Biblicum Franciscanum, 11; Jerusalem: Tip dei PP Franciscani).

Corcoran, T.H.
1962 'Roman Fish Sauces', *The Classical Journal* 58: 204-10.

Costa, P., E. Dies, C. Gomez, P. Guerin and G. Perez
1992 *L'Alt de Benimaquia: Iberos y Fenicios en la Marina Alta* (Valencia: University of Valencia) (Excavations Brochure).

Cotton, M.A.
1979 *The Late Republican Villa at Posto Francolise* (London: The British School at Rome).

Cotton, M.A., P.V. Blanckenberg and J.B. Ward-Perkins
1965 'Francolise (Caserta) Rapporto provvisorio del 1962–64 sugli scavi di due ville romane della Repubblica e del Primo Impero', *Notizie degli scavi di antichita* 19 (Anno 362): 237-52.

Cotton, H.C., and J. Geiger
1989 *Masada*. II. *The Latin and Greek Documents* (Jerusalem: Hebrew University).

Coupry, M.J.
1957 'IX Circonscription', *Gallia* 15: 240-56.
1959 'Circonscription de Bordeaux', *Gallia* 17: 377-409.

Courtois, J.C.
1962 'Contributions à l'étude des civilisations du bronze ancien à Ras Shamra Ugarit', in C.F.A. Schaeffer (ed.) *Ugaritica IV* (Paris: P. Geuthner): 415-75.

Cowper, H.S.
1896 'The Senans of Megalithic Temples of Tripoli', *The Antiquary* 1896: 37-45.
1897 *The Hill of the Graces* (London: Methuen).

Cresswell, R.
1965 'Un pressoir à olives au Liban: Essai de technologie comparé', *L'homme Revue française d'anthropologie* 5: 33-63.

Cross, F.M.
1968 'Jar Inscriptions from Shiqmona', *Israel Exploration Journal* 18: 226-33.
1969 'Judean Stamps', *Eretz-Israel* 9: 20*-27*.

Crowfoot, J.W., K.M. Kenyon and E.L. Sukenik
1942 *Samaria-Sebaste*. I. *The Buildings* (London: Palestine Exploration Fund).

Crowfoot, J.W., G.M. Crowfoot and K.M. Kenyon
1957 *Samaria-Sebaste*. III. *The Objects* (London: Palestine Exploration Fund).

Curtis, R.I.
1983 'In Defense of Garum', *The Classical Journal* 78: 232-40.
1984–86 'Product Identification and Advertising on Roman Commercial Amphorae', *Ancient Society* 15–17: 209-28.

D'Alembert, J. le Rond, and D. Diderot
1762 *Agriculture et économie rustique, Vol. I pt. 1 de recueil de planches sur les arts libéraux et les mécaniques avec leur explication. Encyclopédie ou dictionnaire raisonne des arts et métiers* (Paris).

Dalman, G.
1908 'Die Schalensteine Palästinas in ihrer Beziehung zu alter Kultur und Religion', *Palästina Jahrbuch* 4: 23-53.
1928–42 *Arbeit und Sitte in Palästina* (7 vols.; Gütersloh: Bertelman [repr. Hildesheim: G. Olms, 1964]).

Danby, H. (trans.)
1933 *The Mishnah* (London: Oxford University Press).

Dar, S.
1978 'seqer yishuvim qedumim 'al katef ha-hermon (A Survey of Ancient Settlements on the Slopes of Mt Hermon)', in S. Applebaum (ed.), *Ha-Ḥermon Ve-Margelotav* (The Hermon and its Slopes) (Jerusalem: The Section for Homeland Studies of the Kibbutz Movement): 52-120 (Hebrew).
1980 'Khirbet Jema'in: A Village from the Period of the Monarchy', *Qadmoniot* 13 (51-52): 97-100 (Hebrew).
1982 'Umm Er-Rihan Region', *Excavations and Surveys in Israel* 1: 110-12.
1985 'Khirbet Summaqa 1985', *Excavations and Surveys in Israel* 4: 104-107.
1986 *Landscape and Pattern: An Archaeological Survey of Samaria 800 BCE 636 CE* (Oxford: British Archaeological Reports).

Daremberg, C.V., and E. Saglio
1877–1919 *Dictionnaire des antiquités grecques et romaines* (Paris [repr.; Graz: Anstalt 1962–63]).

Davidson-Weinberg, G.
1988 *Excavations at Jalame* (Columbia: University of Missouri).

Davies, N.D.G.
1907 *The Tomb of Nakht at Thebes* (Tytus Memorial Series, 1; New York: The Egyptian Expedition).
1922 *The Tomb of Puyemre at Thebes*. I. *The Hall of Memories* (Tytus Memorial Series, 2; New York: Metropolitan Museum of Art).
1923 *The Tomb of Two Officials of Thutmoses the Fourth: Nos. 75 and 90* (The Theban Tombs Series; London: Egypt Exploration Society).
1933 *The Tomb of Nefer-Hotep at Thebes* (Egyptian Expedition, 9; New York: Metropolitan Museum of Art).
1943 *The Tomb of Rekhmire at Thebes Vol. II* (Egyptian Expedition, 10; New York: Metropolitan Museum of Art).

Davies, D.
1978 'H. Mesah', *Hadashot Arkhiologiot* 65-66: 5-6 (Hebrew).

Dawkins, R.M.
1902–1903 'Notes from Karpathos', *Annual of the British School at Athens* 9: 176-210.

1904– 1905 'Excavations at Palaikastro IV', *Annual of the British School at Athens* 11: 258-92.

De Azevedo, M.C., C. Caprino, A. Ciasca, E. Coleiro, A. Davico, G. Garbini, F.S. Mallia, G.P. Marchi, P. Mingati, S. Moscati, E. Paribeni, B. Puugliese, M.P. Rossignani, V.S.M. Scrinari and A. Stenico

1965 *Missione archeologica italiana a Malta: Rapporto preliminare della campagna 1964* (Rome: Istituto di Studi del Vicino Oriente, Universita).

1966 *Missione archeologica italiana a Malta: Rapporto preliminare della campagna 1965* (Rome: Istituto di Studi del Vicino Oriente, Universita).

1967 *Missione archeologica italiana a Malta: Rapporto preliminare della campagna 1966* (Rome: Istituto di Studi del Vicino Oriente, Universita).

De Boe, G.

1975 'Villa romana in localita "Posta Crusta". Rapporto provvisorio sulle campagne di scavo 1972 e 1973', *Notizie degli scavi di antichita* 1975: 516-30.

Dechandol, H., M.P. Feuillel and T. Odiot

1983 'Le grand domaine viticole du Molard', *Histoire et archéologie* 78: 56-57.

Della Corte, M.

1921 'Pompei scavi eseguiti da privati nel territorio Pompiano', *Notizie degli scavi di antichita* 1921: 415-67.

Delmas, J.

1988 *Presses et pressoirs Rouergats* (Rodez: Amis du Musée du Rouergue).

Demsky, A.

1972 '"Dark Wine" from Judah', *Israel Exploration Journal* 22: 233-34.

1979 'A Note on Smoked Wine', *Tel Aviv* 6: 163.

Dentzer, J.M.

1985 *Hauran I: Recherches archéologiques sur la Syrie du sud à l'époque hellénistique et romaine* (Paris: Paul Geuther).

De Rossi, G.M.

1979 *Bovillae* (Forma Italiae, 1.15; Florence: Istituto di topografia antica della Universita di Roma).

De Vaux, R.

1951 'La troisième campagne de fouilles à Tell el Farah près Naplouse', *Revue biblique* 58: 393-430, 566-90.

De Vogue, Le cte, C.J.M.

1865–77 *Syrie centrale architecture civile et religieuse de I-er à VII siècle* (2 vols.; Paris: Noblet et Baudry).

Dodwell, C.R., and P. Clemoes

1974 *The Old English Illustrated Hexateuch: British Museum Cotton Claudius B IV* (Copenhagen: Rosenkilde & Bagger).

Dorner, H., and E. Kutsch

1963 'Archäologische Bemerkungen zu Etam', *Zeitschrift des deutschen Palästina Vereins* 79: 8-12.

Dorpfeld, W.

1895 'Die Ausgrabungen am Westabhange der Akropolis II', *Mitteilungen der kaiserlich deutschen archäologischen Institut* 20: 161-206.

Dothan, M.

1956 'The Excavations at Nahariyah: Preliminary Report (Seasons 1954/1955)', *Israel Exploration Journal* 6: 14-25.

1972 'Ashdod: Seven Seasons of Excavations', *Qadmoniot* 5 (17): 2-13 (Hebrew).

Dothan, T., and S. Gitin

1984 'Tel Miqne Ekron 1984', *Excavations and Surveys in Israel* 3: 78-80 and cover.

1986 'Tel Miqneh Ekron 1985/1986 Season', *Excavations and Surveys in Israel* 5: 74-77.

Drachman, A.G.

1932 *Ancient Oil Mills and Presses* (Copenhagen: Levin & Munksgaard).

1963 *The Mechanical Technology of Greek and Roman Antiquity* (Copenhagen: Munksgaard).

Dufkova, M., and J. Pečirka

1970 'Excavations of Farms and Farmhouses in the Chora of Chersonesos in the Crimea (Farms Strshelilski)', *Eirene* 8: 123-74.

Dunand, M., and R. Duru

1957 *Oumm el-Amed* (Paris: Maisonneuve).

Duncan, G.

1926 'New Rock Chambers and Galleries on Ophel', *Palestine Exploration Quarterly Statement*: 7-14.

Du Plat Taylor, J.

1952 'A Late Bronze Age Settlement at Apliki, Cyprus', *The Antiquiaries Journal* 32: 133-66.

1957 *Myrtou Pigadhes* (Oxford: Ashmolean Museum).

Duran, F.

1958 *Le moulin des bouillons* (Museum Brochure).

Duval, N.

1976 'Encore les "Monuments à Auges" d'Afrique Tebessa Khalia, Hr Faraoun (3rd Article)', *Mélanges de l'école française de Rome, Antiquité* 88: 929-59.

Duval, N., and F. Baratte

1973 *Les ruines de Sufetla* (Sbeitla; Tunis: Société Tunisienne de Diffusion).

Dyson, S.L.

1972 'Excavations at Buccino 1971', *American Journal of Archaeology* 76: 159-63.

Edelstein, G., and Y. Gat

1980 'yerushalayim madregot saviv la (Terraces around Jerusalem)', *Teva Ve-Aretz* 22.5: 191-95 (Hebrew).

Eisenberg, E.

1977 'Mavo Modi'in', *Hadashot Arkheologiot* 61–62: 27-28 (Hebrew)

Eitam, D.

1979 'Olive Presses of the Israelite Period', *Tel Aviv* 6: 146-55.

1980 'The Production of Oil and Wine in Mount Ephraim in the Iron Age' (Unpublished MA thesis, Tel Aviv University) (Hebrew).

1986 'Tel Miqne-Ekron Survey of Oil Presses 1985', *Excavations and Surveys in Israel* 5: 72-74.

1987 'Olive-Oil Production during the Biblical Period', in Heltzer and Eitam 1987: 16-36.

Eitam, D., and M. Heltzer

1996 *Olive Oil in Antiquity* (Padua: Sargon).

Eitam, D., and A. Shomroni

1987 'Research of the Oil Industry during the Iron Age at Tel Miqne: A Preliminary Report', in Heltzer and Eitam 1987: 37-56.

Elgavish, J.

1969 'Shiqmona', *Israel Exploration Journal* 19: Notes and News, 247-48.

1970a 'Shiqmona', *Hadashot Arkhiyologiot* 33: 16 (Hebrew).

1970b 'Shiqmona: A Biblical City', *Qadmoniot* 3 (11): 90-93 (Hebrew).

1972 'Shiqmona 1971', *Hadashot Arkhiyologiot* 43: 5-6 (Hebrew).

1978 'Shiqmona, Tel', in M. Avi-Yonah and E. Stern (eds.), *Encyclopedia of Archaeological Excavations in the Holy Land* (4 vols.; Jerusalem: Massada): 1101-109.

Epstein, C.

1976 'Kh el-Hutiya', *Hadashot Arkhiyilogiot* 59–60: 8-9 (Hebrew).

1978 'A New Aspect of Chalcolithic Culture', *Bulletin of the American Schools of Oriental Research* 229: 27-45.

1993 'Oil Production in the Golan Heights during the Chalcolithic Period', *Tel Aviv* 20: 133-46.

Epstein, J.N.

1924 *The Gaonic Commentary on the Order Toharoht Attributed to Rav Hay Gaon* (Berlin: Meqizei Nirdamim [repr. Jerusalem: Devir-Magnes]).

Etienne, R.
1960 *Le quartier nord-est de Volubilis* (Paris: de Boccard).
Eusennat, M.
1957 'L'archéologie marocaine de 1955 à 1957', *Bulletin d'archéologie marocaine* 2: 199-229.
1967 'Circonscription de Provence Cote d'Azur-Corse (Region Sud)', *Gallia* 25: 392-435.
Evans, A.J.
1900–1901 'The Palace of Knossos', *Annual of the British School at Athens* 7: 1-119.
1921–35 *The Palace of Minos at Knossos* (4 vols.; London: Macmillan).
Faccena, D.
1957 'Tivoli (localita Granaraccio) resti della parte rustica di una villa', *Notizie degli scavi di antichita* 1957: 148-53.
Fantar, M.
1984 'A Gamartii avant la conquête romaine', in *Actes du 1ère colloque international sur l'histoire et l'archéologie de l'Afrique du Nord (Perpigan 14–18 Avril 1981)* (Bulletin archéologique du CTHS, NS 17b; Paris): 3-19.
Feig, N.
1984 'Lohamei HaGetaot Area D', *Excavations and Surveys in Israel* 3: 73.
1987–89 'Wine Presses from Tell el-Samaria', *Israel: People and Land* 5–6: 73-91 (Hebrew).
1988–89 'Meron', *Excavations and Surveys in Israel* 7–8: 127-28.
Feliks, J.
1963 *Agriculture in Palestine in the Period of the Mishna and Talmud* (Jerusalem: Magnes Press) (Hebrew).
Felleti Maj, B.M.
1955 'Roma (Via Tiberina) villa rustica', *Notizie degli scavi di antichita* 1955: 206-16.
Fernandez Castro, M.C.
1983 'Fabricas de aceite en el campo hispano-romano', in Blazquez Martinez 1983: 569-99.
Feuille, G.L.
1937–38 'Note sur les vestiges d'un établissement agricole dans les environs de Taenae', *Bulletin archéologique du comité des travaux historiques et scientifiques*: 141-50.
Figuier, L.
1877 *Les merveilles de l'industrie*. IV. *Industries agricoles et alimentaires* (Paris: Furne Jouvet).
*Finet, A.
1974–77 'Le vin à Mari', *Archiv für Orientforschung* 25: 122-31.
Fisher, M.
1985 'Excavations at Horvat Zikhrin', *Qadmoniot* 18 (71-72): 112-21 (Hebrew).
Fitzgerald, G.M.
1939 *A Sixth Century Monastery at Beth-Shan* (Philadelphia: University of Pennsylvania).
Flower, B., and E. Rosenbaum
1958 *The Roman Cookery Book: A Critical Translation of The Art of Cooking by Apicius* (London: Harrap).
Forbes, R.J.
1955–64 *Studies in Ancient Technology* (9 vols.; Leiden: Brill).
Forbes, H.A., and A. Foxhall
1978 ' "The Queen of all Trees": Preliminary Notes on the Archaeology of the Olive', *Expedition* 21: 37-47.
Fraenkel, S.
1898 Review of S. Krauss, *Griechische und lateinische Lehnwörter im Talmud Midrasch und Targum*, I', *Zeitschrift der deutschen morgenländischen Gesellschaft* 25: 290-300.
Frank, T.
1932 'Notes on Cato's de Agricultura', in *Mélanges Gustave Glotz Volume I* (Paris: Universitaires): 377-80.
Frankel, B.
1981 'Notes and Remarks regarding Oil Production in Juda', *Teva VaAretz* 23 (4): 183 (Hebrew).
Frankel, R.
1984 'The History of the Processing of Wine and Oil in Galilee in the Period of the Bible the Mishna and the Talmud' (Doctoral dissertation, Tel Aviv University).
1985 'Western Galilee, Oil-Presses', *Excavations and Surveys in Israel* 4: 110-14.
1986a 'Horvat Din'ila', *Excavations and Surveys in Israel* 5: 21-23.
1986b 'Khirbet el-Quṣeir', *Excavations and Surveys in Israel* 5: 88-91.
1987 'Oil Presses in Western Galilee and the Judea: A Comparison', in Heltzer and Eitam 1987: 63-80.
1988–89 'An Oil Press at Tel Ṣafṣafot', *Tel Aviv* 15–16: 77-91.
1992a 'Some Oil Presses in Western Galilee', *Bulletin of the American Schools of Oriental Research* 286: 39-71.
1992b 'Galilee (Pre-Hellenistic)', in D.N. Freedman (ed.), *The Anchor Bible Dictionary* (New York: Doubleday): II, 879-95.
*Frankel, R., S. Avitsur and E. Ayalon
1994 *History and Technology of Olive Oil in the Holy Land* (Arlington, VA: Olearius).
Frankel, R., and E. Ayalon
1988 *gefen gitot ve-yayin be-'et ha-'atiqa* (The Vine, Wine Presses and Wine in Ancient Times) (Man and his Work, 7; Tel Aviv: Eretz-Israel Museum) (Hebrew)
Frankel, R., J. Patrich and Y. Tsafrir
1990 'The Oil Press at Ḥorvath Beit Loya', in Bottini, Di Signi and Alliata 1990: 287-300.
Frantz, A.
1988 *The Athenian Agora*. XXIV. *Late Antiquity: AD 267–700* (Princeton, NJ: The American School of Classical Studies at Athens).
French, R.K.
1982 *The History and Virtues of Cyder* (New York: St Martins; London: Robert Hale).
Frey, J.B.
1933 'Les juifs Pompei', *Revue biblique* 42: 365-83.
Gal, Z.
1985–86 'Vineyard Cultivation at ʿEmek Ḥarod and its Vicinity 1986 during the Roman Byzantine Period', *Israel: People and Land* 2–3: 129-38 (Hebrew).
Gal, Z., and R. Frankel
1993 'An Olive Press Complex at Ḥurbat Roš Zayit (Ras ez-Zetun) in Lower Galilee', *Zeitschrift des deutschen Palästina Vereins* 109: 36-41.
Gaidukevych, V.F.
1958 'Vinodelye na Bospore', in V.F. Gaidukevych and T.N. Knipovych (eds.), *Bosporskiye Topoda II* (Moscow: Materialy i Issledovaniya po Arkheolgii): 352-457.
Galili, E., and J. Sharvit
1994–95 'Evidence of Olive Oil Production from the Submerged Site at Kfar Samir Israel', *Mitekufat Haeven: Journal of the Israel Prehistoric Society* 26: 122-23.
Galili, E., M. Weinstein-Evron and D. Zohary
1989 'Appearance of Olives in Submerged Neolithic Sites along the Carmel Coast', *Mitekufat Haeven* 22: 95*-97*.
Garcia Diego, J.A.
1983 *Los veintuin libros de los ingenios y de las maquinas. Ms. 3372-76 Biblioteca nacional Madrid* (Ediciones Turner; Madrid: Colegio de ingenieros de caminos canales y puertos).
Gat, Y.
1979 'An Olive Press at the High Commissioner's Palace', *Hadashot Arkheologiot* 72: 29-30 (Hebrew).
1980 'tagliyot arkhiologiot be-talpiot mizrah bi-yerushalaim (Archaeological Discoveries at Talpiot East Jerusalem)', *Qardom* 10-11: 71-72 (Hebrew).
G.B.H.M.S.O (Great Britain, Her Majesty's Stationary Office)
1964 *Weather in the Mediterranean* (London: HMSO, 2nd edn).

Geltzer, M.L.
1963 'Selskaya obshchina i prochiye vidi zemlevladeniya v drevnem Ugarite (The Village Community and Other Forms of Land Tenure in Ancient Ugarit)', *Vestnik Drevnei Istorii* 1963.1: 35ff. Hebrew translation in R. Nadel (trans.) *ha-shilton ha-ʿironi ha-ʿaṣmi be-mizraḥ ha-qadum* (Urban Self-Rule in the Ancient East) (Jerusalem: Hebrew University): 1-53.
1965 'Eshche raz ob obshchinnom samoupravlenii v Ugarite (More as to Community Self-Rule in Ugarit)', *Vestnik Drevnei Istorii* 1965.2: 3ff. Hebrew translation in R. Nadel (trans.), *ha-shilton ha-ʿironi ha-ʿaṣmi be-mizraḥ ha-qadum* (Urban Self-Rule in the Ancient East) (Jerusalem: Hebrew University): 110-39.
Geographical List
1976 *Department of Antiquities of Israel: Geographical List of The Record Files, 1918–1948* (Jerusalem).
Geoponica
1895 *Geoponica sive cassiani bassi scholastici de resticos eclogae* (ed. H. Beckh; Leipzig: Teubner).
Gershuni, L.
1991 'Khʾel Marmitta', *Hadashot Arkhiologiot* 96: 24-25 (Hebrew).
Gianfrotta, P.A.
n.d. *Castrum Novum* (Forma Italiae Regio 7.3; Rome: De Luca).
Gichon, M.
1979–80 'The Upright Screw-Operated Pillar Press in Israel.', *Scripta Classica Israelica* 5: 206-44.
Gilles, K.-J.
1995 *Neurere Forschungen zum römischen Weinbau an Mosel und Rhein* (Trier: Rheinisches Landeshuseum).
Gitin, S.
1987 'Tel Miqne-Ekron in the 7th c. BC: City Plan Development and the Oil Industry', in Heltzer and Eitam 1987: 81-97.
Giveon, R.
1988 'Area I: An Oil Press and Domestic Buildings from the Literary Estate of R. Giveon', in Mazar 1988: 215-18 (Hebrew).
Giveon, R., and M. Linn
1978 'An Oil Press at Mishmar Ha-ʿEmeq', *Hadashot Arkhiologiot* 65–66: 7 (Hebrew).
Gnirs, A.
1908 'Istriche Beispiele für Formen der antik-römischen villa rustica', *Jahrbuch für Altertums Kunde* 2: 124-43.
1914 'Forschungen in Pola und in der Polesana', *Jahreshefte des österreichischen archäologischen Instituts in Wien* 17: 181-84.
1915 'Forschungen über antiken Villenbau in Südistrien', *Jahreshefte des österreichischen archäologischen Instituts in Wien* 18: 101-64.
Goldman, F.
1907 *Der Olbau in Palästina zur Zeit Mišnah* (Pressburg: Adolf Alkalay).
Gonzalez Blanco, A., and J.A. Hernandez Vera
1983 'Mas restos de industria oleicolo romana en la Rioja', in Blazquez Martinez and Rodriguez 1983: 611-16.
Gonzalez Blanco, A., P. Lillo Carpio, A. Guerrero Fuster and S. Ramallo Asensio
1983 'La industria del aceite en la zona de la actual provincia de Murcia durante la epoca romana (Primera aproximacion al tema)', in Blazquez Martinez and Rodriguez 1983: 601-10.
Goodchild, R.G.
1951 'Roman Sites on the Tarhuna Plateau of Tripolitania', *Papers of the British School at Rome* 19 (NS 6): 43-65.
Gophna, R., and M. Kislev
1979 'Tel Ṣaf (1977–1978)', *Revue biblique* 86: 112-14.
Gordon, C.H.
1965 *Ugaritic Textbook* (Rome: Pontifical Institute).
Goudinau, C.
1977 'Circoscription de Cote-d'Azur', *Gallia* 35: 495-510.
1984 'Un contrepoids de pressoir à huile d'Entremont (Bouches-du-Rhone)', *Gallia* 42: 219-21.
Graber, O.
1966 'Qasr el-Sharqui, Preliminary Report', *Annales archéologiques de Syrie* 16: 5-46.
Grace, V.R.
1952 'Timbres amphoriques trouvé à Delos', *Bulletin de Correspondance hellénique* 76: 514-40.
1953 'The Eponyms Named on Rhodian Amphora Stamps', *Hesperia* 22: 116-28.
1956 'The Canaanite Jar', in S.S. Weinberg (ed.), *The Aegean and the Near East: Studies presented to Hetty Goldman* (New York): 80-109.
1961 *Amphoras and the Ancient Wine Trade* (Princeton, NJ: American Schools of Classical Studies).
Grant, E.
1931 *Ain Shems Excavations (Palestine) 1928–31 Part I (Text, Illustrations)* (Haverford, PA: Haverford College).
1932 *Ain Shems Excavations (Palestine) 1928-29-30-31 Part II* (Haverford, PA: Haverford College).
1934 *Rumeilah being Ain Shems Excavations Part III.* (Haverford, PA: Haverford College/J.H. Furst).
Grant E., and G.E. Wright
1938 *Ain Shems Excavations (Palestine) Part IV (Pottery)* (Haverford, PA: Haverford College/J.H. Furst).
1939 *Ain Shems Excavations (Palestine) Part V (Text)* (Haverford, PA: Haverford College).
Greenhot, Z.
1990 'Ḥ. Ḥermeshit', *Hadashot Arkheologiot* 95: 47-49 (Hebrew).
Grimal, P., and T. Monod
1952 'Sur la véritable nature du "Garum" ', *Revue des études anciennes* 54: 27-38.
Gsell, S.
1901 *Les monuments de l'Algérie Vol. II* (Paris: Libraire des écoles d'Athènes et de Rome du Collège de France et de l'Ecole Normale Sup).
Guerin, V.
1868–80 *Description géographique, historique et archéologique de la Palestine* (3 parts, 7 vols.; Paris: Imprimerie National [repr. Amsterdam: Oriental Press, 1969]).
Guillemard, F.H.H.
1888a 'Monoliths in the Island of Cyprus', *The Athenaeum* 3155: 474-75.
1888b 'Monoliths in Cyprus', *The Athenaeum* 3172: 201.
Guidi, G.
1935 'Orfeo liber pater e oceano in mosaici della Tripolitania', *Africa Italiana* 6: 110-55.
Guitart Duran, J.
1970 'Excavacion en la zona sudeste de villa Romana de sentroma (Tiana)', *Pyrenae* 6: 111-65.
Gur, A.
1960 'yeynot eres yisrael biymey qedem (Wines from Eretz Israel in Antiquity)', *Teva Va-Aretz* 3.1: 33-40 (Hebrew).
Gur, A., and H. Speigel
1960 *The Olive* (Tel Aviv: Israel Ministry of Agriculture) (Hebrew).
1963 'The Olive', *Encyclopedia Hebraica* 16: 785-93 (Hebrew).
Güterbock, H.G.
1968 'Oil Plants in Hittite Anatolia', *Journal of the American Oriental Society* 88: 66-71.
Gutman, S., and D. Wagner
1986 'Gamla, 1984–86', *Excavations and Surveys in Israel* 5: 38-41.
Hadjisavvas, S.
1987 'An Introduction to Olive Oil Production in Cyprus', in Heltzer and Eitam 1987: 98-105.

1988 'Olive Oil Production in Ancient Cyprus', *Report of the Department of Antiquities Cyprus* 1988: 111-19.

1990 'Windlass Vs Screw: A Case-Study for the Reconstruction of an Olive Press', *Report of the Department of Antiquities Cyprus* 1990: 181-85 (pl. XXIX).

1991 'Perforated Monoliths: Myths and Reality', in Amouretti, Brun and Eitam 1991: 69-80.

*1992a *Olive Oil Production in Cyprus from the Bronze Age to the Byzantine Period* (Nicosia: Paul Åström).

*1992b 'Olive Oil Production and Divine Protection, in P. Astrom (ed.), *Acta Cypra*. Part 3. *Acts of an International Congress on Cypriote Archaeology held in Goteberg 22–24 August 1991*: 233-49.

Hamel, G.

1983 'Poverty and Charity in Roman Palestine' (Unpublished PhD dissertation; Santa Cruz: University of California).

*Hamilakis, Y.

1996 'Wine, Oil and the Dialectics of Power in Bronze Age Crete: A Review of the Evidence', *Oxford Journal of Archaeology* 15: 1-32.

Hamilton, R.W.

1934–35 'Note on a Chapel and Winepress at 'Ain el Jedide', *Quarterly of the Department of Antiquities in Palestine* 4: 111-17.

Hasluck, F.W.

1914–16 'Stone Cults and Venerated Stones in the Graeco Turkish Area', *Annual of the British School at Athens* 21: 62-83.

Heath, T.H.

1910–11 'Hero of Alexandria', *Encyclopedia Brittanica* (11th edn), 13: 378-79.

*Heine, P.

1982 *Weinstudien* (Wiesbaden: Otto Harrassowitz).

Heltzer, M.

1979 '*Dimtu-gt-pyrgos*: An Essay about the Non-Etymological sense of these Terms', *Journal of Northwest Semitic Languages* 2: 31-35.

1982 *The Internal Organisation of the Kingdom of Ugarit* (Wiesbaden: Ludwig Reichart Verlag).

*1990 'Vineyards and Wine in Ugarit (Property and Distribution)', *Ugarit-Forschungen* 22: 119-35.

1996 'Olive Growing and Olive Oil in Ugarit', in Eitam and Heltzer 1996: 77-89.

Heltzer, M., and D. Eitam

1987 *Olive Oil in Antiquity* (Proceedings of Conference preliminary publication; Haifa: Haifa University, Israel Oil Industry Museum, Dagon Museum) (Hebrew).

Hennessy, J.B.

1967 *The Foreign Relations of Palestine during the Early Bronze Age* (London: Quaritch).

Herzog, Z.

1980 'Excavations at Tel Michal 1978', *Tel Aviv* 7: 111-51.

1982 'Wine Presses at Tel Michal', *Hadashot Arkheologiot* 80–81: 14 (Hebrew).

Herzog, Z., G. Rapp and O. Negbi

1988 *Excavations at Tel Michal, Israel* (Tel Aviv: University of Minnesota, Sonia and Marco Nadler Institute of Archaeology, Tel Aviv University).

Hestrin, R.

1972 *Inscriptions Revealed* (Jerusalem: Israel Museum).

Hestrin, R., and Y. Yeivin

1971 'An Oil-Press and its Operation', *Qadmoniot* 4 (15): 92-95 (Hebrew).

Hirschfeld, Y.

1979 'Ancient Wine Presses in the Area of the Ayalon Park' (Unpublished paper) (Hebrew).

1981 'Ancient Wine Presses in the Area of the Ayalon Park', *Eretz-Israel* 15: 383-90 (Hebrew).

1983 'Ancient Wine Presses in the Park of Ayalon', *Israel Exploration Journal* 33: 207-17.

1985 'Khirbet el-Quneitra: A Byzantine Monastery in the Wilderness of Ziph', *Eretz-Israel* 18: 243-55 (Hebrew).

1987 'The Judean Desert Monasteries in the Byzantine Period: Their Development and Internal Organization in the Light of Archaeological Research' (Unpublished PhD dissertation, Jerusalem, Hebrew University) (Hebrew).

Hirschfeld, Y., and R. Birger Calderon

1991 'Early Roman and Byzantine Estates near Caesarea', *Israel Exploration Journal* 41: 81-111.

Hirschfeld, Y., and A. Kloner

1988–89 'Khirbet el-Qasr: A Byzantine Fort in the Judaean Desert', *Bulletin of the Anglo-Israel Archaeological Society* 8: 5-20.

Hjelmquist, H.

1973 'Some Economic Plants from Ancient Cyprus: Appendix VI', in V. Karageorghis, *Excavations in the Necropolis of Salamis III* (Nicosia: Department of Antiquities for the Republic of Cyprus): 229-55.

Hodder, I., and C. Orton

1976 *Spatial Analysis in Archaeology* (Cambridge: Cambridge University Press).

Hoffner, A.A.

1974 *Alimenta Hethaeorum* (Newhaven: American Oriental Society).

Hogarth, D.G.

1889 *Devia Cypria: Note of an Archaeological Journey in Cyprus in 1888* (London: Henry Frowde).

Hönigsberg, P. von

1962 'Römische Ölmuhlen mahlen noch in Oberägypten', *Mitteilungen des deutschen archäologischen Instituts Abteilung Kairo* 18: 70-79.

Hood, S.

1981 Personal letter received by the author.

Hood, S., and D. Smyth

1981 *Archaeological Survey of the Knossos Area* (London: The British School at Athens/Thames & Hudson).

Horle, J.

1929 *Catos Hausbücher* (Paderborn: Schonigh [repr. Johnson Reprint Corporation, 1968]).

Horowitz, A.

1971 'Climatic and Vegetational Developments in North-Eastern Israel during Upper Pleistocene Holocene Times', *Pollens et Spores* 13.2: 255-78.

1979 *The Quaternary of Israel* (New York: Academic Press).

Houel, J.

1782 *Voyage pittoresque des Isles de Sicile, de Malte et de Lipari* (4 vols.; Paris: Imprimerie de Monsieur).

Hübner, V.

1985 'Die literarische und archäologische Zeugnisse über den vorchristlichen Athos', *Antike Welt* 16: 33-44.

Humbel, X.

1976 *Vieux pressoirs sans frontières* (Paris: Guinegard).

Hutchinson, R.W.

1962 *Prehistoric Crete* (Harmondsworth: Penguin).

Hutteroth, W.D., and K. Abdulfallah

1977 *Historical Geography of Palestine, Transjordan and Southern Syria in the Late 16th Century* (Erlangen: Sebstverlag der Frankischen Geographischen Gesellschaft).

Ilan, Z.

1976 'ḥirbet Jaboris mimṣaey atar qadum ba-midbar shomron (Kh. Jaboris) (Finds from an Ancient Site in the Samarian Desert)', *Nofim* 4: 20-27 (Hebrew).

1982 'seqer ve-ḥafira be-ḥurvat ʿerav(ʿirbin) she-be-ramat idmit (A Survey and Excavation at Ḥ. ʿErav [ʿIrbin] on the Idmit Plateau)', *Nofim* 16: 5-18 (Hebrew)

1984 'Meroth: A Fortified Village of the Roman Period', *Eretz-Israel* 17: 141-46 (Hebrew).

Irelli, G.C.

1965 'S. Sebastiano de Vesuvio Villa Rustica Romano', *Notizie degli scavi di antichita* (Supplemento 8.9): 161-78.

Irimie, C.
1974 *Brukental Museum Sibiu: Museum der bäuerlichen Technik* (Sibiu).

Jashemski, W.F.
1972 'A Vineyard at Pompei. Part II: The Vineyard Complex', *Archaeology* 25: 132-39.
1979 *The Gardens of Pompey Herculaneum and the Villas Destroyed by Vesuvius* (New Rochelle, NY: Caratzas).

Jastrow, M.
1926 *A Dictionary of the Targumim, the Talmud Babli and Yerushalmi and the Midrashic Literature* (New York: Choreb; London: Shapiro Valentine).

Jean, M.
1984 'Une taillerie de meules de moulins au Puget sur Argens', *Moulins de Provence* 5: 18-21.

Jones, R.F.J., S.J. Keay, J.M. Nolla and J. Tarrus
1982 'The Late Roman Villa of Vilauba and its Context: A First Report on Field-Work and Excavations in Catalunya, North-East Spain 1978–81', *Antiquaries Journal* 62: 266-82.

Jungst, E., and P. Thielscher
1954 'Catos Keltern und Kollergange', *Bonner Jahrbücher* 154: 32-93.
1957 'Catos Keltern und Kollergange', *Bonner Jahrbücher* 157: 53-126.

Kahane, P.
1951 'Rishpon (Appolonia) 1951', *Bulletin of the Department of Antiquities of the State of Israel* 3: 41-43 (Hebrew).

Kallai, Z.
1965 'Remains of the Roman Road along the Mevo Beitar Highway', *Israel Exploration Journal* 15: 195-203.
1969 'Bet Ḥoron Taḥton', *Hadashot Arkhiologiot* 31-32: 13 (Hebrew).

Kaplan, J.
1978 'Tel Aviv', in Avi-Yonah and Stern 1978: 1159-170.

Kaufmann, K.M.
1910 *Die Menasstadt I* (Leipzig: Karl W. Hiersman).

Kelm, G.L., and A. Mazar
1983 'Tel Batash (Timnah)', *Israel Exploration Journal* 33: 126, pl. 16c.
1987 '7th Century BCE Oil Presses at Tel Batash, Biblical Timnah', in Heltzer and Eitam 1987: 121-25.

Kelso, J.L.
1968 'The Excavations of Bethel (1934–1960)', *Annual of the American Schools of Oriental Research* (Cambridge): 39.

Kempers, A.J.B.
1962 *Oliemolens* (Arnhem: Het Nederlands Openlucht-museum).

Kenyon, K.M.
1976 'Discussion', *Phil. Trans. R. Soc. London* B.275: 194

Kestenbaum, H.
1983 'Fiq', *Excavations and Surveys in Israel* 2: 32-33.

Kislev, M.E.
1994–95 'Wild Olive Stones at Submerged Chalcolithic Kfar Samir, Haifa, Israel', *Mitekufat Haeven: Journal of the Israel Prehistoric Society* 26: 134

Kittel, R. (ed.)
1973 *Biblia Hebraica* (Stuttgart: Würtembergische Bibelanstalt).

Kjaer, H.
1929 'The Excavations of Shiloh', *Journal of the Palestine Oriental Society* 10: 87-174.

Kloner, A.
1971 'Kh. el-Qasr', *Hadashot Arkhiologiot* 38: 22-23 (Hebrew).
1974 'Bet Shemesh (An Oil Press)', *Hadashot Arkhiologiot* 51–52: 26 (Hebrew).
1981 'The Underground City of Maresha', *The Eighth Archaeological Congress in Israel: Abstracts of Lectures* 7 (Jerusalem: Israel Exploration Society, Israel Department of Antiquities) (Hebrew).
1986 'Horvat Benaya', *Excavations and Surveys in Israel* 5: 12-15.
1989 'Underground Olive Presses with Uprights from Judea', *Niqrot Zurim* 15: 66-73 (Hebrew).

Kloner, A., and Y. Hirschfeld
1987 'Khirbet e-Qasr: A Byzantine Fort with an Olive Press in the Judean Desert', *Eretz-Israel* 19: 132-41 (Hebrew).

Kloner, A., and Y. Tepper
1987 *The Hiding Complexes in the Judean Shephelah* (Tel Aviv: Hakibbutz Hameuchad Israel Exploration Society) (Hebrew).

Kloner, A., and N. Sagiv
1987 'The Technology of Oil Production in the Hellenistic Period', in Heltzer and Eitam 1987: 133-38.
1989 'Maresha: Olive Oil Production in the Hellenistic Period', *Niqrot Zurim* 15: 17-65 (Hebrew).
1989–90 'Maresha in the Shephela: Olive-Oil Production in the Hellenistic Period', *Israel Land and Nature* 15: 58-65.

Kochavi, M. (ed.)
1972 *Judaea Samaria and the Golan Archaeological Survey 1967–1968* (Jerusalem: The Archeological Survey of Israel, Carta).

Kohut, A. (ed.)
1926 *Aruch completum lexicon targumicis talmudicis et midrasschicis auctore Nathane filio Jechielis* (Vindobona: Menorah).

*Kopaka, K., and L. Platon
1993 'Installations minoennes de traitement des products liquides', *Bulletin de correspondance hellénique* 117: 35-101.

Kotzia, N.C.
1955 'Anaskafe tes Basilikis toi Laireotikoi Olimpoi', *Praktika tes Athenes Archaeologikas Heterias* 1952: 92-128.

Krauss, S.
1910–12 *Talmudische Archäologie* (3 vols.; Leipzig: Gustav Fock).

Kuhnen, H.P.
1987 *Nordwest-Palästina in hellenisch-römischer Zeit* (Weinheim: Acta Humaniora VCH).

Kutcher, Y.
1961 *Words and their History* (Jerusalem: Kiriath Sepher) (Hebrew).

Lamon, R.S., and M. Shipton
1939 *Megiddo I: Seasons 1925–34, Strata I–IV* (Chicago: Oriental Institute, University of Chicago).

Lancha, J.
1981 *Recueil des mosaïques de la Gaulle III.2* (Paris: CNRS).

Lander, Y.
1990 *Mount Nafḥa Map (196) 1201* (Jerusalem: Archaeological Survey of Israel).

Laporte, J.P.
1978 'La Tudicula: Machine antique à écraser les olives et les massues de bronze d'Afrique du nord', *Bulletin archéologique du comité des travaux historiques et scientifiques* 10–11, 1974–75: 167-74.
1983 'Fermes huileries et pressoirs de Grande Kabylie', *Bulletin archéologique du comité des travaux historiques et scientifiques* (NS) 19: 127-40.

Lapp, P.W.
1963 'Ptolemaic Stamped Handles from Judah', *Bulletin of the American Schools of Oriental Research* 172: 22-35.
1964 'The 1963 Excavations at Taᶜannek', *Bulletin of the American Schools of Oriental Research* 173: 4-44.
1969 'The 1968 Excavations at Taᶜannek', *Bulletin of the American Schools of Oriental Research* 195: 2-49.

Lassus, J.
1957 'Cheregas (dept. d'Alger): une huilerie', *Libyca* 5: 307-309.

Latour, P.
1901 'Recherches sur la marche du progrès dans les pressoirs', *Société d'histoire d'archéologie et de littérature de l'arrondissement de Beaune mémoires* 1900: 127-43.

Lauffer, S.
1971 *Diokletians Preisedict* (Berlin: W. de Gruyter).
Laufner, R., K.H. Faas and H. Cuppers
1987 *2000 Jahre Weinkultur an Mosel Saar Ruwer* (Trier: Rheinisches Landesmuseum).
La Vega
1783 'Descrizione del ritrovamente e ristaurazione di un antico Molino da Olia', in *Memoria sulla economica olearia antica e moderna* (Napoli): 53-71.
Lefebure, M.
1910 'Egypte gréco-romaine. II: Crocodopolis (suite et Théadelphie)', *Annales du service des antiquités de l'Egypte* 10: 115-72.
1923–24 *Le tombeau de Petosiris* (Cairo: Service des antiquités).
Lehavy, Y.M.
1974 'Excavations at Neolithic Dhali-Agridi', in L.E. Stager, *American Expedition to Idalion Cyprus.* (Cambridge, MA: American Schools of Oriental Research): 95-102, Pl. I (p. 85).
Lemaire, A.
1977 *Inscriptions hébraïques.* I. *Tome Les Ostraca* (Paris: Cerf).
1980 'A Note on Inscription XXX from Lachish', *Tel Aviv* 7: 92-94.
Leveau, P.
1984 *Caesarea de Maurétanie: Un ville romaine et ses campagnes* (Rome: Ecole Française de Rome).
Lewis, C.T. and C. Short
1969 *A Latin Dictionary* (Oxford: Oxford University Press).
Lieberman, S.
1955–88 *Tosefta Ki-Fshutah* (10 vols.; New York: The Jewish Seminary of America).
Liphschitz, N., R. Gophna, M. Hartman and G. Biger
1991 'Beginning of Olive (*Olea Europaea*) Cultivation in the Old World: A Reassessment', *Journal of Archaeological Science* 18: 441-53.
Litchfield, C.
1982 'Extracting Oil with a Rotary Hand Quern', *Transactions of the International Molonogical Society* 5: 340-42.
Liverani, P.
1987 'Termini Muti de Centuriazione o Contrappesi di Torchi?', *Mélanges de l'Ecole française de Rome. Antiquité* 99: 111-27.
Livneh, M.
1966 'Tel Barom', *Hadashot Arkhiologiot* 20: 4-5 (Hebrew).
Llopis y Llopis, S.
1948 'Pie de pressa de aceite romana de Santa Cruz de Moya (Cuenca)', *Archivo Espanol de Arqueologia* 70: 298-99.
Loffreda, S.
1980 *A Visit to Caphernaum* (Jerusalem: Franciscan Printing Press).
Loud, G.
1948 *Megiddo II: Seasons 1935–39* (Chicago: Oriental Institute, University of Chicago).
Löw, I.
1928 *Die Flora der Juden* (4 vols.; Leipzig [repr. Hildesheim: G. Olms, 1967]).
Lutz, H.F.
1922 *Viticulture and Brewing in the Ancient Orient* (Leipzig: J.C. Hinrich).
Macalister, R.A.S.
1912 *The Excavations of Gezer* (3 vols.; London: Palestine Exploration Fund).
Mader, A.E.
1918 *Altchristliche Basiliken und Localtradition in Sudjudaa* (Studien sur Geschichte und Kultur des Altertums 8.5-6; Paderborn: Ferdinand Schoningh).
Magen, I.
1978 'Qalandia: A Farm', *Hadashot Arkhiologiot* 67–68: 47-48 (Hebrew).
1982a *Qedem Museum: The Archeological Discoveries at Qedumim Samaria* (Qedumim, Samaria: Regional Municipality Samaria).
1982b 'Qedumim', *Excavations and Surveys in Israel* 1: 96-100.
1990 'Mount Gerizim: A Temple City', *Qadmoniot* 23 (91–92): 70-96 (Hebrew).
Maier, F.G., and M.L.V. Wartburg
1983 'Excavations at Kouklia (Palaepaphos): Twelfth Preliminary Report, Seasons 1981 and 1982', *Report of the Department of Antiquities Cyprus 1983*: 300-14.
Maiuri, A., and H.J. Bejen
1965 *Ercolano Pompei e Stili Pompeiani* (Rome).
Major, J.K.
1985 *Animal Powered Machinery* (Aylesbury [Shire Album 128]).
Malul, M.
1996 'ZE/IRTU (SE/IRDU): The Olive Tree and its Products in Ancient Mesopotamia', in Heltzer and Eitam 1996: 92-99.
Manacorda, D.
1982 *I Volusi Saturnini* (Bari).
Mandelkern, S.
1969 *Veteris Testamente Concordantiae* (Jerusalem: Schocken).
Mari, Z.
1983 *Tibur Pars Tertia* (Forma Italiae 1.17; Rome: Leo Solchki).
Marinatos, S., and M. Hirmer
1960 *Crete and Mycenae* (London: Thames & Hudson).
Matijasic, R.
1982 'Roman Architecture in the Territory of Colonia Iulia Pola', *American Journal of Archaeology* 86: 53-64.
Mattingly, D.J.
1985 'Olive Oil in Roman Tripolitania', in D.J. Buck and D.J. Mattingly (eds.), *Town and Country in Roman Tripolitania* (Oxford: British Archaeological Reports): 27-46.
1988a 'Megalithic Madness and Measurement or How Many Olives Could an Olive Press Press?', *Oxford Journal of Archaeology* 7: 177-95.
1988b 'The Olive Boom, Oil Surpluses, Wealth and Power in Roman Tripolitania', *Libyan Studies* 19: 21-41.
1990 'Paintings, Presses and Perfume Production at Pompeii', *Oxford Journal of Archaeology* 9: 71-90.
1994 'Regional Variation in Roman Oleoculture: Some Problems of Comparability: Landuse in the Roman Empire', *Analecta Romana, Instituti Danici Supplementum* 22: 91-106.
Mattingly, D.J., and M. Zenati
1984 'The Excavation of Building Lm 4E: The Olive Press', in G.W.W. Barker and G.D.B. Jones, *The UNESCO Libyan Valleys Survey.* VI. *Investigations of a Romano Libyan Farm,* Part I: *Libyan Studies* 15: 1ff., 13-18.
Mau, A.
1896 'Ausgraben von Boscoreale', *Mitteilungen des kaiserlich deutschen archäologischen Instituts römische Abteilung* 11: 131-40.
1907 *Pompei Life and Art* (trans. F.W. Kelsey; London: Macmillan).
Maurin, L.
1964 'Établissement vinicole à Allas-les-Mines (Dordogne)', *Gallia* 22: 209-21.
1967 'Thurburbo Majus et la Paix Vandale', *Les Cahiers de Tunisie* 15: 225-54.
Mayerson, P.
1985 'The Wine and Vineyards of Gaza in the Byzantine Period', *Bulletin of the American Schools of Oriental Research* 257: 75-80.
*1992 'The Gaza Wine Jar (Gazition) and the Lost Ashkelon Jar (Askalonium)', *Israel Exploration Journal* 42: 76-80.

*1993 'The Use of Ascalon Wine in the Medical Writers of the Fourth to the Seventh Centuries', *Israel Exploration Journal* 43: 169-73.

Mazar, B. (ed.)
1988 *Geva: Archaeological Discoveries at Tell Abu-Shusha, Mishmar Ha-ʿEmeq* (Tel Aviv: Israel Exploration Society, Hakibbutz Hameuchad) (Hebrew).

Mazar, B., A. Biran, M. Dothan and I. Dunayevsky
1964 'ʿEin Gev Excavations in 1961', *Israel Exploration Journal* 14: 1-49.

Mazor, G.
1981 'The Wine Presses of the Negev', *Qadmoniot* 14 (53–54): 51-60 (Hebrew).

McCown, C.C.
1947 *Tel En-Nasbeh Vol. I* (Berkeley: The Palestine Institute of Pacific Schools of Religion; New Haven: The Amercan Schools Of Oriental Research.

*McGovern, P.E., S.T. Fleming and S.H. Katz
1996 *The Origins and Ancient History of Wine* (Philadelphia: The University of Pennsylvania. Museum Gordon and Breach).

Meister, A.L.F.
1763 *De Torculario Catonis Vasis Quadranis* (Goettenberg: Victorini Bossigelii).

Melena, J.L.
1983 'Olive Oil and Other Sorts of Oil in the Mycenaean Tablets', *Minos* NS 18: 89-123.

Mercando, L.
1979 'Cesano di Senigallia (Ancona)', *Notizie degli Scavi di Antichita* 1979: 110-31.

Meron, Y.
1985 'Maresha Cave Complexes', *Niqrot Zurim* 11–12: 108-12 (Hebrew).

Meunier, J.
1941 'L'huilerie romaine de Kherbet-Agoub (Perigotville)', *Bulletin de la Societé Historique et Géographique de la Region de Setif* 2: 35-41.

Miége, J.L. (ed.)
1983 *L'huile d'olive Méditerranée* (Maison de la Méditerranée Mémoires et Documents, 2; Aix en Provence: Institut de Recherches Méditerranéenes, Université de Provence [CNRS]).

Miguelez Ramos, C.
1989 'La agricultura tradicional en Ibiza: Introduccion al estudio de la cultura material', *Etnografia Española* 7: 7-58.

Mingazzini, P.
1940 'Petralia Sottana (Palermo)', *Notizie degli Scavi di Antichita* 1940: 227-33.

Monneret de Villard, U.
1927 '*Il Monastero Di S. Simeons presso Aswan*. I. *Descrizione Archeologica* (Milano: Liberia Pontificia Arcivescovile S. Giuseppe).

Montet, P.
1913 'La fabrication du vin dans les tombeaux anterieurs au Nouvel Empire', *Recueil de Travaux Relatifs à la Philologie et à l'Archéologie Egyptiennes et Assyriennes* 35: 117-24.

Moreno Navarro, I., E. Aquilar, J. Agudo, A. Cholan, E. Fernandez de Paz, J.M. Montes de Oca, A. Melgai, A. Morena and J. Moeno
1981 'El cultivo la vina: La fabricacion de aguardiente y la colonia agricola de Galeon. Estudia etnologico de la evolucion y crisis de les actividades econimicas tradicionales de la Sierra (Sevilla)', *Etnografia Española* 2: 187-254.

Moritz, L.A.
1958 *Grain-Mills and Flour in Classical Antiquity* (Oxford: Oxford University Press).

*Morizot, P.
1993 'L'aures et L'olivier', *Antiquites Africaines* 29: 177-240.

*Muheisen, A., and F. Villeneuve
1990 'Khirbet adh-Dhaih: A Nabataean Site in Wadi el La'ban', *Annual of the Department of Antiquities of Jordan* 34: 5-18 (Arabic).

Müllinen, E.
1908 'Beiträge zur Kenntnis des Karmels', *Zeitschrift des deutschen Palästina Vereins* 31: 1-258.

Na'aman, N.
1986 'Hezekiah's Fortified Cities and the LMLK Stamps', *Bulletin of the American Schools of Oriental Research* 261: 5-21.

Nagatsune, O.
1974 *Seiju Roku: On Oil Manufacturing* (trans. E. Arigu; ed. C. Litchfield; New Brunswick, NJ: Olearius).

Naveh, Y.
1981 ' "Belonging to Makhbiram" or "Belonging to Food-Servers" ', *Eretz-Israel* 15: 301-302 (Hebrew).

Neef, R.
1990 'Introduction, Development and Environmental Implications of Olive Culture: The Evidence from Jordan', in S. Bottema, G. Entjes-Nieborg and W. Van Zeist (eds.), *Man's Role in the Making of the Eastern Mediterranean Landscape* (Rotterdam: Balkena): 295-300.

Negev, A.
1978 'Subeita', in M. Avi-Yonah and E. Stern 1978: 1116-24.

Nelson, H.H.
1930 *Medinat Habu*. I. *Earlier Records of Ramses III* (Chicago: Oriental Institute of Chicago Publications, 8).

Nestle, E.
1903 'Miscelleni', *Zeitschrift für Alttestamentliche Wissenschaft* 23: 337-46.

Netzer, E.
1989 'Water Channels and a Royal Estate from the Hasmonean Period in the Western Plains of Jericho', in D. Amit, Y. Hirschfeld and J. Patrich (eds.), *The Aqueducts of Ancient Palestine* (Jerusalem: Yad Izhak Ben-Zvi): 273-82 (Hebrew).

Newberry, P.E.
1893 *Beni Hassan I* (London: Egypt Exploration Fund).
1894 *Beni Hassan II* (London: Egypt Exploration Fund).
1894–95 *El Bersheh I* (London: Egypt Exploration Fund).

Neyses, A.
1979 'Drei neuentdeckte gallo-römische Weinkelterhäuser im Moselgebiet', *Antike Welt* 10: 56-59.

Nix, L., and W. Schmidt (trans.)
1900 *Herons von Alexandria Mechanik und Katoptrik* (Leipzig: Teubner).

Oates, D.
1952–53 'The Tripolitanian Gebel: Settlement of the Roman Period around Gasr Ed-Dauun', *Papers of the British School at Rome* 20-21: 81-117.

Ohata, K.
1966 *Tel Zeror I* (Tokyo: The Society for Near Eastern Studies in Japan).

Ohnefasch-Richter, M.A.
1913 *Grieschische Gebräuche auf Cypern* (Berlin: Dietrich Reimar [Ernst Vohsen]).

Olami, Y.
1981 *Daliya Map (31) 15–22* (Jerusalem: Archaeological Survey of Israel).

Oliver-Smith, P., and W. Widrig
1982 'Roma. Loc. Tor Bella, Monica Villa rustica romana, Relazione preliminare sulle campagne di Scavo 1976 e 1977 nell'agro romano', *Notizie degli scavi di anticha*: 99-114.

Oren, E.
1970 'French Hill in Jerusalem', *Hadashot Arkhiologiot* 34–35: 19 (Hebrew).

1971 'French Hill', *Hadashot Arkhiologiot* 38: 16-19 (Hebrew).

Orion, E.
1982 *Wine and Oil Presses in the Negev Heights* (Sdeh Boqer) (Hebrew).

Ottosen, M.
1980 'Sarafand/Sarepta and its Phoenecian Background', *Qadmoniot* 8 (51–52): 122-26 (Hebrew).

Osten, V.
1929 *Exploration in Central Anatolia: Season of 1926* (Chicago: University of Chicago Press).

Pala, C.
1970 *Nomentum* (Forma Italiae 1.12; Roma: De Luca).

Palladius
1898 *Agricultura* (ed. J.C. Schmitt; Leipzig: Teubner).

Pallotino, M.
1937 'Capena Resti di costruzioni romane e medioevali in localita "Montecanino"', *Notizie degli Scavi di Antichita* 1937: 7-28.

*Palmer, R.
1994 *Wine in the Mycenaean Palace Economy* (Aegaeum, 10; Liège: Université de Liège University of Texas at Austin).

Parain, C.
1963 'Typologie des pressoirs préindustriels et aires de diffusion des types successifs en Europe occidentale', in *VIe Congrés international des sciences anthropologiques et ethnologiques Paris 30 juillet 6 aout, 1960. II. Ethnologie (premier volume)* (Paris: Musée de l'Homme): 605.
1979 *Outils ethnies et développement historique* (Paris: Editions sociales).

Paskvalin, V.
1974 'Anticki Torkular v Bihovu Kod Trebinya', *Gasnik Zemaljskog Muzeja Bosne i Hercegovine u Sarajevu Arheologija* 29: 289-93.

Paton, W.R., and J.L. Myres
1898 'On Some Karian and Hellenic Oil Presses', *Journal of Hellenic Studies* 18: 209-17.

Patrich, J., and Y. Tsafrir
1985 'A Byzantine Church and Agricultural Installations at Khirbet Beit Loya', *Qadmoniot* 18 (71–72) 106-12 (Hebrew).

Paul, S.M.
1975 'Classification of Wine in Mesopotanian and Rabbinic Sources', *Israel Exploration Journal* 25: 42.

Peacocke, P.S., and D.F. Williams
1986 *Amphorae and the Roman Economy* (London: Longman).

Peleg, Y.
1981 'How Ancient Olive Presses Worked', *Land and Nature* 6: 98-103.

Peltenberg, E.J.
1978 'The Sotira Culture: Regional Diversity and Cultural Unity in Late Neolithic Cyprus', *Levant* 10: 55-74.

Pesce, G.
1936 'Regione III (Lucania et Bruth)', *Notizie Degli Scavi di Antichita* 1936: 67-76.

Pilar Pascual, M. de, and E.J. Moreno Arrastio
1980 'Prensas de aceite Romanas en la rioja', *Archivo español de arqueologia* 53: 199-209.

Pinkerfeld, Y.
1961 'Montfort Plans', in M. Yedaya and A. Gil (eds.), *ma'aravo shel ha-galil (Western Galilee)* (Sulam Sur and Ga'aton): 102 (Hebrew).

Pliny
1949–63 *Naturalis Historiae* (10 vols.; trans. H. Rackman; London: Loeb).

Plommer, H.
1973 *Vitruvius and Later Roman Building Manuals* (Cambridge: Cambridge University Press).

Polge, H.
1967 'Généalogie du pressoir', *Bulletin de la societé artistique littéraire et scientifique du Gers*: 139-59.

Ponsich, M.
1970 *Recherches archéologiques à Tanger et dans sa region* (Paris: CNRS).
1974 *Implantation rurale antique sur le Bas-Guadalquivir*, II (Madrid: Casa de Velasquez).
1979 *Implantation rurale antique sur le Bas-Guadalquivir*, III (Madrid: Casa de Velasquez).

Porat, Y., S. Dar and S. Applebaum
1985 *The Antiquities of ʿEmeq Hefer* (Tel Aviv: Hakibbutz Hameuchad) (Hebrew).

Prausnitz, M.
1975a 'Romema (Haifa)', *Hadashot Arkhiologiot* 53: 5-6 (Hebrew).
1975b 'Romema (Haifa)', *Hadashot Arkhiologiot* 56: 15 (Hebrew).

Prickett, J.L.
1980 'A Scientific and Technological Study of Topics Associated with the Grape in Greek and Roman Antiquity' (Doctoral disseration, University of Kentucky; Ann Arbor: University Microfilms International).

Pritchard, J.B.
1959 *Hebrew Inscriptions and Stamps from Gibeon* (Philadelphia: University of Pennsylvania).
1964 *Winery, Defences and Soundings at Giveon* (Pennsylvania: University Museum, University of Pennsylvania).
1978 *Recovering Sarepta: A Phoenecian City* (Princeton, NJ: Princeton University Press).

Pritchett, W.K.
1956 'The Attic Stelai (Part II)', *Hesperia* 25: 178-317.

Provera, M.
1982 'La Coltura Della Vite Nela Tradizione Biblica ed Orientale', *Bibbia e oriente* 24: 97-106.

Quinion, M.B.
1982 *Cidermaking* (Haverfordwest: Shire Album 95).

*Rahmani, L.
1991 'Two Byzantine Winepresses in Jerusalem', *'Atiqot* 20: 95-110.

Rainey, A.F.
1966 'Gath of the Philistines. I. Location', *Christian News from Israel* 7 (2–3): 30-38.
1982 'Wine from Royal Vineyards', *Bulletin of the American Schools of Oriental Research* 245: 57-62.

Ramsay, M.W., and G.L. Bell
1909 *The Thousand and One Churches* (London: Hodder & Stoughton).

Rapin, C.
1987 'La trésorerie hellénistique d'Aï Khanoum', *Revue archéologique* 1: 48-70.

Raveh, K.
1990 'Dor', *Hadashot Arkhiologiot* 95: 28 (Hebrew).

*Reeves, J.C.
1992 'The Feast of the First Fruit of Wine and the Ancient Canaanite Calendar', *Vetus Testamentum* 42: 350-61.

Reisner, G.A., C.S. Fisher and D.G. Lyon
1924 *Harvard Excavations at Samaria 1908–1910* (Cambridge, MA: Harvard University Press).

Reggiani, A.M.
1978 'La Villa Rustica Pilella Ne l'Ager Tiburtinus', *Archeologia Classica* 30: 219-25.

Reich, R.
1991 'A Note on the Roman Mosaic at Magdala on the Sea of Galilee', *Liber Annus* 41: 455-58.

Remesal Rodriguez, J.
1986 *La annona militaris y la exportacion de aceite betico a Germania* (Madrid: Universidade Complutense).
1996 'The Production and Commerce of Baetican Olive Oil

during the Roman Empire', in Eitam and Heltzer 1996: 101-12.

Renan, E.
1874 *Mission en Phénicie* (Paris: Teubner).

Renfrew, C.
1972 *The Emergence of Civilisation: The Cyclades and the Aegean in the Third Millenium* BC (London: Methuen).

Rey-Coquais, J.P.
1977 'Inscriptions grecques et Latines découvertes dans les fouilles de Tyr (1963–1974) I. Inscriptions de la Necropolis', *Bulletin de Musée de Beyrouth* 29.

*Ricci, C.
1924 *La coltura della vite e la fabbricazione del vino nell'eqitto greco-romana* ('Aegyptus'; Studi Della Scuola Papirologica, 4.I-R; Milan Accademia Scientifico-Letteraria in Milano [repr. Milan: Ciasalpino-Goliardica, 1972]).

Riley, J.A.
1979 'The Coarse Pottery from Benghazi', in J.A. Lloyd (ed.), *Excavations at Sidi Khrabish Benghazi (Berenice)*, II (Tripoli: Socialistic Peoples Libyan Arab Jamhiriya. Secretariat of Education, Department of Antiquities): 91-497.

Rivals, C.
1975–76 'Les moulins à vent des plaines septentrionales: 1976 hegemonie du mouluin sur pivot. II. Les moulins à huile du nord de la France', *Ethnologie française* 5–6: 163-80.

Robinson, D.M., and J.W. Graham
1938 *Excavations at Olynthus.* VIII. *The Hellenic House* (Baltimore: John Hopkins).

Robinson, G.B.
1913 'The Mosaic Olive Press at Moreshet Gath', *Journal of Biblical Literature* 32: 54-56.

Röhricht, R.
1893 *Regesta Regni Hierosolymitani (MXCCVII–MCCXCI)* (Innsbruck: Wagner).

Roll, I., and E. Ayalon
1981 'Two Large Wine Presses in the Red Soil Regions of Israel', *Palestine Exploration Quarterly* 130: 111-25.

Ronen, A., and Y. Olami
1978 *ʿAtlit Map (1)* (Jerusalem: Archaeological Survey of Israel).
1983 *Archaeological Survey of Israel: Map of Haifa-East (23) 15-24* (Jerusalem: Archaeological Survey of Israel).

Rosen, B.
1986–87 'Wine and Oil Allocations in the Samaria Ostraca', *Tel Aviv* 13-14: 39-45.

Rossiter, J.J.
1978 *Roman Farming in Italy* (Oxford: Bar Int. 52).
1981 'Wine and Oil Processing at Roman Farms in Italy', *Phoenix* 35: 345-61.

Rostovtzeff, M.
1957 *The Social and Economic History of the Roman Empire* (rev. P.M. Frazer; Oxford: Clarendon Press).

Roure, A., P. Castanyer, J.M. Nolla, S.J. Keay and J.I. Tarrus
1988 *La Villa Romana de Vilauba (camos): Estudi d'un Assentament Rural (Campanyes 1979–85)* (Serie Monografica, 8; Girona, Catalunye: Centre d'Investigacions Arqueologiques de Girona).

Rozier, (l'Abbe), M.
1776 *Vues économiques sur les moulins et pressoirs connus en France et d'Italie* (Paris: Ruault]).

Ruggiero, M.
1881 *Degli scavi di Stabia da 1749 al 1782* (Napoli: Tipographia deli Academia Reale della Scienze).

Runnels, C.N., and J. Hansen
1986 'The Olive in the Prehistoric Aegean: The Evidence for Domestication in the Early Bronze Age', *Oxford Journal of Archaeology* 5: 299-308.

Saarisalo, A.
1927 *The Boundary between Issacher and Naphtali* (Helsinki: Suomlaisen Tiedeakatemian Toimituksia).

Sacket, J.R.
1982 'Approaches to Style in Lithic Archaeology', *Journal of Anthropological Archaeology* 1: 60-112.
1985 'Style and Ethnicity in the Kalahari', *American Antiquity* 50: 154-59.

Saez Fernandez, P.
1987 *Agricultura romana de la Betica* (Sevilla: Monografias del Departamento de Historia Antigua de la Universidad Sevilla).

Safrai, Z.
1996 'The Economic Implication of Olive Oil Production in the Mishnah and Talmud Periods', in Eitam and Heltzer 1996: 119-23.

Safrai, Z., and M. Linn
1988 'Excavations and Surveys in the Mishmar Ha-Emeq Area', in Mazar 1988: 167-214 (Hebrew).

Saidah, R.
1968–69 'Archaeology in the Lebanon', *Berytus* 18: 119-42.

Saladin, M.H.
1887 'Rapport sur la mission faite en Tunisie de novembre 1882 à avril 1883', *Archives des missions scientifiques* (3rd series) 13: 1-225.

Saller, S.J.
1941 *The Memorial of Moses on Mt Nebo* (Publications of the Studium Biblicum Franciscanum, 1; 3 vols.; Jerusalem: Franciscan Press).
1946 *Discoveries at St John's En Karim* (Publications of the Studium Biblicum Franciscanum, 3; Jerusalem: Franciscan Press).
1957 *Excavations at Bethany (1949–1952)* (Publications of the Studium Biblicum Franciscanum, 12; Jerusalem: Franciscan Press).

Saller, S.J., and B. Bagatti
1949 *The Town of Nebo* (Publications of the Studium Biblicum Franciscanum, 7; Jerusalem: Franciscan Press).

Saller, S., and E. Testa
1961 *The Archaeological Setting of the Shrine of Bethphage* (Studium Biblicum Franciscanum Smaller Series, 1; Jerusalem: Franciscan Press).

Sasson, V.
1981 '*smn rhṣ* in the Samaria Ostraca', *Journal of Semitic Studies* 26: 1-5.

Scale, G.
1936 'Scalea-Troamenti varii', *Notizie degli scavi di anticha* 1936: 67-74.

Schapiro, N.
1932 'Die Weintechnologie in der Mišna und im Talmud', *Zeitschrift für Semitistik und verwandte Gebiete* 8: 53-60.

Schafer-Schuchardt, H.
1988 *L'olivia: La grande storia di un piccolo frutto* (Favia-Bari).

Schedule of Monuments
1964 *Schedule of Monuments and Historical Sites: Reshumot, Yalqut haPirsumim 1091*, 18.5: 1349-561.

Scheltema, H.J., and D. Holwerda
1957 *Basilicorum Libri*, LX (Series B volumen III Scholia in Libr. XV-XX; Groningen: J.B. Wolters; The Hague: Martinus Nijhoff).

Schick, C.
1887 'Artuf und seine Umgebung', *Zeitschrift des deutschen Palästina Vereins* 10: 131-59.
1893 'The Ruins of Jubeiah', *Palestine Exploration Fund Quarterly Statement* 1893: 201-203.
1899 'Ancient Rock Cut Wine Presses at 'Ain Karim', *Palestine Exploration Fund Quarterly Statement* 1899: 41-42.

Schneider, A.M.
1937 *The Church of the Multiplying of the Loaves and Fishes* (London: A. Ousley).

Sciallano, M., and P. Sibella
1991 *Amphores, comment les identifier* (Aix-en-Provence: Chaudoreille, Edisud).

Sellin, E.
1904 *Tell Taʿannek* (Vienna: Denkschriften der kaiserlichen Akademie der Wissenschaften, 50).

Seltman, C.T.
1957 *Wine in the Ancient World* (London: Routledge & Kegan Paul).

Serra Rafols, J. de C.
1952 *La villa romana de la Dehesa de la Cocosa* (Institucion de Servicios Culturales; Revista de Estudios Extremos; Anejos; Badajoz: Diputacion Provincial de Badajoz).

Serrano, R.G.
1969 'Notas historicas sobre la elaboracion de aceite de oliva en la provincia de Jean', *Etnologia y Tradiciones Populares* 1969: 229-33.

Serrano Ramos, E., and A. de Luque Morano
1976 'Memoria de las excavaciones de Manguarra y San Jose (Cartama Malaga)', *Noticiaro arqueologico hispanico* 4: 491-546.
1979 'Una villa romano en Cartama (Malaga)', *Mainake* 1: 147-64.
1980 'Memoria de la segunda y tercera campana de excavaciones en la Romana de Manguarra y San Jose Cartama (Malaga)', *Noticario arqueologia hispanico* 8: 255-390.

Shambun, A., and A. Stross
1991 'ʿEin Fattir', *Hadashot Arkhiologiot* 96: 26-27 (Hebrew).

Shaw, J.W.
1977 'Excavations at Kommos (Crete) during 1976', *Hesperia* 46: 199-240.

Shennan, S.
1989 'Introduction: Archaeological Approaches to Cultural Identity', in S. Shennan (ed.), *Archaeological Approaches to Cultural Identity* (London: Unwin Hyman): 1-32.

Shiloh, Y., and A. Hurwitz
1975 'Stone Dressing Techniques and Ashlar Quarries of the Iron Age', *Qadmoniot* 8 (30–31): 68-71 (Hebrew).

Shtal, A.
1981 'hafaqat shemen zayit be-yehuda (Olive Oil Production in Juda)', *Teva Vaaretz* 23 (2): 80-81 (Hebrew).

Siegelman, A., and M. Linn
1981 'Mishmar Ha-ʿEmeq, Tel Shosh', *Hadashot Arkhiologiot* 76: 15-16 (Hebrew).

Sinclair, L.A.
1960 'An Archaeological Study of Gibeah (Tell el Ful)', *Annual of the American Schools of Oriental Research*: 34-35.

Singer, I.
1987 'Oil in Anatolia according to Hittite Texts', in Heltzer and Eitam 1987: 183-86.

Sleeswyk, A.W.
1980 'Phoenecian Joints, Coagmenta Punicana', *Nautical Archaeology* 9: 243-44.

Smith, D.J.
1956 *Report of Durham University Exploration Expedition to French Morocco* (Durham: Durham University Exploration Society).

Soden, W., von
1959 *Akkadisches Handwörterbuch* (Wiesbaden: Otto Harrassowitz).

Sodini, J.P., G. Tate, B. Bauani, S. Bauani, J.L. Biscop and D. Orssaud
1980 'Dehes (Syrie de Nord) Campagnes I–III 1976–1978: Recherches sur l'habitat rural', *Syrie* 67: 1-305.

Sogliano, A.
1897 'Boscoreale Villa Romana in Contrado Detta Giuliana', *Notizie degli scava di antichita communicate alla R. accademia dei Lincei*: 391-402.

Sordinas, A.
1971 *Old Olive Oil Mills and Presses on the Island of Corfu, Greece* (Occasional Papers, 5; Memphis: Memphis State University Anthropological Research Centre).

South, A.K.
1980 'Kalavasos-Ayios Dhimitros. A Summary Report', *Report of the Department of Antiquities, Cyprus*: 22-53.

*Sparkes, B.A.
1976 'Treading the Grapes', *Bulletin Antieke Beschaving* 51: 47-64.

Sparkes, B.A., and L. Talcott
1974 *Pots and Pans of Classical Athens* (Princeton, NJ: American Schools of Classical Studies at Athens).

Spiro, M.
1978 *Critical Corpus of the Mosaic Pavements on the Greek Mainland Fourth/Sixth Centuries with Architectural Surveys* (New York: Gorland).

Stafford, H.
1755 *A Treatise on Cyder-Making Founded on Long Practice and Experience* (London).

Stager, L.E.
1983 'The Finest Olive Oil in Samaria', *Journal of Semitic Studies* 28: 241-45.
1985 'The Firstfruits of Civilisation', in J.N. Tubb (ed.), *Palestine in the Bronze and Iron Ages: Papers in Honour of Olga Tufnell* (London: Institute of Archaeology): 172-88.
1990 'Shemer's Estate', *Bulletin of the American Schools of Oriental Research* 277–78: 93-107.

Stager, L.E., and S.R. Wolff
1981 'Production and Commerce in Temple Courtyards: An Olive Press in the Sacred Precinct at Tel Dan', *Bulletin of the American Schools of Oriental Research* 243: 95-102.

Steinzaltz, A (ed.)
1969 *Babylonian Talmud Tractate Shabbat Vol. 2* (Jerusalem: Israel Institute for Talmudic Publications) (Hebrew).

Stern, E.
1978 *Excavations at Tel Mevorakh 1973–1976* (Qedem, 9; Jerusalem: Hebrew University).

Stern, H.
1958 'Les mosaïques de l'eglise de Sainte Constance à Rome', *Dumbarton Oaks Papers*, 12: 159-218.

Stol, M.
1985 'Remarks on the Cultivation of Sesame and the Extraction of its Oil.', *Bulletin on Sumerian Agriculture* 2: 119-26.

Sukenik, E.L.
1942 'A Stamp of a Jewish Wine Merchant from the Vicinity of Jerusalem', *Kedem Studies in Jewish Archaeology* 1: 20–23 (Hebrew).

Talgam, R., and Z. Weiss
1988 ' "The Dionysus Cycle" in the Sepphoris Mosaic', *Qadmoniot* 21 (83–84): 93-99 (Hebrew).

Tallon, M.
1972 'Une inscription du Liban Nord: Note sur les vestiges archéologiques mentionnés dans le texte', *Mélanges de l'Université Saint-Joseph* 47: 109-19.

Tchalenko, G.
1953 *Villages antiques de la Syrie du Nord* (Institut français d'archéologie de Beyrouth bibliotheque archéologique et historique, 50; Paris: P. Geuthner).

Tchernia, A.
1986 *Le vin de l'Italie romaine* (Rome: Ecole française de Rome: Palais Farnese).

Tepper, Y.
1987 'The Oil Presses at the Maresha Region', in Heltzer and Eitam 1987: 25*-42* (Hebrew).
1988 'seqer batey ha-bad be-meʿarot she-liyad maresha (A Survey of Oil-Presses in Caves near Maresha)', in Y. Tepper and Y. Shahar, *meʿarot maresha ve-meʿarot ha-mistor-meḥqarim* (The Caves of Maresha and Hiding Complexes Researches) (Tel Aviv: The Department of Homeland Studies of the Union of Kibbutz Movements): 69-116 (Hebrew).

Thomson, H.A.
1937 'Buildings on the West Side of the Agora', *Hesperia* 6: 1-226.

Thomson, H.A., and R.E. Wycherley
1972 *The Athenian Agora*. XIV. *The Agora of Athens* (Princeton, NJ: The American School of Classical Studies at Athens).

Thompson, H.O.
1972 'A Tomb at Khirbet Yajuz', *Annual of the Department of Antiquities of Jordan* 17: 37-41, 126-37.

Thouvenot, R.
1941 'La maison d'Orphée à Volubilis', *Publications du service des antiquités du Maroc* (Rabat) 6.
1954a 'Elements de pressoir à huile trouvés à Sale', *Publications du service des antiquités du Maroc* (Rabat) 10: 227-30.
1954b 'Le site de Julia Valentia Banasa', *Publications du service des antiquités du Maroc* (Rabat) 2: 1-54.
1958 'Maisons de Volubilis: Le palais dit de Gordien et la maison à la mosaïque de Venus', *Publications du service des antiquités du Maroc* (Rabat) 12.

Tufnell, O.
1953 *Lachish III (Tell ed-Duweir): The Iron Age* (London: London University).
1958 *Lachish IV (Tell ed-Duweir): The Bronze Age* (Oxford: Oxford University Press).

Tylor, J.J., and F.L. Griffith
1894 *The Tomb of Paheri at El Kab* (London: 11th Memoir of the Egypt Exploration Fund [Joint volume with Naville, E. Ahnas el Medineh]).

*Tyree, E.L., and E. Stefanoubaki
1996 'The Olive Pit and Roman Oil Making', *Biblical Archaeologist* 59.3: 171-78.

Tzaferis, V.
1984 'The Excavations at Kursi-Gergasa', *ʿAtiqot* 16.

Tzaferis, V., and T. Shai
1972 'Rama', *Hadashot Arkhiologiot* 44: 7 (Hebrew).
1976 'Excavations at Kafr er-Ramah', *Qadmoniot* 9 (34–35): 83-85 (Hebrew).

Urmann, D.
1974 'batei gitot le-yiṣur devash anavim be-golan (Wine-Presses for the Production of Grape Syrup in the Golan)', *Teva Vaaretz* 16: 173-76.

Ussishkin, D.
1977 'The Destruction of Lachish by Sennacharib and the Dating of the Royal Storage Jars', *Tel Aviv* 4: 28-60.
1978 *Excavations at Tel Lachish 1973–1977 Preliminary Report* (Reprint series no. 3 from Tel Aviv, 5; Tel Aviv: Tel Aviv University).

Varro
1979 'Marcus Terentius Varro: Rerum Rusticarum', in W.D. Hooper (trans.), *Cato and Varro* (London: Loeb): 161-528.

Viana, A.
1959 'Notas Historicas, Arqueologicas e Etnograficas do Baixo Alentejo', *Arquivo de Beja* 16: 3-43.

Vickery, K.F.
1936 *Food in Early Greece* (Illinois Studies in the Social Sciences, 10.3; Urbaña, IL: University of Illinois).

Vilbosh, N.
1947 'le-taʿasiyat shemen zayit biymey qedem (Olive Oil Production in Ancient Times)', in S. Avitzur (ed.), *haroshet ha-maʿase* (Industry and Practice) (Tel Aviv: Milo): 79-81 (Hebrew).

Vincent, L.H., and F.M. Abel
1932 *Emmaus: Sa basilique et son histoire* (Paris: Ernest Leroux).

Vincze, I.
1959 'Ungarische Weinkelter', *Acta Ethnographica Academiae Scientiarum Hungaricae* 8 (fasc. 1-2): 99-129.

Vindry, G.
1981 'L'huilerie romaine du Candeou à Peymeinade: Prospections et sondages dans les terroirs agricoles', *Histoire et Archéologie* 57: 71-74.

Violet, E.
1938 'Le pressoir à grand point', *Annales d'Ige*: 161-72.

Vito, F.
1984 'Kh Bata', *Hadashot Arkhiologiot* 85: 10-11 (Hebrew).

Vitruvius
1962 *De Architectura* (trans. F. Granger; 2 vols.; Cambridge: Loeb).

Wace, A.J.B.
1921–23 'Excavations at Mycenae', *Annual of the British School at Athens* 25: 9-147.

Waele, J.A. de
1980 'Agrigento gli scavi sulla Rupe Atenea (1970–1975)', *Notizie degli scavi di anticha*: 395-452.

Waetzoldt, H.
1985 'Olpflanzen und Pflanzenole in 3. Jahrtausend', *Bulletin on Sumerian Agriculture* 2: 77-96.

Wagner, D.
1987 'Oil Production at Gamla', in Heltzer and Eitam 1987: 187-91.

Warmington, B.H.
1960 *Carthage* (Middlesex: Pelican).

Warner, G.
1912 *Queen Mary's Psalter: Miniatures and Drawings by an English Artist of the 14th Century Reproduced from Royal MS 2BVII in the British Museum* (London: British Museum/Oxford University).

Warren, P.A.
1968 'A Textile Town 4500 Years Ago', *The Illustrated London News* 17 February 1968: 25-27.
1972 *Myrtos: An Early Bronze Age Settlement in Crete* (Athens: The British School of Archaeology at Athens).
1980 Personal letter to author.

Wartburg, M.L.V., and F.G. Maier
1989 'Excavations at Kouklia (Palaepaphos) 15th Preliminary Report: Seasons 1987 and 1988', *Report of the Department of Antiquities, Cyprus*: 177-88.

Waterman, L.
1937 *Preliminary Report of the University of Michigan Excavations at Sephoris Palestine in 1931* (Ann Arbor: University of Michigan).

Weinfeld, M.
1996 'The Use of Oil in the Cult of Ancient Israel', in Eitam and Heltzer 1996: 125-27.

Weissner, P.
1983 'Style and Social Information in Kalahari San Projectile Points', *American Antiquity* 50: 160-66.
1985 'Style or Isochrestic Variation? A Reply to Sacket', *American Antiquity* 50: 160-66.

White, K.D.
1970a *Roman Farming* (London: Thames & Hudson).
1970b *A Bibliography of Roman Agriculture* (Reading: University of Reading).
1975 *Farm Equipment of the Roman World* (Cambridge: Cambridge University Press).

Wilkinson, J.G.
1837 *Manners and Customs of the Ancient Egyptians* (London: J. Murray).

Williams-Davies, J.
1984 *Cider Making in Wales* (National Museum of Wales, Welsh Folk Museum).

Winkler, A.J.
1949 'Grapes and Wine', *Economic Botany* 3: 46-70.

Woolley, C.L.
1914–52 *Carchemish* (3 vols.; Oxford: British Museum).

Worlidge, J.
1689 *The Second Parts of Systema Agriculturae or the*

Mystery of Husbandry and Vinetum Britannicum or Treatise of Cyder & ct.

Wright, G.E.

1965 *Shechem: The Biography of a Biblical City* (London: Duckworth).

Wulff, H.E.

1966 *The Traditional Crafts of Persia* (Cambridge, MA: London: MIT Press).

Yaakobi, H., and Y. Meron

1985 'Hiding Complexes at Spot Height 270', *Niqrot Zurim* 11–12: 54-61 (Hebrew).

Yadin, Y.

1972 *Hazor: The Schweich Lectures of the British Academy, 1970* (London: British Academy/Oxford University Press).

1981 'Was the Temple Scroll a Product of a Sect?', in B. Mazar (ed.), *Thirty Years of Archaeology in Eretz Israel: Papers in Honour of J. Abiram* (Jerusalem: The Israel Exploration Society): 152-71 (Hebrew).

1983 *The Temple Scroll* (English Translation; Jerusalem: Israel Exploration Society).

Yadin, Y., R. Amiran, T. Dothan, I. Dunayevsky and J. Perrot

1959 *Hazor*, II (Jerusalem: Magnes Press).

Yadin, Y., Y. Aharoni, R. Amiran, T. Dothan, M. Dothan, I. Dunayevsky and J. Perrot

1961–89 *Hazor*, III–IV (Jerusalem: Magnes).

Yadin, Y., and J. Naveh

1989 *Masada*. I. *The Aramaic and Hebrew Ostraca and Jar Inscriptions* (Jerusalem: Hebrew University).

Yankovskaya, N.B.

1963 'Obshchinnoye samoupravleniya v Ugarite (gapantii i struktura) (Self-Rule of Communities at Ugarit [responsibility and stucture])', *Vestnik Drevnei Istorii*. Hebrew translation in A. Nadel (trans.), *ha-shilton ha-ʿironi ha-ʿaṣmi be-mizraḥ ha-qadum* (City Self-Rule in the Ancient East) (Jerusalem: Hebrew University): 35ff., 54-109 (Hebrew).

Yeivin, S.

1978 'el-Areini, Tel', in Avi-Yonah and Stern 1978: 89-97.

Yeivin, Z.

1966 'Two Ancient Oil Presses', *Atiqot* (Hebrew Series) 3: 52-62.

1982 'Korazin', *Excavations and Surveys in Israel* 1: 64-67.

1984 'Korazim 183–184', *Excavations and Surveys in Israel* 3: 66-71.

Yeivin, Z., and G. Edelstein

1970 'Excavations at Tirat Yehuda', *ʿAtiqot* (Hebrew Series) 6: 50-67 (Hebrew).

Yeroulanou, A., N. Belesioti and E. Georgiadis

1977 *Traditional Methods of Cultivation in Greece* (Athens: Benaki Museum/Athens Publishing Centre).

Yogev, O.

1982 'Aderet', *Excavations and Surveys in Israel* 1: 1.

Young, J.H.

1963 'A Migrant City in the Peloponnesus', *Expedition* 5: 2-12.

Younger, W.

1966 *Gods, Men and Wine* (London: The Wine and Food Society).

*Zapetal, V.

1920 *Der Wein in der Bibel* (Biblische Studien, 20.1; Freiburg: Herder).

Zayadine, F.

1977–78 'Excavations on the Upper Citadel of Amman Area A (1975 and 1977)', *Annual of the Department of Antiquities of Jordan* 22: 20-45, 193-96.

Zemer, A.

1977 *Storage Jars in Ancient Trade* (Haifa: The National Maritime Museum, Haifa).

Zohary, D., and F. Spiegel-Roy

1975 'Beginnings of Fruit Growing in the Old World', *Science* 187 (4171): 319-27.

Zori, N.

1977 *The Land of Issacher: Archaeological Survey*. (Jerusalem: Israel Exploration Society).

Zuck, Z.

1982a 'niquiy gat be-shilat (Excavating a Wine Press at Shilat)', *Nofim* 16: 62-64 (Hebrew).

1982b 'Shilat', *Hadashot Arkhiologiot* 78–79: 51 (Hebrew).

1982c *Archaeological Survey in Western Samaria* (Yarkon Field School) (Hebrew).

1984 'Yaar Qula: Winepresses', *Excavations and Surveys in Israel* 3: 107-108.

Index

Abdulfallah, K. 37
Abu Sinan North 54
Accadian, language 37, 186
Achaya, K.T. 75
Acre *see* Akko
Adams W.Y. 58, 154
Adderet 100
Aegean 28, 29, 39, 44, 45, 50, 59, 91, 104, 106, 170-73, 176, 180 *see also* Greece
Afeq Antipatris 49, 53, 58
Africa, North *see* North Africa
Africa, Roman Province 98
Agmati, Zachariah ben Judah 190
Agrigento, Sicily 175
Aharoni, Y. 45, 198, 199
Ahituv, S. 198
Ahlström, G.W. 52, 55
Aix en Provence 40
ʿAjlun 101, 161
Akerraz, A. 98
Akko (Acre) 136, 205
Albright, W.F 63, 99, 100
Alexandre-Bidon, D. 159
Algeria 58, 73, 94, 95, 103-105, 125, 161, 166, 171, 175, 179
Allas la Mines, France 157
Alpago-Novello, L. 120
Altbauer, M. 198
ʿAlya, Kh.84 139
Amineum 206
Amiran, R. 35, 39, 68
Ammon, Ammonite 201
Amorgos (island), Amorgos Weight 103-104, 119
Amouretti, M.C. 39, 40, 44, 46, 61, 104
ʿAmqa Central 118
ʿAmqa North 118
ʿAmqa South 52
ʿAmud, el, Libya 95
Anatolia 27, 35, 39, 72, 104, 105, 114, 119, 171, 173, 180 *see also* Turkey
Anderson, J.G.C. 119
Apicius 39, 45, 205
Aplici, Cyprus 67
Aquae Sirenses, Algeria 97
Arabic language 44, 47, 87, 88, 113, 185-88, 193, 199
Arad letters 45, 198
ʿAreini, Tel 57
Arginunta, Kalymnos, Arginunta Press, Arginunta Screw Weight 110-112, 119, 120, 121, 160, 163
Argolid 105
Arsuf (Rishpon, Apollonia), Arsuf Press 153
Ashan 199
Ashdod 199
Ashqelon, Ashqelon Weight 110, 158
Aswan (Simons Kloster), Egypt 58, 155
Athens, the Agora, Greece 90, 119, 156
Athos, Greece 59
Austria 123
ʿAvdon, Tel 66
ʿAvedat 150
ʿAvedat South-West 151
ʿAvedat West 152
Avi-Yonah, M. 165
Aviam, M. 44
Avigad, N. 198, 199
Avitsur, S. 47, 50, 134
Ayali, M. 194
Ayalon Screw Press 28, 141-45, 149, 151, 153, 165, 167-69, 173
Ayalon, E. 44, 54, 134
Azzefoun, Algeria 58, 155
ʿAzzun, Kh. (Tabsur) 153

Babylonian, Old 199
Baird, J.A. 42
Balat 51
Balata (Shechem) 68, 165
Banat Barr, Kh. 68
Banaqfur, Syria 89
Banjole, Yugoslavia 157
Bara, el-, Syria 89
Barbariga, Yugoslavia 93, 175
Barnett, R.D. 116, 166
Barom, Tel 100
Baruch, U. 38
Basch, L. 165
Bat el-Jebel East 58
Bata, Ḥ. 115, 134
Batash, Tel (Timna) 63, 64, 68, 100
Beaune, France 50
Beaujolais district, France 161
Beer Sheva, Tel 54
Beḥyo Syria, Beḥyo Weight 89, 114, 118, 119
Beida, el-, Libya, el-BeidaWeight 105
Beitin (Bethel) 58
Bekkouch, Leveau Site 180, Algeria 97
Bel, A. 50
Ben David, C. 49
Ben Yaʿakob, Y. 41, 139
Benaya, Ḥ. 144
Beni Hassan, Egypt 58
Benimaquia-Denia,Spain 157, 158
Bet Guvrin 101, 102, 161
Bet Ha-ʿEmeq, Tel, Bet ha-ʿEmeq Screw Weight 82, 83, 114, 116, 123, 140, 144, 165, 173
Bet Hashitta 146
Bet Loya, Ḥ. 83, 114, 118, 167, 167, 173, 178
Bet Mirsham, Tel Bet Mirsham Press 27, 62, 63, 64, 66, 68, 77, 100, 164, 165, 167, 179
Bet Natif, Ḥ. 127
Bet Sheʿarim, H. 190
Bet Shemesh, Tel 58, 63, 64, 100

Bet Ṣur (Burj el-Sur) 127
Betica 39
Bettir, Syria 175
Biblical references, Hebrew Bible 27, 29, 35, 37, 38, 41, 44-46, 51, 56, 159, 185-86, 188, 191, 195, 198, 199 *see also* Septuagint
Biblical references, New Testament 41, 56, 159
Billiard, R. 39
Binford, L.R. 177
Bir Sgaoun, Algeria 94
Black Sea 35
Blazquez Martinez, J.M. 40
Blümner, H. 42
Boardman, J. 44, 56
Borowski, O. 185
Bosco Tre Case, Italy 78, 83
Boscoreale, Italy 73, 74, 92
Bourgogne region, France 108, 109
Bouzon, H. 92
Bovillae, Italy 137
Bowyer, P.R. 159
Box Press 93, 110, 163
Brand, Y. 193
Brøndsted, J. 74, 148
Brown, J.P. 35, 39, 198, 206
Brun, J.P. 25, 30, 39, 40, 47, 49, 72, 74, 86, 93, 120, 163
Burj el-Sur *see* Bet Ṣur
Burj, Kh. el- north 51
Butler, H.C. 58

Cadiz, Spain 74
Caesarea, Algeria *see* Mauritanian Caesarea
California 42
Callot, O. 25, 30, 49, 72, 88, 89, 118, 146, 147, 155
Camino de Pago, Medrano, Spain 173
Campania, Italy, Campanian Platform Presses 50, 74, 91, 92, 162, 170, 171, 175
Camps-Fabrer, H. 25, 39, 49
Capena, Monte Canino, Italy 92
Capua, Italy 120
Carinola, Italy 120
Carmel, Mt 28, 36, 56, 65, 84, 85, 113, 116, 139, 140, 143, 145, 146, 160, 167-69, 179, 187, 201
Carmi, I. 36, 56
Carra de Vaux 88, 125
Carthage, Carthaginian 27, 98, 105, 171 *see also* Punic
Casanova, A. 50
Caspian Sea 35
Casse-Cou Press 92, 162
Cato 'the Censor', Cato's Press 26-29, 35, 37, 39, 41-49, 74, 86, 91-93, 97, 107, 138, 148, 162, 165, 171, 172, 174, 178, 194, 199, 206
Cesnola, L.P. di 90
Chalonnais district, France 161
Champagne region, France 108
Champvallon France 78
Chateau de Vinzelles France 78
Cheillé, France 90, 162
Chersonesos, Crimea, Russia 84, 91, 104, 156
Chevallier, R. 40
Chios (island) 39, 73, 75, 171
Christofle 95, 96, 100, 103, 156
City of Menes, Egypt 58, 155
Cohen, A. 44
Collumella 27, 29, 35, 37, 39, 42, 43-49, 54, 74, 92, 139, 140, 147, 148, 174, 175, 178, 192, 194, 195, 202, 204, 206
Conder, C.R. 26, 127
Constantine, Algeria 103
Coon, C.S. 50
Corcoran, T.H. 204
Corfu (island) 50, 123
Corsica (island) 42, 50, 92, 162, 172
Cotta, Morocco, Cotta Weight 98, 105, 111, 121
Cotton, H.C. 206
Courtois, J.C. 66
Cowper, H.S. 95
Creswell, R. 188
Crete (island) 36, 59, 62, 67, 90, 156, 175
Crimea, Russia 28, 84, 91, 104, 155, 157, 171
Crocodopolis-Theadelfia Egypt 58, 155
Cross, F.M. 198, 199
Crusaders 136, 137, 146
Ctesibius 87
Curtis, R.I. 205
Cyclades (islands) 59, 104
Cypro-Syllabic 35
Cyprus (island) 27, 28, 39, 49, 59, 62, 66, 67, 73, 83, 90-91, 93, 100, 101, 103, 104, 110, 119-16, 137, 156, 164, 170-72, 175
Cyrillus 189
Czechoslovakia 26, 46

Dabussiya 123
Dalman, G. 44, 46, 47, 50, 63, 78, 101, 126, 183, 184, 185, 193, 195, 204
Dalmatia 148
Dan, Tel 62, 65, 100, 199
Danby, H. 187, 202
Danube Basin 35
Danun, Sheikh 118
Dar Qita, Syria 119
Dar, S. 44, 56
De Vogue, Le cte. C.J.M. 88
Dead Sea Sect 36, 37
Dehas Syria 90
Deir Mišmiš, Syria 89
Deir Siman, Ḡ Siman Syria 58
Deir, el- 72
Delos (island), Delos Weight 90, 104, 119, 156
Demsky, A. 199, 202
DeVaux, R. 58
Devon district, Great Britain 162
Dhioros, Cyprus 156
Diderot, D. 109
Dijesta 188
Dinᶜila, Kh., Dinᶜila Screw Weight 72, 82, 85, 88, 113-116, 119, 121, 144, 163, 165, 173, 179, 180
Diocletian 39, 47, 192
Donzere, France 157, 158
Dothan, M. 199
Drachman, A.G. 25, 49, 74, 86-88, 91, 97, 103, 107, 108, 113, 119, 124, 125, 148, 156, 165, 172
Dressel, H. 39
Duhdah, Kh. el- 114, 173
Dukas, Kh. 126, 127
Duma, Kh. (Deir, el-) 144

Echnertach, Luxemburg 108
Edam, Holland 126
Egypt, Alexandria Museum Goblet 154
Egypt, Egyptian 26, 28, 35, 37, 39, 42, 50, 55, 58, 73, 137, 154-156, 158, 177
ᶜEin el-Judeida, ᶜEin el Judeida Screw Weight 113, 114, 146, 169, 178
ᶜEin Fattir 127
ᶜEin Gev 199
ᶜEin Hod 110
ᶜEin Nashut, ᶜEin Nashut Press 130-35, 167, 179, 195
Eitam, D. 40, 49, 57, 63
ᶜEizariya, el- 178
Ekron (*see also* Miqne) 64
Elche, Spain 96, 111, 175, 176
Elgavish, J. 146
ᶜen Hadda 116
England 126, 145, 159, 161, 162 *see also* Great Britain
Entremot, France 100, 103, 104
Epstein, C. 36, 57
Epstein, J.N. 189
Evans, A.J. 67

Fakhura 131
Fassuta 68, 84, 174
Fayum, Egypt, Fayum Press 124, 162
Feast of the first fruit of oil 37
Feliks, J. 39
Fenis, Italy, Fenis Press 111, 115
Fernandez Castro, M.C. 49
Fez, Morocco 83, 96, 105, 108, 109, 163, 175
Finet, A. 39
Fisher, C.S. 198
Flower, B. 39, 45, 206
Forbes, H.A. 57, 74, 75, 105
Foxhall, A. 57, 74, 75, 105
France 27, 28, 30, 35, 47, 49, 50, 59, 73, 87, 93-94, 100, 103, 104, 106, 108, 109, 120-121, 125, 126, 157, 158, 161-63, 170-75
Francolise Posto, Italy, Francolise Posto Weight 92, 105, 111, 120-21, 173, 174
Frank, T. 39
Frankel, B. 49, 126
Frankel, R. 39, 49, 54, 62, 83, 118, 128, 134, 165, 168, 178
French R.K. 159
French, language 189
Frey, J.B. 205

Gades, Spain 178
Gaidukevych, V.F. 103, 155, 156
Galilee, Galilean 28, 69, 84-86, 89-91, 95, 97, 100, 116, 121, 122, 128-30, 145, 152, 153, 165-67, 169, 171, 179, 180, 194
Galilee, Lower 29, 81, 84, 113, 116, 132, 135, 141, 143-46, 153, 160, 167-69, 177, 183, 186
Galilee, Sea of 101, 133, 134, 169, 179
Galilee, Upper 36, 76, 80, 83-85, 100, 101, 113, 117, 141, 145, 146, 153, 160, 161, 165, 167-69, 177
Galilee, Western 29, 52, 73, 98, 131, 134, 136, 137, 141, 145, 153, 161, 167, 171, 179, 180, 186
Galili, E. 36, 38, 56
Gallego Gongora Diaz 42
Gamla (Salem Tel el-) 76, 134, 169
Gammarth, Tunisia 103
Gat Carmel 199
Gaza amphora 106, 152, 206
Gaza Wine 152, 206
Geiger, J. 206
Geltzer, M.L. 185
Genoese Press 123, 178
Geoponica 29, 39, 42, 205
Germany, German 42, 93, 94, 104, 108, 123, 124, 157, 158, 171, 189
Gesenius, W. 185, 186
Gezer, Tel 49, 65, 68, 100, 170
Gibeon, *see* Jib-el
Gichon, M. 78, 127
Givᶜat Hayeᶜur 169
Golan 28, 29, 56, 57, 84, 100, 101, 117, 132, 139, 141, 145, 161, 169, 176
Goldman, F. 49, 189, 192
Gordon, C.H. 186, 198, 199
Goudineau, C. 103
Govit, Ḥ. 101
Grace, V.R. 39
Grand Point Press 108, 109, 161, 162
Grande Kabylie, Algeria 155
Granvas Lozere, France 159
Great Britain 123, 124, 162, 172 *see also* England
Greece, Greek 27, 29, 35, 37, 42, 44, 59, 75, 75, 93, 94, 104, 1052, 119, 156-58, 180 *see also* Aegean
Greek, language 30, 35, 37, 39, 43, 46, 47, 148, 188, 190, 193-97, 198, 200-206
Grimal, P. 205
Guadalquivir, Spain 72
Guerin, V. 26
Guillemard, F.H.H. 90
Gur, A. 200
Gush Ḥalav, Gischalla 194
Gütterbock, H.G. 39

Hadjisavvas, S. 49, 67, 90, 103
Haga Treada, Greece 45
Haifa 40, 56, 57
Haifa, Romema (Rushmiya, Kh.) 100
Halies, Greece 90, 105
Ḥaluṣa, Ḥorevot 150
Ḥamad Kh., Ḥamad Press 141, 144
Hamel, G. 39
Ḥanita, Ḥanita Screw Press 28, 141, 144, 145, 165, 167, 173
Hansen, J. 36, 38, 44, 45
Ḥasun, Kh. 112
Hauran 127
Ḥazor, Tel 198, 199
Heath, T.H. 87
Hebron, Hebron district 42, 51, 198, 199
Helbon 199
Heltzer, M. 39, 40, 45, 185
Henchir Choud et-Battal, Tunisia 95, 105, 108
Hennesey, J.B. 35
Herault, France 125
Herculaneum, Italy 50
Ḥermeshit, H. 83
Hermon, Mt 202
Hermopolis Tomb of Peosris, Egypt 154
Hero of Alexandria 27-29, 49, 87, 88, 92, 107, 111, 113, 116, 122-26, 138, 148, 162, 165, 172, 173, 178, 180
Herod, King 206

Hestrin, R. 199, 206
Hirmer, M. 67
Hirschfeld, Y. 49, 187
Hittite 35, 39, 44
Hjelmquist, H. 35, 38
Hoffner, A.A 39
Hogarth, D.G. 90
Holland 126
Holwerda, D. 188
Honigsberg, P. von 42
Horle, J. 49, 74
Horowitz, A. 38
Ḥudash, Kh. 68, 146
Huevar-Huelva, Spain 47, 96, 175
Humbel, X. 50, 123, 124, 178
Hungary 50, 108-111, 123, 124, 161-63
Hurvitz, A. 52
Hutchinson, R.W. 59
Hutteroth, W.D. 37

Ibiza 50, 96, 162, 172, 175
Iksal, Iksal Screw Weight 117, 160
Ilan, Z. 165
India, Indian 75
Isfahan, Iran, Persia 87, 97, 172
Italy, Italian 27, 35, 39, 42, 49, 59, 73, 74, 88, 91-3, 97, 105, 110, 111, 119-20, 121, 123, 125, 137, 138, 148, 156, 158, 171, 172, 174, 175, 178, 201, 202

Japan 125
Jashemki, W.F. 38
Jastrow, M. 187, 188, 191-93, 200-202, 204, 205
Jericho 150
Jerusalem 29, 63, 113, 114 118, 145, 150, 169, 178, 179, 199
Jerusalem, Notre Dame 114
Jewish 86, 168, 178, 205
Jezreel Valley 70
Jib, el- (Gibeon) 56, 199
Jordan Valley 100
Jordan, country 139, 140, 161
Josephus Flavius 113, 165, 194
Judaea, Judaean 29, 44, 70, 76, 84-86, 130, 143, 144, 161, 164, 165, 167, 168, 177, 179, 182, 199 (Juda),
Judeida, el- 136
Judur, Kh. 146
Jungst, E. 46, 49, 86, 87
Justinian 188

Kabri, el- 113
Kafr Kama 134
Kafr Nabo, Syria 89, 119
Kafr Rut, Kh. 127
Kafr Yasif 79
Kalavasos, Cyprus 67
Kalymnos (island) 90, 110, 119, 119
Karim, el- 55
Karkara, Ḥ. 77-79, 100, 118, 165, 175, 190
Karpathos (island) 123
Karrantinnaya Slobodka, Crimea, Russia 155
Kasfa, Kh., Kasfa Screw Weight 114, 112, 118-20, 144, 169, 173
Kefar Ḥananya 130
Kenyon, K.M. 36
Kerje, el- 47
Kfar Hay, Lebanon 101, 161
Kibbutz Erez 106
Kislev, M.E. 36, 56
Kitchener, H.H. 26, 127
Kjeir, el-, Syria 89
Kloner, A. 74, 127
Knidos, Turkey 39
Knossos, Crete 67, 156
Kohut, A. 190, 192, 193
Kommos, Crete 67
Korazim 132
Kouklia Styllarka, Cyprus 90, 104
Kouris Valley, Cyprus 90
Krauss, S. 39, 44, 49, 187, 189-91, 197, 200, 202
Ksegbe, G. Barisa, Syria 119
Kunya,Turkey 58
Kutcher, Y. 198, 204

La Roquebrussanne, Le Grand Loou I, France 157
La Vega 49
Lakhish (Lachish) 62, 198, 199, 202
Lambesc, France 157
Languedoc region, France, Languedoc Press 108, 109, 110, 114, 115
Lania, Cyprus 110
Laodicea, Syria 194
Laporte, J.P. 74, 155
Lapp, P.W. 58, 199
Latium 178
Latour, P. 50
Lauffer, S. 39, 47, 195
Laureatic Olympus, Greece 90, 119
Lebanon 75, 101, 102, 118, 127, 140, 164, 171, 173, 199
Lemaire, A. 198
Lenoir, M. 98
Lepcis Major (Leptis Magna), Libya 39
Lesbos (island) 39, 119
Leveau, P. 96
Libya, Libyan 35, 37, 94, 95, 103, 171, 177
Lieberman, S. 187, 189, 191, 202
Liman West 152
Lion-head spouts 154, 155, 158
Liphschitz, N. 36, 38, 57
Liverani, P. 49, 119, 173
lmlk seal impressions 199
Loche, France 78
Losnich, Germany 157
Löw, I. 200
Lutz, H.F. 39
Luvim, Ḥ., Luvim Screw Weight 116, 135, 140, 142, 167, 173
Luxemburg 108
LXX *see* Septuagint

Ma Ougelmine, el-, Algeria 58, 171
Maconnais district, France 161
Madaure, Algeria, Madaure Oil Separator 96, 104, 105, 108, 175
Madfane, Medfane, Matfana 78
Madrid 40
Mago 39, 43, 139, 171, 202
Maier, F.G. 137
Malta 73, 84, 91, 104, 165, 171, 175,
Malul, M. 37, 39, 43, 98
Mana East 136
Mandaean 186

Mandelkern, S. 186, 198
Manguarra y San Jose, Spain 94, 97
Manot, H., Lower, Manot Press 136, 137, 146, 161
Mansur el-Aqqab 146
Maresha, Tel, Maresha Press 27, 28, 62, 70, 71, 74, 76, 77, 84, 85, 127, 164, 165, 167
Mari 39
Mari Kopetra, Cyprus 90, 91
Mariffi, Italy 95
Marinatos, S. 67
Marj el-Qital, Marj el-Qital Press 146
Maroni, Cyprus 67, 99
Marseilles 94, 105
Marus, Ḥ. 80
Masada (Golan) 42, 51
Masada Inscriptions 206
Mattingly, D.J. 25, 38, 39, 45, 49, 103, 179
Mauritania 27, 98
Mauritanian Caesarea 96, 171, 179
Mavo Modiʿim 82
Mayerson, P. 152, 206
Mazar, B. 199
Mazor, G. 25, 150
Medinat Habu reliefs, Egypt 177
Megiddo, Tel 51, 57, 63, 66
Meister, A.L.F. 26, 49, 97
Melena, J.L. 39, 44, 45
Mesopotamia 39
Methana, Greece, Methana Weight 28, 59, 105, 175
Miʿilya, Miʿilya Screw Weight 47, 82, 83, 84, 117, 117, 165, 173
Michal, Tel 54, 149
Midrasa, Ḥ., Midrasa Screw Weight 117
Miguelez Ramos, C. 50
Mikonos (island) 156
Miqne, Tel (Ekron) 62, 63, 64, 68, 100
Mirmiki, Crimea, Russia 156
Mishkena, H., Mishkena Press 135, 136, 167
Monasteries 178
Monod, T. 205
Monte Cupellazo, Italy 59
Montet, P. 58
Moritz, L.A. 68, 72
Morroco, Morrocan 35, 72, 103, 105, 107, 109, 111, 161, 171, 175
Moṣah 199
Mosel region, Germany 157
Mubarak (Mevorakh, Tel) 63
Myres, J.L. 50, 103
Myrtou, Cyprus 67

Naʿaman, N. 199
Nabataean 150
Nablus 47
Nadur, Malta 98
Nagatsune, O. 125
Nahariya 54
Najjara, Kh. el- 82
Naṣba, Tel el- 63, 199
Naveh, Y. 199, 206
Neef, R. 36, 38
Negev, Negev Highlands 28, 139, 140, 144, 152, 158, 188
Nelson, H.H. 177
Nes ʿAmin East 54
Netzer, E. 150
Neumage, Germany 157
Nicosia, Pasydy Plot, Cyprus 59
Nimrud, Iraq 116
Nix, L. 88
Normandy 145, 159
North Africa 27-29, 35, 39, 45, 49, 50, 59, 73-75, 83, 84, 88, 91, 94-98, 103, 104, 106, 121, 155-57, 161, 163, 170-72, 174, 176, 179
Numidia 27, 98

Oates, D. 103
Olami, Y. 52, 55
Ollioles, France 100
Olynthos, Greece 73, 74, 90, 156
Oshrat North 146
Oud el-Htab, Tunisia 104
Oued Athmenia, Kharba, Algeria 96, 97

Pachna Ayios Stefanos, Cyprus 90
Pachna Sykes, Cyprus 90
Palaestina Secunda 80, 113, 165
Palaikastro, Crete 67
Palladius 29, 37, 39, 41, 42, 43, 59, 154, 158, 202-205
Palmer, R. 39
Parain, C. 50
Paris 40
Parod 152
Partizani, Crimea, Russia 155
Paton, W.R. 50, 103, 118
Patrasia, Crimea, Russia 156
Patrich, J. 39, 83, 165, 178
Paul, S.M. 39, 186, 200
Peacocke, P.S. 39, 152, 206
Peleg, Y. 126
Peltenberg, E.J. 59
Persia 50, 87, 97
Petralia, Sicily 59
Phaestos, Crete 59, 67
Philistia 65
Philonianum 206
Phocaea, Phocaean 94, 105
Phoenicia, Phoenician, 27, 29, 80, 81, 85, 92, 106, 113, 127, 145, 157, 158, 165-67 169, 171, 173, 177-79, 198
Phoenician Joints 86, 165
Phrygia, Phrygian 202
Piesport, Germany 104, 157
Pindakos, Chios 73-75, 90, 100
Pisgat Zeʿev, Deir Ghazali 114, 118
Pliny the Elder 27-29, 35, 37, 39, 41-49, 86, 87, 93, 107, 108, 110, 122, 125, 137-39, 148, 162, 171, 172, 178, 189, 195, 199, 200, 203-206
Plommer, H. 42, 154
Plugata 202
Pola,Yugoslavia 93
Polge, H. 50
Pompeii 26, 38, 50, 73, 74, 91, 92, 124, 137, 159
Pompeian Donkey Mill 72
Ponteves, France, Ponteves Screw Weight 94, 95, 120, 121, 156
Pontus, Turkey 105, 114, 119-21, 173, 174
Portugal 42, 110, 121, 163, 172, 173
Posta Crusta Foggia, Italy 157
Praesos, Crete 59, 90
Pricket, J.L. 43

Pritchard, J.B. 56, 118, 199
Provence, France 27, 30, 39, 49, 93, 94, 109, 120
Ptolemy Eugertes II 87
Puget sur Argens, France 118
Punic 39, 43, 104, 105, 139, 171, 202

Qala 68
Qalandiya 54, 100
Qashish, Tel 57
Qasr, Kh. el- 127
Qastra, H. (Kafr Samir, Kh.) 80
Qat Kh. el- (Bir el-Qat), Qat Screw Mortice 153, 169
Qedumim, Qedumim Press-Bed 84, 85, 114, 153, 167
Qiri, Tel 68
Qiriat el- Mekhayyat (Mt Nebo), Jordan 140
Quinion, M.B. 159, 162
Quneitra, Kh. el- 127
Qurnat el-Harmiya 63
Quseir, Kh. el- (Rajmi, Kh.) 80, 82, 83, 84, 91, 98, 114, 117, 134
Quza, el- 80

Rabat, Morroco 97, 105, 114, 121
Rafat, Kh. 129
Rainey, A.F. 185, 199
Rama, el-, Rama Press 135, 136, 144, 146, 167
Ramat Rahel 199
Ras el-Hammam, Libya 96, 175
Ras Shamra *see* Ugarit
Rasm el Beida 127
Rasm Harbush 36, 175
Reeves, J.C. 36, 37
Rehovot, Dueran, Kh. 139, 152
Reisner, G.A. 198
Remesal Rodrigues, J. 39, 40, 45
Renfrew, C. 38, 45, 59
Rey-Coquais, J.P. 205
Rhine Valley 124
Ricci, C. 39
Riley, J.A. 152, 206
Rome, Villa Rondanini *see* Villa Rondanini
Rosen, B. 45
Rosenbaum, E. 39, 45, 206
Rosh Pina 114
Rosh Zayit, H., Rosh Zayit Press 27, 65, 66, 68, 81, 100, 165, 167, 175
Rossiter, J.J. 25, 91, 92, 93
Rozier, M. 108
Ruggiero, M. 49, 91
Rumania 110, 123, 125, 126,163
Runnels, C.N. 36, 38, 44, 45

Sacket, J.R. 177
Safrai, Z. 39
Ṣafṣafot, Tel 123, 128-30
Sagiv, N. 74
Saint Laure de la Plaine, Layon, France 91, 162
Saint Romain en Gal, France 154
Saladin, M.H. 95
Salamis, Cyprus 49, 90
Salone, Yugoslavia 74, 93
Samaria, Samarian, region 29, 44, 84-86, 100, 113, 114, 116, 131, 132, 161, 164, 165, 167, 168, 177, 179, 180
Samaria, town *see* Sebastia-Samaria
Samaria ostraca 45, 46, 186, 198, 199
Samaria Screw Weight 111-13, 115, 117, 118, 119, 120, 121, 137, 146, 160, 167, 172, 173, 178, 180
Samaritan 86, 168, 178
San Paul Milqi, Malta 84, 98, 166, 172
Sanary Saint Ternide, France 92
Santa Lucia di Mercurio, Corsica 46
Sarepta, Lebanon, Sarepta Screw Weight 114, 118, 120, 121, 173
Sarfud, Syria, Sarfud Screw Weight 89, 119, 120, 121
Sasson, V. 186, 199
Schapiro, N. 43
Scheltma, H.J. 188
Sciallana, M. 39, 152, 206
Sea Peoples 177
Sebastia-Samaria 45, 83
Segal, D. 36, 56
Šeiḫ Barakat, Syria 114, 119, 173
Sellin, E. 58
Semana Weight 28, 93, 94, 97, 102-106, 119, 120, 121, 156, 157, 161, 171, 172, 173
Sentroma Tiana, Spain 94
Seneca 35
Septuagint (lxx) 41, 198, 202
Sergible, Syria 89
Sette Finestre, Italy 92, 98, 157
Shafarʿam 190
Sharon (coastal plain) 84, 116, 167, 201
Sharvit, J. 36, 56
Shennan, S. 177
Shephala (western foothills of Judaea) 70, 76, 130
Shiloh, Y. 52
Shiqmona, Tel 66, 68, 100, 146, 198
Shivta, Horevot 150
Shubeika East 52
Shubeika, Kh. el- 79
Shush, Tel 76
Si, Syria 145
Sibella, P. 39, 152, 206
Sicily 137, 175,
Singer, I. 39, 44
Sippori, Sippori Press 130-33, 167, 190
Sirin 80
Siyar el-Ghanam 82, 100, 146, 178
Sleeswyk, A.W. 165
Socoh 199
Soden, W. von 185
Sordinas, A. 50
Spain, Spanish 27, 28, 35, 37, 45, 47, 49, 72-74, 94, 104, 106, 110, 111, 121, 123, 125, 137, 157, 158, 161, 163, 171-75, 178
Sparkes, B.A. 59
Spiegel-Roy, F. 35, 36
Stager, L.E. 35-37, 46, 56-58, 62, 66, 186
Stavros, Greece 42, 59, 156
Steinzaltz, A. 202
Stol, M. 186
Sudan 154
Sugar factory 136,
Sukenik, E.L. 202, 206
Sumaq, H. 76
Ṣur Natan 70
Switzerland 28, 110, 163
Syria, Syrian 7, 28, 30, 49, 62, 66, 67, 72, 73, 75, 83, 84, 88-95, 98, 105, 114, 118, 118, 145, 147, 155, 157, 167, 170-75, 179, 194

Taʿanach region 52
Tabgha, Tabgha Press 71, 133-35, 169, 179
Tafila, el- 47, 63, 101, 102, 161
Taissons press 109, 161, 162
Talmud, Talmudic references including Mishna Tosephta etc. 27, 29, 35, 37, 42-44, 46, 47, 49, 54, 55, 57, 68, 74, 81, 101, 138-40, 145, 147, 148, 178, 186-97, 199-206
Tannim, H. 127
Taourienne, Leveau site 204, Algeria, Taourienne Pier Base 97, 166
Taqle, Syria,Taqle Weight 28, 89, 105, 171
Tarentium 206
Tchalenko, G. 88
Tchernia, A. 39
Tel Aviv, Eretz-Israel Museum 54, 143, 149
Tel Aviv, Hevra Hadasha 149
Teleilat Ghassul 36
Tepper, Y. 127
Testaccio Mound, Rome 39, 45
Tezarin Taghzuth, Morroco 83, 105, 108, 109
Thalesfa east, Leveau Site 20, Algeria 96
Thassos (island) 39
Theilscher, P. 46, 49, 86, 87
Thompson, H.O. 62
Thuberbo Majus, Algeria 96, 174
Tiʿinnik (Taʿanach) 47, 55
Tiberias 202
Tifrit Naʾit el Hady, Algeria 58
Tigzert, Algeria 58
Tipasa, Algeria 155, 175
Tira, Kh. el- 167
Tirat Yehuda 84
Tiritake, Crimea, Russia 155, 156
Tivoli Pier Base 27, 92-94, 137, 171
Tripolitania 94
Tsafrir, Y. 39, 83, 118, 165, 178
Tunis, Tunisia 35, 73, 94-95, 103, 104, 105, 161, 171, 174
Tur, el- (Beth Phage) 145, 169,
Turkey 58, 59, 173, 174 *see also* Anatolia
Turrios Berceo, Spain 121
Tuweiri, el- 118
Tyre 205
Tyre-Qabr Hiram 140

Ugarit, Ugaritic 35-37, 39, 45, 57, 66, 99, 185, 186, 198
Umm el-ʿAmad, Lebanon 70, 77, 81, 171
Urmann, D. 49, 55, 187, 188
Usha, Ḥ. 80, 190
Ussishkin, D. 198, 199
USSR 26

Valencio do Douro, Portugal 42
Var, Provence, France 30, 49, 93, 120
Varro 29, 35, 39, 41, 42, 45, 47, 138, 143, 148, 199
Vathypetro, Crete 67, 175
Vaucluse (Cavallon district), France 59
Verige Bay, Yugoslavia 157, 158
Vesuvius 74, 91
Vickery, K.F. 39
Vilbosh, N. 189
Villa Rondini, Rome 74, 92
Villanueva Ariscal, Seville, Spain 42
Vincze, I. 50, 161
Vintage festival 36
Violet, E. 107, 108
Vitruvius 27, 28, 49, 86, 107, 117
Volubilis, Morroco, Volubilis Weight 72, 97, 98, 105, 111, 121

Waetzoldt, H. 39
Wales 159
Warren, P.A. 59
Wartburg, M.I.V. 137
Weinfeld, M. 39, 44
Weinstein-Evron, M. 38
Weissner, P. 177
Weradim, Ḥ., Weradim Press 133-35, 169, 179
White, K.D. 35, 39, 74, 147
Williams, D.F. 39, 152, 206
Williams-Davies, J. 159
Wolff, S.R. 58
Worlidge, J. 159
Worms Press 124
Wulff, H.E. 50

Yaara West 152
Yadin, Y. 36, 199, 206
Yajuz, Kh. 76
Yanovskaiah, N.B. 185
Yatir Naḥal East 51
Yatir Naḥal West 51
Yeivin, Z. 49, 132, 189
Yerovasa, Cyprus 90
Yirka Central 153
Yirka East 52
Yoqneam, Tel 57, 100
Younger, W. 42
Yugoslavia 26, 27, 84, 90, 91, 93, 94, 123, 157, 158, 171, 175

Zabadi, H., Zabadi Press, Zabadi Weight 27, 71, 77-82, 84, 85, 100, 114, 165, 174, 175, 190, 195
Zapetal, V. 39
Zawit, Ḥ. 84
Zemer, A. 39, 152, 206
Zeror, Tel 63
Zibdiyah, name 199
Ziph 199
Zohary, D. 35, 36, 38